Fractal and Fractional in Geotechnical Engineering

Fractal and Fractional in Geotechnical Engineering

Guest Editors

Shao-Heng He
Zhi Ding
Panpan Guo

MDPI

Basel • Beijing • Wuhan • Barcelona • Belgrade • Novi Sad • Cluj • Manchester

Guest Editors
Shao-Heng He
Department of Civil and Environmental Engineering
The Hong Kong Polytechnic University
Hong Kong
China

Zhi Ding
Department of Civil Engineering
Hangzhou City University
Hangzhou
China

Panpan Guo
School of Civil Engineering
Hefei University of Technology
Hefei
China

Editorial Office
MDPI AG
Grosspeteranlage 5
4052 Basel, Switzerland

This is a reprint of the Special Issue, published open access by the journal *Fractal and Fractional* (ISSN 2504-3110), freely accessible at: https://www.mdpi.com/journal/fractalfract/special_issues/53B94Z7S60.

For citation purposes, cite each article independently as indicated on the article page online and as indicated below:

Lastname, A.A.; Lastname, B.B. Article Title. *Journal Name* **Year**, *Volume Number*, Page Range.

ISBN 978-3-7258-5083-9 (Hbk)
ISBN 978-3-7258-5084-6 (PDF)
https://doi.org/10.3390/books978-3-7258-5084-6

Contents

About the Editors

Shao-Heng He

Shao-Heng He is a postdoctoral researcher in the Department of Civil and Environmental Engineering at The Hong Kong Polytechnic University. His research focuses on the macro- and micro-mechanical properties and constitutive modeling of granular materials, including the liquefaction assessment of foundation soils in ultra-deep sea environments, disaster mechanisms of geotechnical materials under complex principal stress axis rotation paths, and numerical prediction methods for the long-term deformation of foundations subjected to traffic loads. His work has been published in top-tier international geotechnical engineering journals. He has also been invited to serve as a reviewer for over 20 international journals.

Zhi Ding

Zhi Ding is a professor in the Department of Civil Engineering at Hangzhou City University. He has long been engaged in research on intelligent construction and maintenance of shield tunnels. He has led more than ten research projects, including the General Program and Youth Program funded by the National Natural Science Foundation of China, the Central Government Guided Local Science and Technology Development Project, the Key Program and Youth Program funded by the Zhejiang Provincial Natural Science Foundation, and the Major Science and Technology Innovation Project of Hangzhou City.

Panpan Guo

Panpan Guo is a lecturer in the School of Civil Engineering at Hefei University of Technology. He is engaged in teaching and research in the fields of intelligent construction and disaster prevention and the control of tunnels and underground engineering, foundation treatment, damage and protection of underground engineering from explosion and intrusion, and resource utilization of solid waste materials. He has led research projects funded by the National Natural Science Foundation of China, Anhui Provincial Natural Science Foundation, and the Open Fund of National Key Laboratories.

Preface

This reprint focuses on exploring and applying the critical roles of fractal and fractional theories in geotechnical engineering design. It addresses the application of these concepts in geotechnical engineering and construction materials, covering topics such as rock fracture behavior, elastic wave propagation, slurry seepage theory, single-particle breakage, artificial frozen soil, tunnel settlement prediction, and more.

Shao-Heng He, Zhi Ding, and Panpan Guo
Guest Editors

fractal and fractional

Editorial

Fractal and Fractional Theories in Advancing Geotechnical Engineering Practices

Shao-Heng He [1,2], Zhi Ding [1,*] and Panpan Guo [3,*]

1 Department of Civil Engineering, Hangzhou City University, Hangzhou 310015, China; shaoheng.he@polyu.edu.hk
2 Department of Civil and Environmental Engineering, The Hong Kong Polytechnic University, Hong Kong
3 School of Civil Engineering, Hefei University of Technology, Hefei 230009, China
* Correspondence: dingz@zucc.edu.cn (Z.D.); guopanpan@hfut.edu.cn (P.G.)

Fractal and fractional theories have developed over several decades and have gradually grown in popularity, with significant applications in geotechnical engineering [1,2]. The contours and pore structures of geotechnical materials often exhibit complex morphologies and scales that fractal theory can effectively describe. Experimental techniques such as scanning electron microscopy (SEM), computerized tomography (CT) scanning [3,4], and mercury intrusion porosimetry (MIP) can be used to obtain pore structure data, which can then be quantitatively characterized based on fractal dimensions and other parameters to capture the complexity and connectivity of the pores. For example, because fractal theory can accurately represent the irregular geometry of natural soils and matches well with their complex structures, it has been applied in predictive models of the soil–water characteristic curve (SWCC) for unsaturated soils [5]. Moreover, fractal theory can be applied to study the distribution and complexity of particle contact networks [6], aiding in understanding the interaction mechanisms between particles and their influence on the macroscopic mechanical behavior of geotechnical materials. Furthermore, traditional soil constitutive models often fall short in describing the complex mechanical behaviors of geotechnical materials. In contrast, fractional-order constitutive models can more precisely capture the nonlinear and memory effects inherent in these materials [7]. For instance, fractional constitutive models can effectively describe the mechanical behavior of geotechnical materials under long-term loading [8], thereby providing more reliable tools for long-term stability analysis in geotechnical engineering.

This Special Issue, "Fractal and Fractional in Geotechnical Engineering", focuses on exploring and applying the critical roles of fractal and fractional theories in geotechnical engineering design. This collection features 13 papers addressing the application of fractal and fractional concepts in geotechnical engineering and construction materials, covering topics such as rock fracture behavior, elastic wave propagation, slurry seepage theory, single-particle breakage, artificial frozen soil, and tunnel settlement prediction. An overview of these papers is provided below.

Shi and Xiao [9] investigated how fractal dimensions affect the mechanical properties and fracture behavior of multi-mineral rocks using three numerical models, i.e., digital texture, Voronoi polygon, and Weibull distribution. Particle flow simulations and uniaxial compression tests were performed on 2D and 3D models, respectively. The results indicated that the Weibull model had the highest fractal dimension and complexity, while the Voronoi model had the lowest, as well as a more regular structure. Compressive strength increased roughly linearly with fractal dimension in all models. Additionally, 3D models showed higher strength than 2D models, but the relationship between fractal dimension and strength followed a similar trend in both cases. This highlights the importance of fractal dimensions in describing rock geometry and mechanics.

Wang and Zhang [10] conducted a numerical study on elastic wave propagation in two-dimensional fractional Brownian fields, examining the effects of the Hurst exponent and standard deviation on wave behavior. Their results demonstrate that higher standard deviation and lower Hurst exponent increase wavefront roughness, cause asynchronous arrivals, and amplify energy attenuation due to stronger scattering and greater modulus variability. This study enhances our understanding of wave propagation in fractal heterogeneous media, with significant implications for seismic exploration and subsurface imaging in complex geological formations.

Gong et al. [11] developed a theoretical penetration grouting model for Bingham fluids based on fractal theory, incorporating the influence of pore structure on slurry infiltration. The model was validated experimentally using a custom apparatus simulating constant flow rate penetration. The results showed strong agreement between theoretical predictions and experimental data, demonstrating the model's effectiveness in describing slurry pressure distribution during grouting. This work provides valuable guidance for grouting design and related engineering applications.

Li et al. [12] developed a comprehensive framework integrating experiments and numerical simulations to study single-particle breakage using 3D particle reconstruction based on a vision foundation model. Their calibrated discrete element method simulations accurately replicated particle breakage behavior. The study revealed strong correlations between 3D fractal dimensions and particle size, crushing strength, and morphology, highlighting fractal dimension as a valuable descriptor of particle properties. This framework advances the understanding of particle breakage mechanics and provides a robust tool for future research.

Sun et al. [13] analyzed highway slope surface displacement using multifractal detrended fluctuation analysis (MF-DFA) and developed a particle swarm optimization (PSO)-optimized long short-term memory (LSTM) model for prediction. Their study revealed multifractal characteristics in the displacement data and classified the slope warning levels. The PSO-LSTM model demonstrated high predictive accuracy. These findings support routine monitoring and provide a strong foundation for improved slope safety management.

Kong et al. [14] investigated the microstructural and fractal characteristics of frozen–thawed sandy soft soil using nuclear magnetic resonance and uniaxial compression tests. Their study demonstrated that freezing temperature and sand content significantly affect pore structure complexity and soil strength, with the fractal dimension serving as a key indicator. Lower freezing temperatures increase fractal dimension and soil strength, while higher sand content enlarges pores and reduces strength. These findings offer valuable insights into freeze–thaw effects on soil stability in soft soil regions.

Liang et al. [15] employed DEM numerical simulations to investigate soil arching mechanisms in granular materials with varying relative densities. They identified three zones based on particle displacement and found that soil arching strength and force chain development are strongly influenced by particle assembly porosity. Denser samples exhibited stronger arching and more pronounced force chain evolution, particularly in the intermediate displacement zone. After reaching peak arching, force chains degraded, leading to a reduction in arching. Ultimately, soil arching at the limit state was found to be independent of relative density. This study highlights the crucial role of contact force chains in soil arching behavior.

Yu and Yin [16] investigated the convergence of stiffness operators and viscoelastic properties in fractal ladder and tree structures. They demonstrated that sequences of stiffness operators converge and that finite-level fractal structures beyond the third hierarchy exhibit behavior similar to infinite-level fractals, simplifying analysis. The study revealed characteristic ultra-long creep times and relaxation tailing effects, underscoring the effectiveness of fractal models in capturing complex viscoelastic behavior. Their work bridges continuous and fractal models, providing valuable insights for materials science and mechanical engineering.

Wang and Wang [17] conducted in situ industrial CT scans to study crack evolution in coal under varying confining pressures, using fractal theory to quantitatively describe crack development. They identified consistent crack evolution stages across pressures and found that fractal dimension changes correlate well with crack dynamics, showing an initial slight decrease followed by significant growth. Higher confining pressures limit crack development and enhance the mechanical strength of coal, primarily leading to shear failure. This study offers a reliable fractal-based method of assessing coal fracture evolution and predicting instability.

Yang et al. [18] applied multifractal theory and multifractal detrended fluctuation analysis (MF-DFA) to characterize tunnel deformation data. They integrated Mann–Kendall analysis to establish dual early warning criteria and employed a particle swarm optimization–long short-term memory (PSO-LSTM) model to predict tunnel settlement. The results showed consistent Class II warning levels across sections and demonstrated that the PSO-LSTM model delivers accurate and stable settlement predictions. This study supports enhanced monitoring and disaster preparedness through quantitative analysis.

Zhang et al. [19] investigated the optimization of peripheral hole charging structures and blasting parameters for extra-long hard rock tunnels to enhance the effectiveness of smooth blasting. By employing laser profiling and multifractal detrended fluctuation analysis (MF-DFA), they compared bidirectional shaped charge blasting with spaced decoupled charge blasting. The results demonstrated that bidirectional shaped charge blasting significantly improved smooth blasting performance, yielding flatter tunnel contours and more uniform overbreak and underbreak. Both blasting methods exhibited multifractal characteristics in overbreak data, with quantitative analyses closely aligning with actual conditions.

Zhang et al. [20] developed an innovative energy calculation method to identify the key factors contributing to rock bursts by comparing elastoplastic and purely elastic models. Their study reveals that a deviatoric stress field, which induces a butterfly-shaped plastic zone, is the primary driver of significant energy release and rock burst occurrence, with trigger stress playing a secondary role. Laboratory experiments employing acoustic emission monitoring validate the butterfly-shaped failure pattern and reveal fractal characteristics in the spatial distribution of events prior to failure. These findings offer a clearer and more quantifiable understanding of the mechanisms underlying rock bursts.

Li et al. [21] developed a dilatancy equation that accounts for the non-coaxiality between the stress and strain rate in granular soils by introducing a novel non-coaxial coefficient grounded in potential theory and material fabric characteristics. This approach establishes a link between plastic strain and the microstructural fabric, enabling the equation to reduce to the classical critical state theory under isotropic conditions while effectively capturing non-coaxial effects under anisotropic conditions. Validation through simple shear tests demonstrates that the equation reliably predicts dilatancy behavior, particularly under principal stress rotation and varying stress ratios.

This collection of 13 papers is expected to inspire new ideas in advancing research on fractal and fractional theories, as well as promoting their broader application in geotechnical engineering. This will facilitate a better characterization of the complex microstructures and mechanical properties of soils, rocks, and structures such as tunnels, ultimately supporting the development of improved design methods and practical tools.

Author Contributions: S.-H.H., Z.D., and P.G. have all been directly and fundamentally involved in writing this editorial, summarizing the collected papers, and approving its publication. All authors have read and agreed to the published version of the manuscript.

Conflicts of Interest: The authors declare no conflicts of interest.

References

1. Wang, L.; Tang, S. Investigation and Application of Fractals in Civil Engineering Materials. *Fractal Fract.* **2023**, *7*, 369. [CrossRef]
2. He, S.-H.; Ding, Z.; Hu, H.-B.; Gao, M. Effect of Grain Size on Microscopic Pore Structure and Fractal Characteristics of Carbonate-Based Sand and Silicate-Based Sand. *Fractal Fract.* **2021**, *5*, 152. [CrossRef]
3. He, S.-H.; Yin, Z.-Y.; Ding, Z.; Li, R.-D. Particle morphology and principal stress direction dependent strength anisotropy through torsional shear testing. *Can. Geotech. J.* **2025**, *62*, 1–23. [CrossRef]
4. He, S.-H.; Yin, Z.-Y.; Ibraim, E.; Ding, Z. Face mask chips-reinforced sands under monotonic and cyclic torsional shearing. *Géotechnique* **2025**. [CrossRef]
5. Zhao, J.; Li, S.; Wang, C.; You, T.; Liu, X.; Zhao, Y. A universal soil–water characteristic curve model based on the particle size distribution and fractal theory. *J. Hydrol.* **2023**, *622*, 129691. [CrossRef]
6. Yang, Y.; Cheng, Y.M. A fractal model of contact force distribution and the unified coordination distribution for crushable granular materials under confined compression. *Powder Technol.* **2015**, *279*, 1–9. [CrossRef]
7. Sun, Y.; Gao, Y.; Song, S. Effect of integrating memory on the performance of the fractional plasticity model for geomaterials. *Acta Mech. Sin.* **2018**, *34*, 896–901. [CrossRef]
8. Sun, Y.; Xiao, Y. Fractional order model for granular soils under drained cyclic loading. *Int. J. Numer. Anal. Methods Geomech.* **2017**, *41*, 555–577. [CrossRef]
9. Shi, R.; Xiao, H. Particle Flow Simulation of the Mechanical Properties and Fracture Behavior of Multi-Mineral Rock Models with Different Fractal Dimensions. *Fractal Fract.* **2025**, *9*, 13. [CrossRef]
10. Wang, S.; Zhang, Z. Wavefield Evolution and Arrival Behavior of Elastic Wave Propagation in Two-Dimensional Fractional Brownian Fields. *Fractal Fract.* **2024**, *8*, 750. [CrossRef]
11. Gong, J.; Xu, C.; Wang, Y.; Gao, S.; Chen, H.; Jia, L. Penetration Grouting Mechanism of Bingham Fluid in Porous Media Based on Fractal Theory. *Fractal Fract.* **2024**, *8*, 739. [CrossRef]
12. Li, R.; Gao, X.; He, S.-H.; Ru, D.; Ding, Z. Fractal Analysis of Particle Size and Morphology in Single-Particle Breakage Based on 3D Images. *Fractal Fract.* **2024**, *8*, 614. [CrossRef]
13. Sun, X.; Su, Y.; Yang, C.; Tan, J.; Liu, D. Research on Slope Early Warning and Displacement Prediction Based on Multifractal Characterization. *Fractal Fract.* **2024**, *8*, 522. [CrossRef]
14. Kong, B.; Yan, Y.; He, H.; Yu, J.; Zou, B.; Chen, Q. Strength and Fractal Characteristics of Artificial Frozen–Thawed Sandy Soft Soil. *Fractal Fract.* **2024**, *8*, 393. [CrossRef]
15. Liang, L.; Cheng, Y.P.; Fan, X.; Ding, Z.; Xu, C. Multi-Scale Research on the Mechanisms of Soil Arching Development and Degradation in Granular Materials with Different Relative Density. *Fractal Fract.* **2024**, *8*, 247. [CrossRef]
16. Yu, X.; Yin, Y. Fractal Operators and Convergence Analysis in Fractional Viscoelastic Theory. *Fractal Fract.* **2024**, *8*, 200. [CrossRef]
17. Wang, A.; Wang, L. A Fractal-Based Quantitative Method for the Study of Fracture Evolution of Coal under Different Confining Pressures. *Fractal Fract.* **2024**, *8*, 159. [CrossRef]
18. Yang, C.; Huang, R.; Liu, D.; Qiu, W.; Zhang, R.; Tang, Y. Analysis and Warning Prediction of Tunnel Deformation Based on Multifractal Theory. *Fractal Fract.* **2024**, *8*, 108. [CrossRef]
19. Zhang, W.; Liu, D.; Tang, Y.; Qiu, W.; Zhang, R. Multifractal Characteristics of Smooth Blasting Overbreak in Extra-Long Hard Rock Tunnel. *Fractal Fract.* **2023**, *7*, 842. [CrossRef]
20. Zhang, W.; Feng, J.; Ren, J.; Ma, J.; Shi, J.; Zhang, J. The Formative Factors of a Rock Burst Based on Energy Calculations and the Experimental Verification of Butterfly-Shaped Plastic Zones. *Fractal Fract.* **2023**, *7*, 829. [CrossRef]
21. Li, X.; Zhu, H.; Yuan, Q. Dilatancy Equation Based on the Property-Dependent Plastic Potential Theory for Geomaterials. *Fractal Fract.* **2023**, *7*, 824. [CrossRef]

fractal and fractional

MDPI

Article

Particle Flow Simulation of the Mechanical Properties and Fracture Behavior of Multi-Mineral Rock Models with Different Fractal Dimensions

Run Shi [1] and Huaiguang Xiao [2,*]

1 Department of Civil Engineering, Monash University, Clayton, VIC 3800, Australia; run.shi@monash.edu
2 School of Civil Engineering, Southeast University, Nanjing 211189, China
* Correspondence: huaiguangxiao@seu.edu.cn

Abstract: To study the effects of rock models with different fractal dimensions on their mechanical properties and fracture behavior, three representative numerical rock models, including the digital texture model, the Voronoi polygon model, and the Weibull distribution model, are established in this paper. These models are used to simulate the structure of multi-mineral rocks and to investigate the influence of fractal dimensions on the mechanical properties and fracture behavior of rocks. Uniaxial compression numerical tests are carried out on 2D and 3D intact rocks under different fractal dimensions using the particle flow simulation method. The relationship between fractal dimensions and uniaxial compression strength and fracture behavior was analyzed. The results show that the fractal dimension of the Weibull distribution model is the largest, followed by the digital texture model, and the fractal dimension of the Voronoi polygon model is the smallest. With the increase in fractal dimension, the uniaxial compressive strength of intact rocks increases significantly, and their relationship is approximately linear. The influence of fractal dimension on rock strength shows a similar trend in both the 2D and 3D models. This study provides a new perspective for the application of fractal dimensions in multi-mineral rock models.

Keywords: fractal dimension; rock; failure mode; grain-based model; PFC

1. Introduction

Mechanical and failure characteristics are not only important physical properties of rocks, but also play a key role in engineering applications such as rock breakage [1], slope protection [2], and deep-ground energy extraction [3]. From a macroscopic point of view, the mechanical properties and failure modes of rocks are mainly influenced by external loads and the intrinsic properties of rocks. Regarding the influence of external loads on mechanics and failure modes, scholars have carried out a large number of physical experiments and numerical simulations under different confining pressure conditions [4–6] and have obtained beneficial results. The study of the intrinsic properties of rocks mainly focuses on the physical properties and geometric characteristics of rocks.

The physical properties of rocks include density, elasticity modulus, Poisson's ratio, tensile strength, compressive strength, shear strength, cohesion, internal friction angle, fracture toughness, and other parameters of intact rock specimens. With the advancement of computational technology, studies on the effect of rock mesoscopic geometric features on mechanics and failure behavior have gradually increased over the last two decades [7–11]. These geometric features mainly include the fracture structure and mineral components of rocks. For rocks with very low porosity, like granite, the geometric features are mainly controlled by the properties of mineral grains, which include the percentage of the components, the grain size [12,13], shape [14], orientation [14], and heterogeneity [15,16].

To investigate the influence of structural complexity on the deformation and failure behavior of rock materials, many researchers have adopted numerical simulation methods to analyze the influence of mineral grains. Currently, there are three typical numerical models for describing rock mineral grains: digital image (texture) mapping models, Voronoi polygon models, and Weibull distribution models. The digital image mapping model can reproduce complex particle characteristics and interlocking relationships of rocks by directly mapping the texture pixels of digital rock images to the particles in the numerical model [17]. The Voronoi polygon model can generate rock models composed of regular geometric lines by controlling mineral geometric parameters such as grain size [12] and shape [14]. This kind of model has advantages in computational efficiency, but its regular polygonal boundary makes it difficult to accurately describe the natural boundary characteristics of rock mineral grains. The Weibull distribution model adopts a random method to generate rock models with high heterogeneity and randomness [18]. Although the reconstruction method of Weibull models is simple, they cannot describe the geometric characteristics of rock mineral grains. These three models play a critical role in analyzing rock mechanics and fracture behavior, but a quantitative method to assess the influence of different numerical models on simulation results is still lacking.

Digital image mapping models, Voronoi polygon models, and Weibull distribution models have different spatial complexities. In order to quantitatively investigate the influence of different numerical models on the numerical results, this paper presented the fractal dimension to describe the three models quantitatively and studied the rock mechanics and fracture behavior with different fractal dimensions.

2. Materials and Methods

2.1. Rock Samples

In this study, granite was selected as the rock material due to its distinct texture where each mineral grain is large, densely packed, and displayed in specific colors, without significant pore or fracture features. The cylindrical sample (as shown in Figure 1a) has a diameter of 5 cm and a height of 10 cm, and the digital rock sample (as shown in Figure 1b) has a length of 10 cm and a width of 5 cm. The pixel resolution of this rectangular model is 500×250 pixels, with each pixel representing a region of 0.2 mm $\times$ 0.2 mm. To calibrate the numerical simulation results, the uniaxial compression physical experiment described in this paper was conducted using a cylindrical rock sample of Φ5 cm $\times$ 10 cm, while the numerical simulation experiment was performed using the 5 cm $\times$ 10 cm rectangular digital rock model. Both samples were obtained from the same parent rock, sharing identical geological conditions, similar textural structures, and material properties, resulting in nearly identical mechanical and deformation characteristics. The granite mainly consists of four minerals: K-feldspar, plagioclase, quartz and biotite, as shown in Figure 1c.

2.2. Numerical Modeling Methods of PFC with Different Fractal Dimensions

The internal structure of rocks is complex and variable, with remarkable fractal features. The different mineral components and their spatial distribution affect the mechanical behavior and fractal dimensions of rocks. Different numerical modeling methods, such as those based on digital rock models, Voronoi polygons, and Weibull distributions, are able to generate granular structures with different fractal dimensions, demonstrating the complex distribution characteristics of rock mineral components. In this section, three particle flow modeling methods that can express the characteristics of different particle fractal dimensions are presented (as shown in Figures 2 and 3).

(a) (b) (c)

Figure 1. Rock samples: (**a**) cylinder sample; (**b**) digital rock model; (**c**) mineral compositions of rock samples.

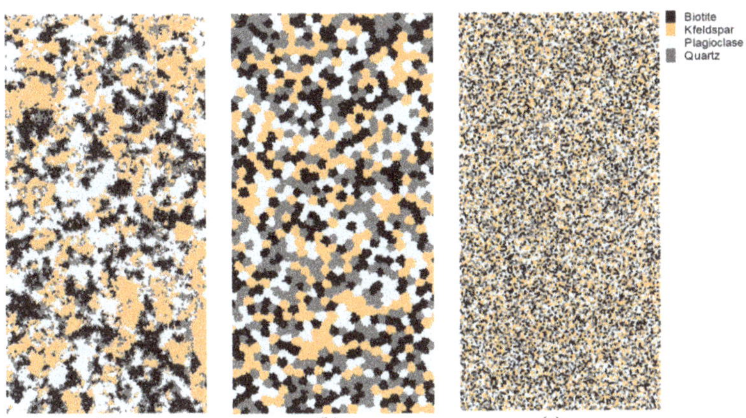

(a) (b) (c)

Figure 2. 2D numerical model of PFC for multi-mineral rocks: (**a**) texture model, (**b**) Voronoi model, (**c**) Weibull distribution model.

(a) (b) (c)

Figure 3. 3D numerical model of PFC for multi-mineral rocks: (**a**) texture model, (**b**) Voronoi model, (**c**) Weibull distribution model.

2.2.1. Construction of Multi-Mineral Particle Flow Numerical Models with Rock Texture

Based on the spatial location mapping method, the steps of constructing a multi-mineral particle flow numerical model from a digital rock model are as follows: (1) Segment the digital rock image (as shown in Figure 1b) to extract the mineral component types and their corresponding spatial coordinates. (2) Create an initial numerical model of particle flow with the same sizes as the digital rock model, and determine the spatial location of each particle in the numerical model. (3) Calculate the coordinates of the nearest pixel in the digital model corresponding to each particle in the numerical model. (4) Map the mineral attributes of the corresponding pixel to the particle in the particle flow model. (5) Traverse through all the balls in the numerical model, classify the components according to the attributes assigned to them. (6) Generate the multi-mineral particle flow numerical model with the rock texture characteristics (as shown in Figure 2a).

The key to constructing a three-dimensional multi-mineral particle flow numerical model with rock texture characteristics is to obtain a three-dimensional multi-component digital rock model. Since the densities of the mineral components of the rock are very close to each other, it is difficult for the traditional X-ray method to effectively distinguish the mineral components, as well as 3D reconstruction. To solve this problem, Xiao et al. proposed a solution based on deep learning and solid texture synthesis technology [19], which is able to generate a 3D multi-component digital rock model from a single 2D digital rock image. In this paper, this method was adopted to successfully construct a 3D multi-mineral digital rock model. Then, a 3D numerical model of multi-mineral components with rock texture features was generated based on the spatial coordinate mapping method (similar to the above-mentioned 2D spatial mapping), as shown in Figure 3a.

2.2.2. Construction of Multi-Mineral Particle Flow Numerical Models with Voronoi Polygons

The steps for constructing a multi-mineral particle flow numerical model with Voronoi polygons are as follows: (1) Define the geometric dimensions of the numerical model, and determine the types of mineral components and their volume percentages in the model to ensure that the percentage of each mineral matches the distribution in the natural rock. (2) Distribute seed points of Voronoi polygonsrandomly or according to a distribution rule throughout the whole model space, based on the number and volume percentage of the mineral components. Importantly, these seed points represent the central positions of each mineral particle. (3) Divide the whole model space into multiple non-overlapping polygons (for 2D models) or polyhedra (for 3D models) using Voronoi polygons. Each Voronoi cell represents the spatial position and shape of a mineral particle, and the boundaries between the cells simulate the contact and interaction between particles, reflecting the fracture distribution of mineral particles in rocks. (4) Assign each cell to the corresponding mineral component type according to the predefined mineral component ratio. Through the above steps, two-dimensional or three-dimensional multi-mineral particle flow numerical models based on Voronoi polygons can be generated, as shown in Figures 2b and 3b.

2.2.3. Construction of Multi-Mineral Particle Flow Numerical Models with Weibull Distribution

The steps for constructing a multi-mineral particle flow numerical model conforming to the Weibull distribution are as follows: (1) Generate a two-dimensional image matrix that serves as the total area of the model, with all the pixel points initially set to zero. (2) Determine the types of mineral components in the model and their volume percentages in the model, and calculate the number of pixels required for each component based on these percentages. (3) Generate random numbers that conform to the Weibull distribution using function for each mineral component. These random numbers represent the properties of particles, such as intensity and size. The Weibull distribution is controlled by the shape parameter (k) and the scale parameter (λ), where the smaller the shape parameter k is, the greater the difference in the properties of the particles, and the larger the value of k is, the more uniform the particles tend to be. The scale parameter λ determines the average size of the particles. The larger the value of λ, the larger the average size of the generated particles. (4) Randomly order the pixel indexes in the image matrix using a function to ensure the random distribution of particles in the image. According to the proportion of each mineral component, randomly selected pixel points are filled with the corresponding Weibull distribution values, so that the distribution of each group of particles in the image is random without overlaps. (5) Save the generated image matrix as a 2D color image, ensuring a clear visual difference in different mineral components. (6) Based on the spatial position mapping method described in Section 2.2.1, a two-dimensional or three-dimensional Weibull distribution numerical model can be further generated, as shown in Figures 2c and 3c.

2.3. Contact Model of PFC-GBM

In PFC, a linearpbond is the most commonly used to generate a bond particle model (BPM). Therefore, in this paper, the linearpbond is also chosen as a bond contact model for balls, including both intra- and inter-component contacts of mineral components. A linearpbond can be imagined as a set of elastic springs uniformly distributed in a rectangular cross-section on the contact plane, and centered on the point of contact between the two particles, as shown in Figure 4a. The bond strength of the linearpbond model depends mainly on the tensile strength, the cohesive force, and the angle of internal friction. When the bond breaks, the contact will be transformed from peak strength to residual strength, and the residual strength mainly depends on the friction coefficient and particle size, as shown in Figure 4b.

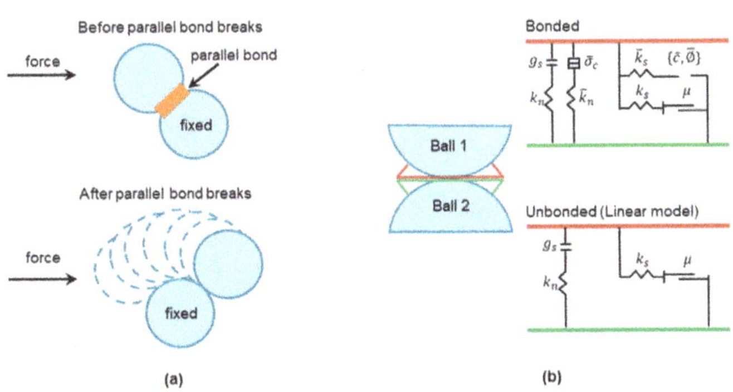

Figure 4. Bonding model between balls in PFC (adapted from [20,21]): (**a**) particle movements after breakage of parallel bond (**b**) linear parallel bond model.

2.4. The Fractal Dimension of Complex Mineral Particles Calculated by the Box-Counting Method

The box-counting method is one of the commonly used techniques for calculating the fractal dimension of complex geometries and has advantages in characterizing the complexity of 2D and 3D particles. First, the structure of the rock is meshed layer by layer. In 2D, the plane is divided into a grid of size $\varepsilon \times \varepsilon$; in 3D, the whole volume is divided into a cubic grid of $\varepsilon \times \varepsilon \times \varepsilon$. Then, at each level of meshing, the number of grids $N(\varepsilon)$ that are filled or occupied in the rock structure is counted. For 2D structures, the grids occupied by particles are counted; for 3D structures, the cubic grids occupied by mineral particles are counted. Third, the grid size ε is reduced to a finer scale, and the above statistical process is repeated to continue calculating N(ε) at different scales. As the grid shrinks, it typically occupies more box counts, indicating that the structure shows more detail and complexity at finer scales. Fourth, the fractal dimension D of the rock structure is calculated by linearly fitting logN(ε) and log (1/ε). The calculation formula is as follows:

$$D = \lim_{\varepsilon \to 0} \frac{\log N(\varepsilon)}{\log(1/\varepsilon)} \tag{1}$$

where N (ε) is the number of grids occupied when the grid side length is ε; ε is the side length of the grid.

3. Calibration of Numerical Results

Based on the PFC-GBM modeling method in Section 2.3, the heterogeneous material structure and contacts of the numerical models were constructed. The micro-parameters under uniaxial compression test conditions are summarized in Table 1 and applied to the failure numerical simulation. Young's modulus, cohesion strength, tensile strength, and friction angle of parallel bonds are assigned different values across different mineral components. The calibrated micro-parameters of different minerals are consistent with the mineral properties proposed by Bass [22]. For instance, quartz has the highest strength, modulus, and cohesion in both Bass's study and Table 1. In addition, the contact parameters between mineral components are lower than those within each mineral because interfaces between minerals are more prone to instability failure than the intact mineral particles.

Table 1. Micro-parameters of grain-based model.

Micro-Parameters	K-Feldspar	Plagioclase	Quartz	Biotite
Mineral grains				
Volume composite (%)	32.86	27.42	19.65	20.07
Particles				
Minimum particle radius of 2D (mm)	0.2	0.2	0.2	0.2
Minimum particle radius of 3D (mm)	0.6	0.6	0.6	0.6
Ratio of maximum to minimum particle radius	1.5	1.5	1.5	1.5
Young Modulus (GPa)	16.00	14.00	20.00	10.00
Stiffness ratio	1.50	1.50	1.50	1.50
Friction coefficient	1.20	1.20	1.20	1.20
Transgranular contacts				
Young Modulus (GPa)	16.00	14.00	20.00	10.00
Cohesion strength (MPa)	98.00	92.00	107.00	66.00
Tension strength (MPa)	98.00	92.00	107.00	66.00
Stiffness ratio	1.50	1.50	1.50	1.50
Friction angle (°)	42.00	40.00	48.00	31.00
Intergranular contacts between different minerals				
Young Modulus (GPa)	6.00			
Cohesion strength (MPa)	30.00			
Tension strength (MPa)	30.00			
Stiffness ratio	1.5			
Friction angle (°)	19.00			

Using the above-mentioned model and physical parameters, the stress–strain curves and fracture patterns of rocks under uniaxial compression conditions were obtained via numerical simulation, as shown in Figure 5. The experimental and numerical stress–strain curves show that the elastic moduli are 20.45 GPa and 20.34 GPa, respectively, with a deviation of 0.54%. The uniaxial compressive strengths are 130.18 MPa and 135.92 MPa, respectively, with a deviation of 4.41%. However, the stress–strain relations have a large difference at the initial stage, which is mainly due to micropores or microcracks in natural rocks, leading to the compression-density stage. In contrast, the rock in numerical simulation is assumed to be intact without micropores or microcracks, accounting for this difference.

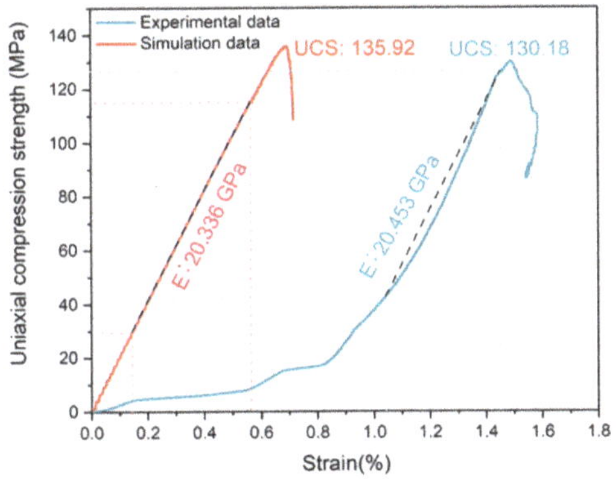

Figure 5. Calibration of deformation and failure for experiment and numerical results.

Regarding the final failure patterns, both physical experiments and numerical simulations are dominated by a single main crack extending from the upper left to thelower right, as shown in Figure 6. Moreover, the failure patterns of the two also exhibit high consistency in both failure mode and location, validating the reliability of the PFC-GBM model and its ability to simulate rock physical processes.

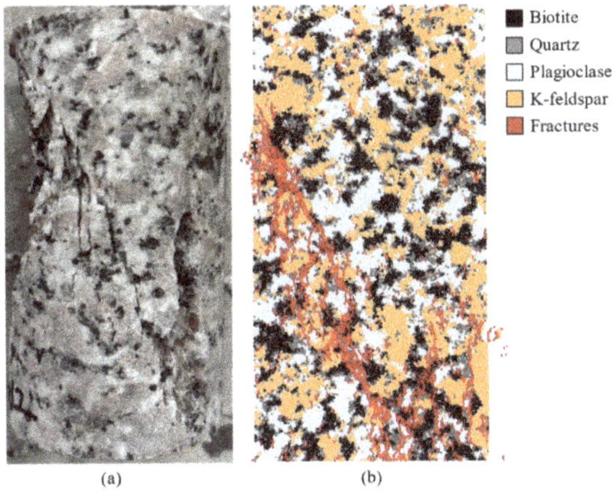

Figure 6. Calibration of failure modes: (**a**) experiment result, (**b**) numerical result.

4. Results and Discussion

4.1. Fractal Dimensions of Different Particle Models

Based on the three types of particle flow model generation methods described in Section 2.2, the multi-mineral component rock models with rock texture, Voronoi polygon, and Weibull distribution were constructed, as shown in Figure 7.

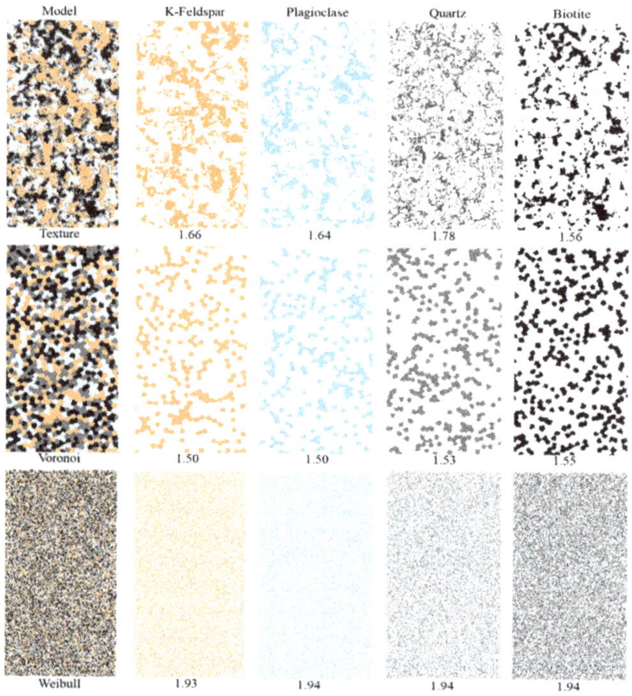

Figure 7. Fractal dimensions of minerals in the 2D texture model, Voronoi model, and Weibull model.

By comparing the texture model in Figure 7 with the digital rock model in Figure 1b, it is found that the texture-based multi-mineral component rock model can retain the mineral particle morphology of the original rock and its complex mosaic relationship very well, and almost completely reproduce the texture characteristics of the rock. The model not only captures the geometric features of the original rocks but also reflects the characteristics of their fractal dimensions. The fractal dimensions of K-feldspar, plagioclase, quartz, and biotite in the texture model are 1.66, 1.64, 1.78, and 1.56, respectively. Quartz has the highest fractal dimension, which indicates the most complex grain morphology, whereas biotite has the lowest fractal dimension, indicating the most regular grain morphology. K-feldspar and plagioclase have similar fractal dimensions, which are between quartz and mica, indicating that their grain complexity is intermediate.

In contrast, there is a significant difference between the Voronoi model in Figure 7 and the digital rock model in Figure 1b. The Voronoi model is constructed using Voronoi polygons with regular boundary shapes, which simplify the description of mineral grain morphology. The contact relationships between its particles are also simplified by the boundaries of the Voronoi polygons, so it cannot reflect the complex mosaic relationship between mineral grains of the original rock. The fractal dimensions of K-feldspar, plagioclase, quartz, and biotite in the Voronoi model were calculated to be 1.50, 1.50, 1.53, and

1.55, respectively, using the box-counting method in Section 2.4. The fractal dimensions for the four mineral components are very close to each other, with the difference between the maximum and minimum values of only 0.05. This similarity is attributed to the fact that all the mineral components are based on Voronoi polygons with relatively regular shapes.

The Weibull model in Figure 7 is quite different from the digital rock model in Figure 1b. The mineral percentages in the Weibull model are randomly distributed in two dimensions according to the proportions from the original rock, showing a high degree of randomness. Using the box-counting method described in Section 2.4, the fractal dimensions of K-feldspar, plagioclase, quartz, and biotite in the Weibull model are 1.93, 1.94, 1.94, and 1.94, respectively. The fractal dimensions of the four mineral components are almost equal and have large values, which shows the high complexity of their grain morphology. This result mainly stems from the random distribution of the mineral components in the model, resulting in their extremely complex morphology.

In summary, the mineral proportions of the three models are consistent with those of the original rocks, but the shapes of the mineral components show significant differences. In terms of the average fractal dimension of the three models, the Weibull model has the largest fractal dimension of 1.94, the Voronoi model has the smallest fractal dimension of 1.52, and the texture model has a fractal dimension between the two models at 1.66. In addition, the fractal dimensions of the different minerals in the texture model have a large difference, while the fractal dimensions of the minerals in the Weibull and the Voronoi models are relatively close. Therefore, the models generated based on the parameterization methods, such as the Weibull distribution and Voronoi polygon, have some limitations in reflecting the intergranular complexity compared to the texture models, as well as in describing the complexity of the mineral shapes.

For 3D models, the fractal dimension is typically between 2 and 3, with larger values indicating higher complexity. Similarly to 2D models, the 3D Weibull model has the largest fractal dimension, followed by the texture model, and the Voronoi model has the smallest. Unlike the 2D models, the difference in fractal dimension between the mineral components is smaller in 3D models. This may be due to the high computational time cost of the 3D models, which have a lower pixel resolution than the 2D models. When the pixel resolution is low, the difference in the fractal dimension between the texture model and the Voronoi model is smaller in 3D than in 2D (as shown in Figure 8).

4.2. Mechanical Properties and Fracture Behavior of 2D Rock Models with Different Fractal Dimensions

Among the three models, the stress-strain curves of the texture model show moderate peak intensities, with peak stresses ranging from 134.80 MPa to 148.99 MPa for the five parallel samples with an error rate of 10.53%. The strains corresponding to the peak stresses range from 0.66% to 0.74% with an error rate of 12.12%. In the elastic deformation stage, the slopes of the curves are consistent across all samples, indicating that the elastic modulus in the texture model is relatively consistent. The stress-strain curves of the Voronoi model exhibit the highest peak strengths ranging from 161.89 MPa to 177.26 MPa with an error rate of 9.49%. The strains corresponding to the peak stresses ranged from 0.79% to 0.87% with an error rate of 10.13%. Similarly, the slopes of the curves in the elastic deformation stage remain consistent for all samples, indicating that the elastic modulus in the model also has a high degree of consistency. In contrast, the stress-strain curves of the Weibull model exhibit the lowest peak intensities ranging from 101.41 MPa to 116.95 MPa, with an error rate of 15.32%. The strains corresponding to the peak stresses range from 0.64% to 0.75% with an error rate of 17.19%. Nevertheless, the slopes of all the samples remain consistent during the elastic deformation stage, indicating that the elastic modulus of the model has a minimal variation among the samples.

All three models exhibit typical brittle failure characteristics with a rapid decrease in stress after the peak value, indicating that all three models rapidly lose their load-bearing capacity after reaching peak stress. The Voronoi model exhibits the highest peak strength and elastic modulus, with the textured model in the middle, while the Weibull model has the lowest strength and elastic modulus. This suggests that the regular geometry in the Voronoi model contributes to the overall load-bearing capacity of the model, while the randomly distributed mineral components in the Weibull model result in weaker mechanical properties (as shown in Figure 9).

Figure 8. Fractal dimensions of minerals in the 3D texture model, Voronoi model, and Weibull model.

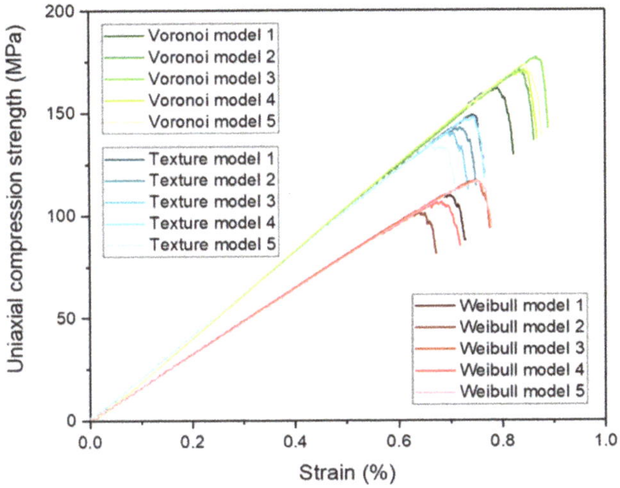

Figure 9. Stress–strain of 2D models with different fractal dimensions.

In the textural model (as shown in Figure 10), the fracture zones are highly correlated with the distribution of mineral components, and the fracture paths mainly propagate along the boundaries of mineral grains. The fracture shows obvious nonlinearity and is accompanied by relatively large number of fine branching cracks, indicating that the fracture paths in this model have a high degree of complexity. The fracture features in each parallel sample maintain a relatively consistent complexity, with the fracture paths propagating mainly along the grain boundaries, and the number of cracks being high and interconnected. Despite the similarity in the overall complexity of the fracture paths, the final fracture morphology in the different parallel samples is variable.

In the Voronoi model (as shown in Figure 10), the facture paths exhibit strong geometric constraints and mainly propagate along the boundaries of the Voronoi units. Since the Voronoi cells have regular shapes, the facture bands are mostly straight lines, and the overall fracture paths are relatively simple with fewer complex branching cracks. The regularity of the fracture paths remains relatively consistent across each parallel sample, and the cracks develop along the polygonal boundaries. However, the fracture distribution locations and final fracture morphology vary considerably among different samples.

In the Weibull model (as shown in Figure 10), the fracture paths present significant randomness. The fractures mainly propagate along the randomly distributed mineral components, forming multiple fracture zones that are separated from each other. The morphology of the fracture zones shows a high degree of randomness and irregularity. The randomness of the fracture regions in each parallel sample remain relatively consistent, and the fracture paths are more dispersed, with no obvious regularity in their direction and expansion mode. The fracture zones between different samples show a certain degree of randomness consistency, but the specific location and morphology of the fractures are significantly different.

All three models suggest that the distribution of mineral components plays a key role in determining where fracture occurs, and the fracture path tends to propagate along the boundaries of mineral grains. In terms of the complexity of the fracture paths, the texture model has the highest complexity, with more fracture branches and nonlinear paths; the fracture paths of the Voronoi model are relatively regular, mainly expanding along polygonal boundaries. The fracture paths of the Weibull model are the most random, with a highly irregular morphology and a variety of directional changes. This suggests that the texture model can better reproduce the complex rupture patterns within the rock, while the Voronoi and Weibull models show distinct characteristics in terms of regularity and randomness.

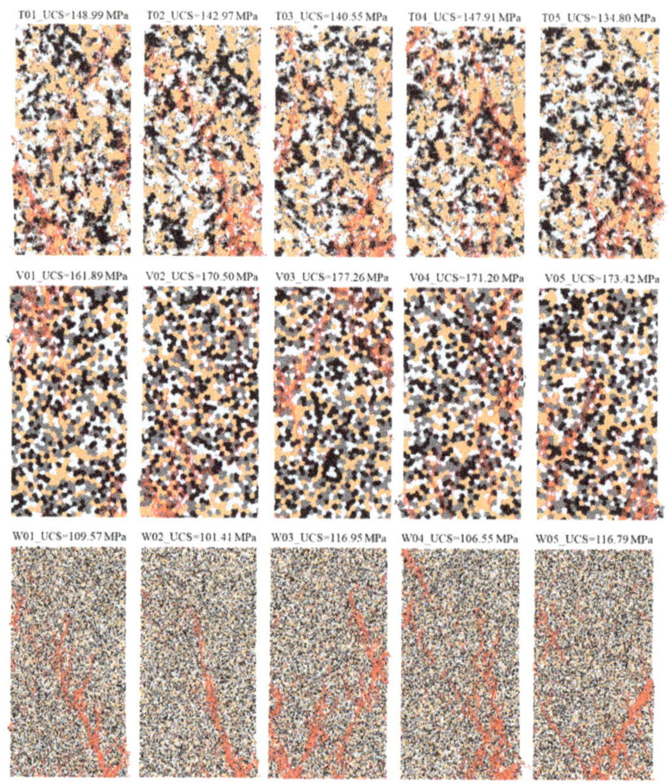

Figure 10. Failure modes of 2D models with different fractal dimensions.

The final crack numbers of the different texture models show some variability, as shown in Figure 11. The texture model has a relatively high number of cracks, ranging from 4031 to 5144 in the five parallel samples. Although the number of cracks varies between samples, the overall difference is small, and the crack number stays at a high level with fewer fluctuations. The number of cracks in the Voronoi model is relatively moderate, with the number of cracks in the five parallel samples concentrating ranging from 3806 to 4766. The number of cracks in the Voronoi model is overall more uniform compared to the texture model, with the number of cracks slightly lower than the texture model. The Weibull model shows significant fluctuations in the number of cracks, with the number of cracks in the five parallel samples ranging from 2200 to 4943. Compared to the other models, the crack number of the Weibull model varies the most between samples, reflecting the influence of the random distribution structure on the crack generation and expansion process. This large range of variation suggests that the Weibull model is highly influenced by random factors in the crack development process, resulting in a high degree of instability in the crack distribution.

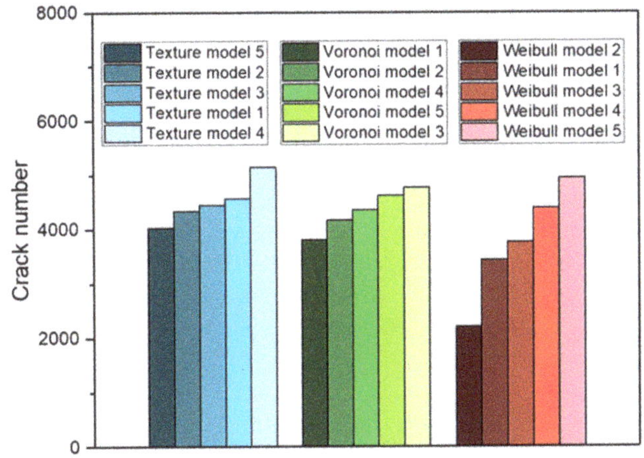

Figure 11. The crack number in 2D models with different fractal dimensions.

4.3. Mechanical Properties and Fracture Behavior of 3D Rock Models with Different Fractal Dimensions

The stress-strain curves of the three models reflect their different mechanical properties, as shown in Figure 12. The stress-strain curves of the textured model exhibit moderate peak stresses ranging from 199.56 MPa to 210.24 MPa with a maximum to minimum error of 5.35%. The strains of peak stresses for this model ranged from 1.06% to 1.12% with an error of 5.66%. The stress-strain curve for the Voronoi model showed the peak stresses ranging from 223.93 MPa to 232.09 MPa with an error of 3.64%. The strain level of this model at the peak stresses is 1.04% to 1.09% with an error of 4.81%. The stress-strain curve of the Weibull model exhibits relatively low peak stresses ranging from 179.60 MPa to 182.30 MPa with an error of only 1.50% showing a relatively small fluctuation. Its peak strain ranges from 1.13% to 1.17% with an error of 3.54%.

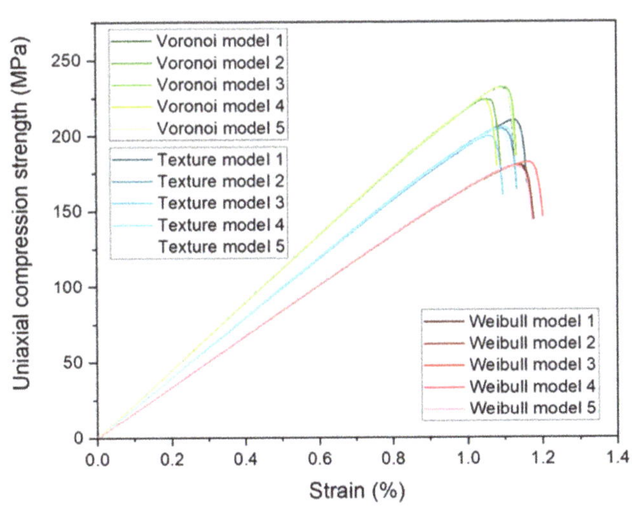

Figure 12. Stress–strain of 3D models with different fractal dimensions.

At the early stage of loading, the stress-strain curves of all three models show linear characteristics, indicating that each model mainly exhibits elastic deformation in the initial stage. With the increase in loading, all three models show significant brittle damage characteristics after reaching the peak strength, which is manifested by the rapid decrease of stress after the peak stress, showing the sudden failure characteristics of the material after losing its load-bearing capacity. However, the three models also show significant differences in strength and deformation characteristics. The Voronoi model shows the highest peak strength and elasticity modulus, the strongest compressive capacity, and the smallest deformation capacity, which is attributed to the limiting effect of its regular geometric structure on crack development. The textured model has medium strength and deformation capacity, with moderate strength consistent with the complex mosaic structure between particles. In contrast, the Weibull model has the lowest strength and elasticity modulus, which reflects the adverse effect of the random distribution of its mineral components on the mechanical properties, resulting in a weaker load-bearing capacity.

For the 3D texture models as shown in Figure 13, the uniaxial compression strengths are between 199.56 MPa and 210.25 MPa, and the number of cracks range from 69,310 to 85,273 with a minor difference. The stress concentration area inside the sample gradually propagates into a fracture zone through the sample, which is closely related to the geometric distribution of the mineral grains and the contact models, showing obvious non-uniformity. Although all samples present distinct main fracture zones, the location and morphology of fracture expansion vary among different samples.

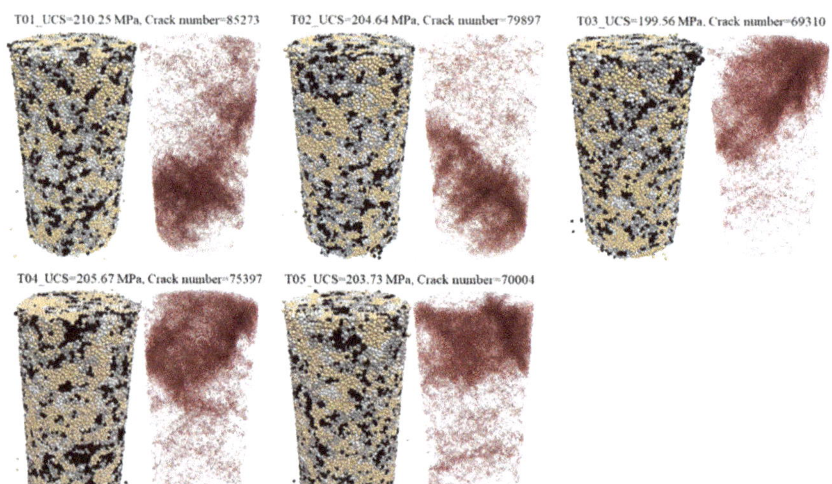

Figure 13. Failure modes of 3D textured rock models.

For the 3D Voronoi models as shown in Figure 14, the average values of uniaxial compression strength and number of cracks are higher by 30.28 MPa and lower by 14,802 compared to the 3D textured models. The differences in uniaxial compression strength and number of cracks among the five Voronoi parallel samples are also small. The distribution of cracks and the final fracture morphology of the five Voronoi parallel models are similar to those of the textured model and will not be repeated here.

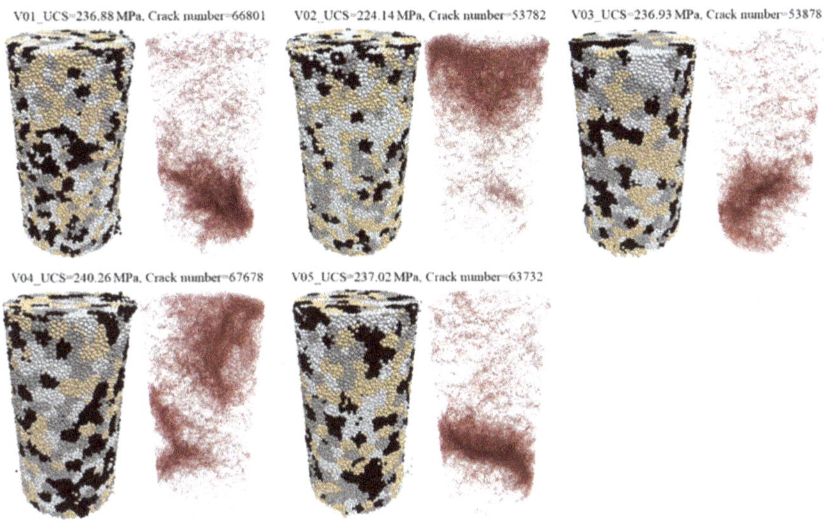

Figure 14. Failure modes of 3D Voronoi rock models.

For the 3D Weibull models as presented in Figure 15, the average uniaxial compressive strength and number of cracks are 24.29 MPa lower and 13,992 higher compared to the 3D textured models. The fracture morphology is similar to the texture model fractures. By comparing the uniaxial compression strength and number of cracks of the three 3D models, it was found that greater uniaxial compression strength corresponds to fewer cracks.

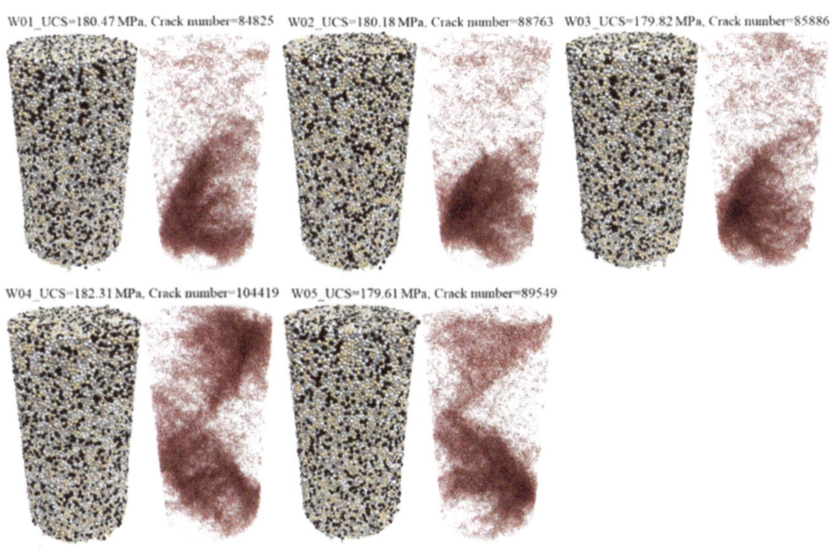

Figure 15. Failure modes of 3D Weibull rock models.

4.4. Uniaxial Compression Strength Relationship Between 2D and 3D Models with Different Fractal Dimensions

To analyze the effect of the 2D and 3D models on the rock strength, the functional relationship between the fractal dimension and uniaxial compression strength is fitted for the 2D and 3D models, respectively, as shown in Figure 16. For all three models with different fractal dimensions, the uniaxial compression strengths of the 3D models are significantly higher than those of 2D models. In terms of uniaxial compression strength, the 3D texture model is higher than the 2D texture model by 61.73 MPa, the 3D Voronoi model is higher than the 2D Voronoi model by 64.19 MPa, and the 3D Weibull model is higher than the 2D Weibull model by 70.22 MPa.

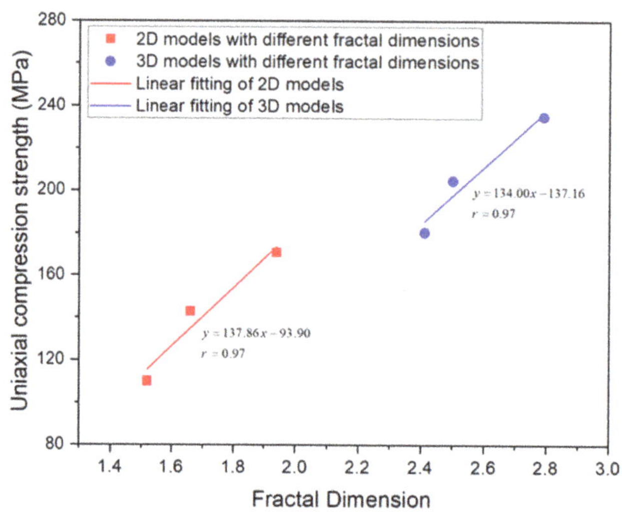

Figure 16. Relationship between fractal dimensions and uniaxial compression strength.

The fractal dimension of the 2D models has a strong linear relationship with the uniaxial compression strength with a linear correlation coefficient of 0.97. Similarly, the linear correlation coefficient in 3D is also 0.97. It is worth noting that the slopes of the 2D and 3D fitted lines are 137.86 and 134.00, respectively, which are very close. These similarities suggest that the effect of fractal dimension on rock strength follows a similar trend in two and three dimensions.

5. Conclusions

Through the uniaxial compression tests of the texture model, the Voronoi model, and the Weibull model, fractal dimensions play an important role in describing the geometrical characteristics, mechanical properties, and fracture behavior of multi-mineral rocks. Based on the above study, the conclusions are as follows:

1. The fractal dimension of the Weibull distribution model is the largest, reflecting the most complex geometric structure; the digital texture model has a moderate fractal dimension, indicating moderate geometric complexity; and the Voronoi polygon model has the smallest fractal dimension, with the regular grain structure.
2. As the fractal dimension increases, the uniaxial compression strength of intact rocks significantly increases, exhibiting an approximately linear relationship. In addition, the fracture propagation paths vary significantly among models with different fractal dimensions.
3. The uniaxial compression strength of 3D models is significantly higher than that of 2D models. Additionally, in the linear fit between fractal dimension and uniaxial compression strength, the slope of the fitted line is similar for both 2D and 3D models,

indicating that the influence of fractal dimension on rock strength follows a similar trend in both conditions.

Author Contributions: Conceptualization, H.X.; methodology, H.X. and R.S.; software, H.X. and R.S.; validation, H.X.; formal analysis, H.X. and R.S.; data curation, R.S.; writing—original draft preparation, H.X. and R.S.; writing—review and editing, R.S.; visualization, R.S.; funding acquisition, H.X. All authors have read and agreed to the published version of the manuscript.

Funding: This work was financially supported by the Natural Science Foundation of Jiangsu Province (BK20230857) and the China Postdoctoral Science Foundation (2023M740610).

Data Availability Statement: The original contributions presented in this study are included in the article, and further inquiries can be directed to the corresponding author.

Acknowledgments: We sincerely thank the editors and reviewers for their detailed comments and efforts toward improving our study.

Conflicts of Interest: The authors declare no conflicts of interest.

References

1. Guo, S.; Zhang, Q.; He, M.; Wang, J.; Liu, J.; Ming, C.; Guo, L.; Fan, L. Numerical investigation on rock fracture induced by a new directional rock-breaking technology. *Eng. Fract. Mech.* **2022**, *268*, 108473. [CrossRef]
2. Brideau, M.-A.; Yan, M.; Stead, D. The role of tectonic damage and brittle rock fracture in the development of large rock slope failures. *Geomorphology* **2009**, *103*, 30–49. [CrossRef]
3. Zhang, L.; He, J.; Wang, H.; Cen, X. Study on heat extraction characteristics in a rock fracture for the application of enhanced geothermal systems. *Geothermics* **2022**, *106*, 102563. [CrossRef]
4. Jiang, C.S.G. Failure characteristics of surrounding rocks along the radial direction of underground excavations: An experimental study. *Eng. Geol.* **2021**, *281*, 106012. [CrossRef]
5. Xie, H.; Lu, J.; Li, C.; Li, M.; Gao, M. Experimental study on the mechanical and failure behaviors of deep rock subjected to true triaxial stress: A review. *J. Min. Sci. Technol.* **2022**, *5*, 915–950. [CrossRef]
6. Guo, S.; Qi, S.; Zhan, Z.; Zhan, Z.; Ma, L.; Gure, E.G.; Zhang, S. Numerical study on the progressive failure of heterogeneous geomaterials under varied confining stresses. *Eng. Geol.* **2020**, *269*, 105556. [CrossRef]
7. Yu, Q.; Zhu, W.; Ranjith, P.G.; Shao, S. Numerical simulation and interpretation of the grain size effect on rock strength. *Geomech. Geophys. Geo-Energy Geo-Resour.* **2018**, *4*, 157–173. [CrossRef]
8. Gao, F.; Stead, D.; Elmo, D. Numerical simulation of microstructure of brittle rock using a grain-breakable distinct element grain-based model. *Comput. Geotech.* **2016**, *78*, 203–217. [CrossRef]
9. Liu, H.; Roquete, M.; Kou, S.; Lindqvist, P. Characterization of rock heterogeneity and numerical verification. *Eng. Geol.* **2004**, *72*, 89–119. [CrossRef]
10. Kong, L.; Shang, J.; Ranjith, P.G. Grain-based DEM modelling of mechanical and coupled hydro-mechanical behaviour of crystalline rocks. *Eng. Geol.* **2024**, *339*, 107649. [CrossRef]
11. De Silva, V.R.S.; Konietzky, H.; Märten, H.; Ranjith, P.G.; Lei, Z.; Xu, T. Grain-scale numerical simulation of crystalline rock fracturing using Soundless Cracking Demolition Agents for in-situ preconditioning. *Comput. Geotech.* **2023**, *155*, 105187. [CrossRef]
12. Peng, J.; Wong, L.N.Y.; Teh, C.I. Influence of grain size on strength of polymineralic crystalline rock: New insights from DEM grain-based modeling. *J. Rock Mech. Geotech. Eng.* **2021**, *13*, 755–766. [CrossRef]
13. Li, X.; Li, H.B.; Zhao, J. Transgranular fracturing of crystalline rocks and its influence on rock strengths: Insights from a grain-scale continuum–discontinuum approach. *Comput. Methods Appl. Mech. Eng.* **2021**, *373*, 113462. [CrossRef]
14. Zhao, X.; Zhao, X.; Elsworth, D.; Hu, W.; Wang, T. A grain texture model to investigate effects of grain shape and orientation on macro-mechanical behavior of crystalline rock. *Int. J. Rock Mech. Min. Sci.* **2021**, *148*, 104971. [CrossRef]
15. Cheng, P.; Li, X.; He, L.; Li, J. Study on the effect of micro-geometric heterogeneity on mechanical properties of brittle rock using a grain-based discrete element method coupling with the cohesive zone model. *Int. J. Rock Mech. Min. Sci.* **2021**, *140*, 104680. [CrossRef]
16. Liu, G.; Cai, M.; Huang, M. Mechanical properties of brittle rock governed by micro-geometric heterogeneity. *Comput. Geotech.* **2018**, *104*, 358–372. [CrossRef]

17. Tan, X.; Konietzky, H.; Chen, W. Numerical Simulation of Heterogeneous Rock Using Discrete Element Model Based on Digital Image Processing. *Rock Mech. Rock Eng.* **2016**, *49*, 4957–4964. [CrossRef]
18. Xu, Y.; Yao, W.; Xia, K. Numerical study on tensile failures of heterogeneous rocks. *J. Rock Mech. Geotech. Eng.* **2020**, *12*, 54–62. [CrossRef]
19. Xiao, H.; He, L.; Zheng, Y.; Yan, S. 3D solid digital and numerical modeling of multimineral heterogeneous rocks based on deep learning. *Geomech. Geophys. Geo-Energy Geo-Resour.* **2022**, *8*, 188. [CrossRef]
20. Bahrani, N.; Purvance, M.; Emam, S. A comparison between 2D and 3D bonded-particle models for rock. In Proceedings of the 47th US Rock Mechanics/Geomechanics Symposium, American Rock Mechanics Association (ARMA), San Francisco, CA, USA, 23–26 June 2013.
21. Itasca Consulting Group Inc. *Particle Flow Code*; Version 5.0; Itasca Consulting Group, Inc.: Minneapolis, MN, USA, 2016.
22. Bass, J.D. Elasticity of minerals, glasses, and melts. In *Mineral Physics & Crystallography: A Handbook of Physical Constants*; Ahrens, T.J., Ed.; AGU: Washington, DC, USA, 1995; pp. 45–63. [CrossRef]

fractal and fractional

MDPI

Article

Wavefield Evolution and Arrival Behavior of Elastic Wave Propagation in Two-Dimensional Fractional Brownian Fields

Shuaifeng Wang [1,2] and Zixin Zhang [1,*]

[1] Department of Geotechnical Engineering, College of Civil Engineering, Tongji University, Shanghai 200092, China; sfwang@tongji.edu.cn
[2] School of Mechanics and Engineering Science, Shanghai University, Shanghai 200444, China
* Correspondence: zxzhang@tongji.edu.cn

Abstract: The fractional Brownian field is often used to reproduce the fractal properties of complex heterogeneous media, which closely represent real-world geological materials. Studying elastic wave transport in this type of heterogeneous media is essential for advancing knowledge in geophysics, seismology, and rock mechanics. In this paper, we numerically investigate the wavefield evolution and arrival behavior of elastic wave propagation in a two-dimensional fractional Brownian field characterized by the standard deviation (σ) and the Hurst exponent (H). Using a high-fidelity finite element model, we quantify the influence of these parameters on wavefront morphology, wave arrival synchronization, and energy decay. Our results reveal that increased matrix heterogeneity with higher σ and lower H values leads to pronounced wavefront roughness, asynchronous arrival phenomena, and increscent energy decay, attributed to enhanced scattering and modulus variability. For smaller H values, rougher modulus distributions scatter wave energy more intensely, producing more coda waves and distorted wavefronts, while smoother fields with larger H fields promote smoother wave propagation. Higher σ amplifies these effects by increasing modulus variability, resulting in more attenuated wave energy and substantial wavefield disturbance. This study contributes to a quantitative understanding of how fractal heterogeneity modulates wave transport and energy attenuation in random media. Our findings hold practical significance for geophysical exploration and seismic tomography, as well as aiding in subsurface imaging and structural evaluation within fractured or stratified rock formations.

Keywords: elastic wave; fractional Brownian field; heterogeneous media; wavefield; arrival behavior

1. Introduction

In geological formations, mechanical properties such as Young's modulus, density, and porosity often display spatial variability [1–3] that follows fractal distributions [4,5], impacting the velocity, attenuation, and scattering of elastic waves [6–8]. The propagation of elastic waves in such heterogeneous media is essential in various geophysical applications, including seismic exploration, subsurface imaging, and hazard assessment [9]. The fractional Brownian field (fBm) is commonly used to model such heterogeneous media due to its ability to represent fractal characteristics observed in natural systems [10,11]. Mandelbrot and Van Ness [10] introduced the concept of fractional Brownian motion, establishing a foundation for modeling self-affine and self-similar structures observed in geological formations. Subsequent studies expanded to identify fractal-based heterogeneity in various geological contexts [4,12]. It has been proven that such fractal characteristics significantly influence wave behavior by inducing scattering, coda waves, and arrival delays [6,13–17]. These findings underscore the importance of fBm-based models for simulating real-world heterogeneity in geological media and understanding how fractal properties affect wave transport.

The investigation on wave propagation through random and fractal media has evolved significantly over the decades. Varadan et al. [6] explored wave scattering and attenuation in random elastic media, establishing theoretical insights into how fractal heterogeneity influences wave transport. Shapiro and Kneib [13] developed numerical models to investigate seismic wave attenuation due to scattering effects in random media. Frankel and Clayton [18] analyzed wave propagation in random velocity models to characterize scattering and energy attenuation, providing insights into high-frequency seismic wave behavior. Aki and Richards [14] quantified the scattering attenuation of wave transport by introducing the quality factor Q, denoted as the ratio of attenuated energies to incident ones. Sahimi and Tajer [4] demonstrated that the elastic moduli and density in rocks are distributed in a self-affine fractal pattern, identifying these properties' influence on seismic wave velocities. This work underscored the importance of fractional Brownian fields in capturing the spatial variability of natural rock properties and highlighted the need for a detailed analysis of wave transport in fractal environments. Allaei and Sahimi [15] analyzed the shape of wavefronts in heterogeneous media, applying theoretical and numerical approaches to quantify how randomness affects wave coherence and transport patterns. Przybilla et al. [19] developed a Monte Carlo scheme to simulate wave transport in 2D random elastic media with spatially distributed scattering coefficients based on the radiative transfer theory. Sahimi et al. [20] emphasized that the distribution patterns of elastic moduli (correlated versus uncorrelated) play a crucial role in the elastic wave transport in heterogeneous solids, leading to the transition between localization and propagation. Garnier and Sølna [21] underlined the relationship between the damping exponent of a traveling wave and the Hurst parameter of random media, which characterizes the correlation properties of the random field. Hamzehpour et al. [22] simulated wave propagation in 2D heterogeneous fractured media where cracks were randomly distributed and the bulk modulus of the matrix was distributed according to fractional Brownian motion. Zhang et al. [23] employed sensitive kernels to represent the impact of the spatial characteristics of media properties on the wavefield, which is beneficial for fetching new tomographic inversion approaches to map small-scale heterogeneity using scattered waves. Wang et al. [17] numerically investigated the competing roles of matrix modulus heterogeneity and fracture networks in wave propagation and quantified the dependence of wave arrival times on the heterogeneity index and fracture stiffness.

Despite these advancements, several unresolved issues and research gaps remain in the study of wave propagation in fractional Brownian fields. It is essential to build more complex models that can capture fractal heterogeneity and anisotropic behavior in geological materials because such material properties impact wave transport characteristics, such as the wavefront profile, arrival behavior, and energy attenuation [1,4,24,25]. Moreover, appropriate indices representing the heterogeneity of fractional Brownian fields should also be included in the model to quantitatively assess their influence on wave transport. This study aims to address these gaps by exploring elastic wave propagation in a 2D fractional Brownian field, which is characterized by Hurst exponent H and standard deviation σ. We systematically investigate the quantitative dependencies of wave propagation behaviors on these fractal heterogeneity indices.

The organization of the remainder of this paper is as follows. In Section 2, we elaborate on the numerical methodology and model setup, detailing how the fractional Brownian field is constructed and the finite element approach is applied. Section 3 presents the results, highlighting how variations in heterogeneity influence wavefield evolution, wavefront roughness, arrival times, and energy decay. Section 4 discusses the implications of these findings for wave transport in fractal media. Finally, Section 5 concludes with a summary of key findings and suggests directions for future research in wave propagation modeling in fractal-based random media.

2. Numerical Methodology

2.1. Model Setup

A squared study domain of size L is built to simulate a plane incident P wave propagating through the heterogeneous media (Figure 1a). The fractional Brownian field (see

Section 2.3) is placed in this $L \times L$ study domain and the elastic wave (see Section 2.2) is excited at the left boundary (red line in Figure 1a) using a five-cycle Hann windowed tone burst with a force-type signal (Figure 1b) of wavelength $\lambda = L/10$. In the study domain, we record the x-direction displacement, aligned with the propagation direction of the plane incident wave, using an array of 301 × 301 arranged receivers. The receivers are spaced at a spatial interval of $\delta = L/300$. To mimic infinite wide-plane wave propagation, we use the following two schemes. First, we generate an auxiliary mirrored domain below the study domain with the same modulus distribution symmetric about the mirror boundary. Meanwhile, two periodic boundaries are separately applied at the bottom of the mirrored domain and the top of the study domain. The random fields in these two areas conjointly constitute a unit cell [26] that is capable of accurately reproducing an infinite plane wave transporting in the x direction and eliminating boundary effects at both the top and bottom of the study domain. Second, we place absorbing layers (see Section 2.2) to the left and right boundaries of the whole model to absorb unwanted reflected waves from these two boundaries. In the middle region (study and mirrored domains), we use a spatial mesh of Delaunay triangular elements with an average element size of $l_{ele} = \lambda/15$ (Figure 1c). For the absorbing layers, mapping elements (Figure 1d) with the same size are generated. To capture the temporal evolution of the wavefield, the time domain is discretized using a timestep $dt = l_{ele}/V_P \times CFL$, where V_P is the velocity of the incident P wave and CFL is the Courant–Friedrichs–Lewy (CFL) number [27] (taking 0.1 in our numerical model).

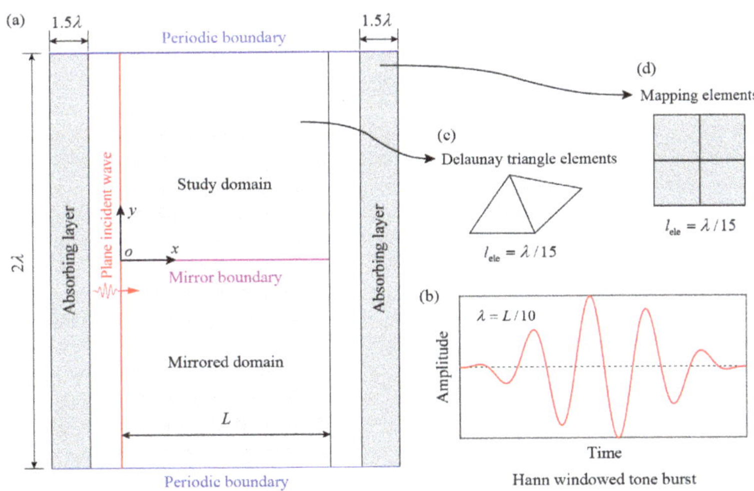

Figure 1. Geometrical model setup for simulating elastic wave transportation in solids: (**a**) overall configuration of numerical model; (**b**) Hann windowed tone burst of wavelength $\lambda = L/10$; (**c**) Delaunay triangular elements in study and mirrored domains; (**d**) mapping elements in absorbing layers.

2.2. *Governing Equations*

The elastic wave propagation in two-dimensional media is governed by the time-domain equation of dynamic equilibrium as

$$\mathbf{M}\frac{\partial^2 \mathbf{u}}{\partial t^2} + \mathbf{C}^* \frac{\partial \mathbf{u}}{\partial t} + \mathbf{K}^* \mathbf{u} = \mathbf{F} \quad (1)$$

where $\mathbf{M}$ is the mass and stiffness matrix, $\mathbf{u}$ stands for the displacement, t is the time, $\mathbf{F}$ represents the external force, and $\mathbf{C}^* = \eta\mathbf{M}$ and $\mathbf{K}^* = \varsigma\mathbf{K}$ are the damping and stiffness matrices, respectively. Herein, η and ς are coefficients representing the mass proportional damping and stiffness reduction, respectively. When waves transport in the study domain, no damping or stiffness reduction is introduced; thus, the damping is set to zero, and the stiffness matrix remains as $\mathbf{K}^* = \mathbf{K}$. To mimic wave transport in a boundless domain, we employ the stiffness reduction method [27] to mitigate a needless reflected wave at the boundaries of the finite-sized model by implementing absorbing layers at the side boundaries. The absorbing layer has a thickness of 1.5λ. In the absorbing layer, the mass proportional damping coefficient $\eta(\chi)$ is derived as $\omega(\chi/(1.5\lambda))^3$, where χ stands for the x-direction distance of a point away from the boundary of the study or mirrored domain; ω denotes the angular frequency of the incident wave. The stiffness reduction coefficient $\varsigma(\chi)$ is derived as $\exp(-\psi(\chi)k\chi)$, where k represents the incident wavenumber; $\psi(\chi) = \psi_{\max}(\chi/(1.5\lambda))^3$ denotes the attenuation factor with $\psi_{\max} = -\ln(\varepsilon/(1.5\lambda k))$ being the maximum attenuation factor. Here, ε is a constant with a small value, usually using 0.01. In this paper, we use the finite element method with the commercial software COMSOL version 5.5 [28] to solve the above elastic wave equations. We conducted a thorough comparison with analytical solutions to assess the validity and precision of our numerical model for simulating wavefield evolution in unbounded heterogeneous media [17,29].

2.3. Generation of Fractional Brownian Fields

Fractional Brownian motion (fBm) is one of the most useful mathematical models for representing the random fractals in nature [10,11,30], which is fundamentally important for understanding wave transport and anomalous diffusion on fractals [31]. Fractional Brownian motion $V_H(t)$ is a single value function of t, of which the increment $V_H(t_{i+1}) - V_H(t_i)$ obeys a Gaussian distribution with the following variance [31]:

$$\left\langle |V_H(t_{i+1}) - V_H(t_i)|^2 \right\rangle \propto |t_{i+1} - t_i|^{2H} \tag{2}$$

where the brackets < > represent the average over all the samples of $V_H(t)$, and H is the Hurst exponent with a value ranging from 0 to 1. Several methods [31,32] have been developed to generate fractional Brownian motion in different dimensions. In this paper, we use the midpoint displacement method [31] to generate 2D fields of fractional Brownian motion due to it being independent of the grid size. The detailed algorithm of generating 2D fractional Brownian fields is referred to in pages 100–101 by Barnsley et al. [31].

It has been found that material properties, e.g., Young's modulus, Poisson's ratio, and density, may be spatially variable in rock masses [4,33], and the distribution usually follows a fBm [4]. Here, we consider the spatial variation in Young's modulus E in our model (Figure 1a) as a 2D fractional Brownian field, having a mean value $\bar{E}$ and a standard deviation σ. We assume that the heterogeneous media are a typical limestone with a material properties list in Table 1. To explore the influence of field heterogeneity on the propagation of elastic waves, two indices, the standard deviation of Young's modulus σ and the Hurst exponent H, are included, taking the values listed in Table 1. The spatial variability of Young's modulus is presented in each numerical element (Figure 2). The probability density of Young's modulus in the random fields is also given in Figure 3. It can be seen that, when σ is smaller, the modulus is more evenly distributed in the field and the magnitudes are more concentrated at the mean value $\bar{E}$. With larger H values, the distribution of E tends to be smoother while the probability density becomes more dispersive with a more obvious divergence to the corresponding conventional Gaussian distribution.

Table 1. Material properties in the fractional Brownian fields.

Properties	Value	Unit
Mean value of Young's modulus, $\bar{E}$	30	GPa
Standard deviation of Young's modulus, σ	1, 2, 4, 8	GPa
Density, ρ	2700	kg/m^3
Poisson's ratio, ν	0.27	--
Hurst exponent, H	0.1, 0.3, 0.5, 0.7, 0.9	--

"--" indicates no unit for this parameter.

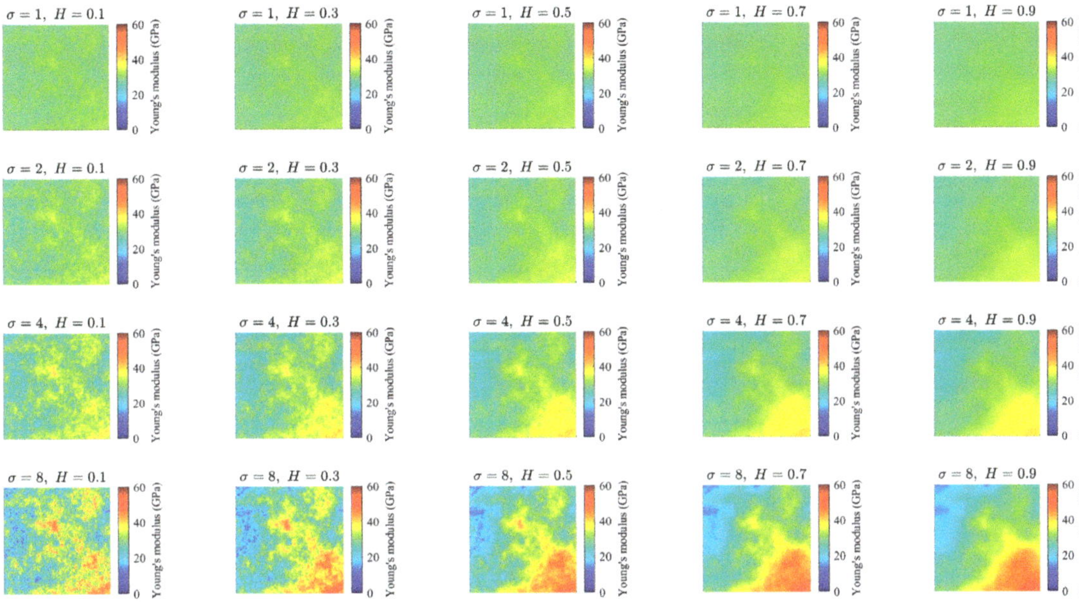

Figure 2. Variability of Young's modulus in the rock matrix characterized by 2D fractional Brownian fields with different values of standard deviation σ and Hurst exponent H.

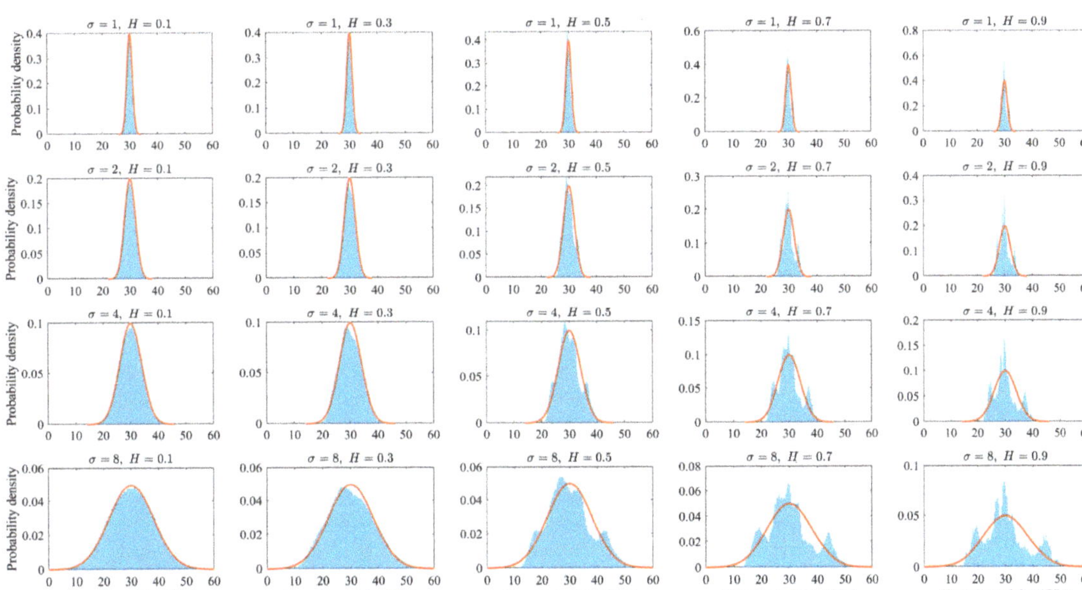

Figure 3. Probability density of Young's modulus in the rock matrix characterized by 2D fractional Brownian fields with different values of standard deviation σ and Hurst exponent H. The red curve in each subplot is the PDF of a Gaussian distribution with the expectation and standard deviation being the same as Young's modulus.

2.4. Characterization of Wave Arrival Behavior

We utilize the wave amplitude to identify the leading front of wave energy to describe the elastic wave arrival behavior, which is called the front of the first arrival wave, abbreviated as FFAW [29]. It is constituted by the points farthest from the source line with the same displacement-to-amplitude ratio, which is denoted as

$$\frac{|U(t_i, x_i, y_i)|}{A(x_i, y_i)} = \frac{|U^0(t^0, x_i, y_i)|}{A^0(x_i, y_i)} \tag{3}$$

where $U(t, x_i, y_i)$ and $U^0(t^0, x_i, y_i)$ represent the displacements of a receiver denoted as (x_i, y_i) at time t_i in the heterogeneous medium and t^0 in the homogeneous medium, respectively; $A^0(x_i, y_i)$ are the wave amplitudes recorded by the receiver (x_i, y_i) in the homogeneous medium while $A(x_i, y_i)$ are the ones in the heterogeneous medium. Specially, t^0 and t_i stand for the times when waves arrive at the receiver (x_i, y_i) in the homogeneous and heterogeneous media, respectively. We can use the same time to derive

$$\frac{|U(t_i, x_j, y_j)|}{A(x_j, y_j)} = \frac{|U(t_i, x_i, y_i)|}{A(x_i, y_i)} \tag{4}$$

and then identify a collection of points to construct the FFAW at a given time t_i in the heterogeneous media. In the same way, we can determine the time at which some point at the right boundary is activated for the first time, i.e., the wave energy reaches the right boundary for the first time, with the corresponding time defined as the breakthrough time $\widetilde{t}_b$. When all the receivers at the right boundary have recorded distinguishable wave signals for the first time, the corresponding time is called the final arrival time $\widetilde{t}_f$.

At a given dimensionless time $\widetilde{t}$, we denote $d(\widetilde{t}, y_i)$ as the distance of a point on the FFAW from the source line (Figure 1a), where y_i is the y-direction coordinate. $d(\widetilde{t}, y_i)$ can be derived from the seismograms recorded by the receiver array. We calculate the roughness $R(\widetilde{t})$ of the FFAW (Barabási and Stanley 1995 [30]) as

$$R(\widetilde{t}) = \sqrt{\frac{1}{N_y}\sum_{i=1}^{N_y}\left[d(\widetilde{t}, y_i) - \overline{d(\widetilde{t})}\right]^2} \tag{5}$$

where N_y is the number of sample points on the y axis (i.e., 301 receivers), and $\overline{d(\widetilde{t})}$ stands for the average value of point-to-source distances of the FFAW at $\widetilde{t}$. We then normalize the roughness by dividing the domain size, namely, $\widetilde{R}= R/L$.

2.5. Quantification of Wave Energy

The decay of elastic waves in random fields is identified by the inverse quality factor Q^{-1} [14] derived as

$$Q^{-1} = -\frac{2\overline{V}}{\omega x}\ln\left(\frac{A}{A^0}\right) \tag{6}$$

where $\overline{V}$ is the mean velocity of elastic waves in heterogeneous media, ω is the angular frequency of incident waves, x is the traveling distance of waves equaling the x-direction distance between the receiver and source line, and A^0 and A are wave amplitudes in homogeneous and heterogeneous media, respectively. The wave velocity is derived from the seismograms, i.e., the normalized x-direction displacement recorded by the receivers at the right boundary of the study domain, which is further averaged over the number of these receivers (301 as indicated in Section 2.1) to determine the mean velocity. We use a fast Fourier transform analysis to transform wave signals recorded by receivers to the frequency domain and average the spectral amplitudes of receivers equidistant from the source line to obtain the mean value, which is then used to derive A. Thus, we can fit a straight line to derive the slope of the curve determined by $\ln(A/A^0)$ versus source-to-receiver distance x.

Consequently, the inverse quality factor Q^{-1} can be calculated by the mean velocity and the slope of $\ln(A/A^0)$ versus x.

3. Results

3.1. Spatiotemporal Wavefield Evolution

Herein, the x-direction displacement is normalized by the displacement amplitude of the incident wave, and the traveling time t of elastic waves in heterogeneous media is normalized to be dimensionless as $\widetilde{t} = t/t_0$, where t_0 is the traveling time of elastic waves propagating from the source line to the right boundary of a homogeneous study domain, taking the derivation of $t_0 = L/V_P$. Figure 4 demonstrates the spatial distribution of the elastic wavefield represented by the normalized x displacement at dimensionless $\widetilde{t} = 1$. The temporal snapshots during wave propagation from $\widetilde{t} = 0$ to 3 are presented as animations in. Figure 5 illustrates the seismograms recorded by the receiver column located at the right side of the study area. When σ is small, e.g., 1, the wavefront keeps a planar shape during the early propagation stage ($\widetilde{t} < 1$ in S1). In cases of smaller H values, the fractional Brownian field has a rougher surface of modulus distribution (Figure 2), inducing a rougher profile of the wavefront (Figure 4) and a more fluctuant waveform (Figure 5) in the later propagation stage ($\widetilde{t} > 1$). Slight coda waves can also be observed behind the frontal wave (Figure 5 and S1). However, due to such a random E distribution, wave energies arrive at the right boundary at almost the same time (Figures 4 and 5). As H increases, larger magnitudes of Young's modulus are more concentrated in the lower right of the study domain (Figure 2), generating larger wave velocity in this region. As a result, waves in the lower region transport faster than those in the upper. Thus, the wavefield becomes inclined in a counterclockwise way at $\widetilde{t} = 1$ (Figure 4) and the following timesteps (S1). In these larger- cases, the distribution of E is smoother and the induced waveform is barely perturbed in the later propagation stage (Figure 5); i.e., no obvious coda wave is detected.

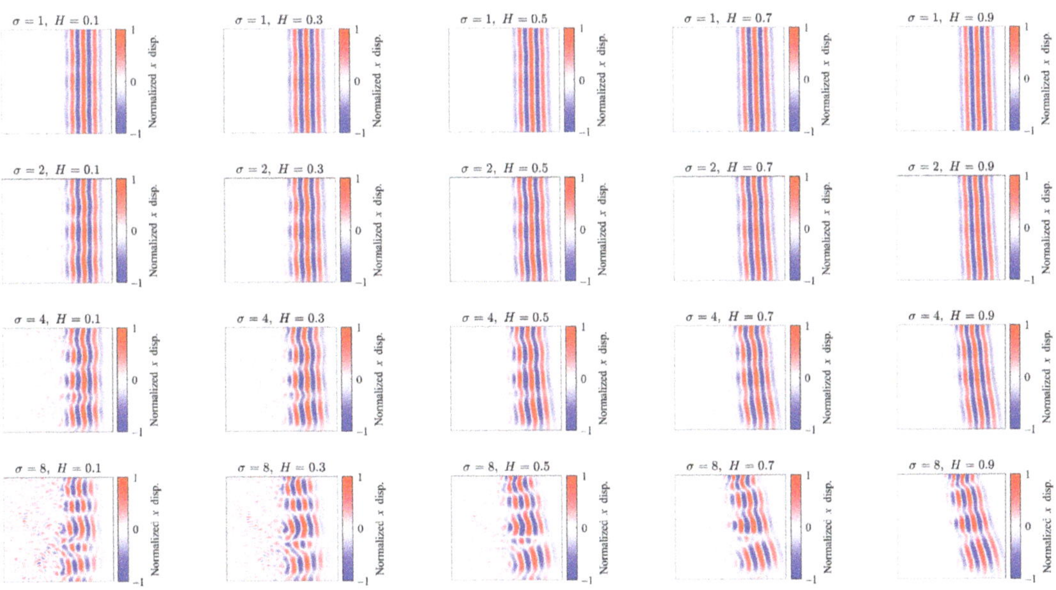

Figure 4. Spatial evolution of elastic wavefield according to normalized x displacement at dimensionless time $\widetilde{t} = 1$.

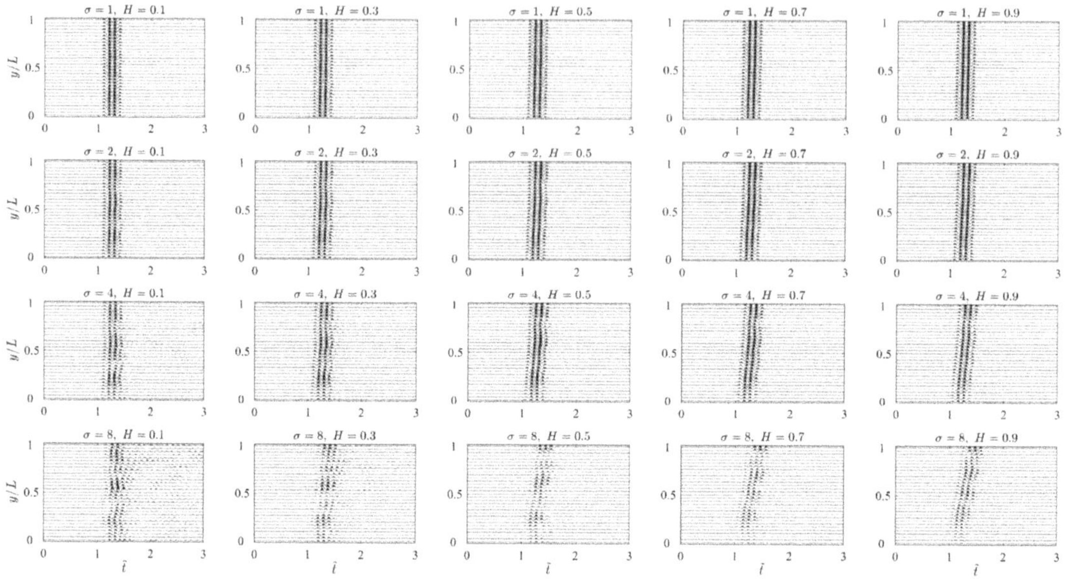

Figure 5. Seismograms according to normalized x-direction displacements recorded by the receiver array at the right boundary of the study domain.

For σ = 2, the waveform is distorted even in the early propagation stage (Figure 5 and $\tilde{t}$ < 0.5 in S1) and more wave energies are scattered by the matrix heterogeneity (Figures 4 and 5). The rough pattern of the wavefront and the asynchronous arrival phenomenon are more recognizable due to the stronger deviation of E. The distortion of the waveform and roughness of the wavefront become more obvious with decreasing H, whereas the skewness of the wavefront (i.e., asynchronous arrival phenomenon) is more apparent with increasing H. As σ becomes larger ($\geq$4), the wavefield tends to be rougher and more waves are scattered into coda (Figures 4 and 5). The dependence of the wavefront and arrival behavior on the Hurst exponent remain the same as the cases of smaller σ: small H generates coda waves and distorts the wavefield; large H separates fast and slow waves and thus causes different arrival times.

As Figures 4 and 5 illustrate, the wavefield evolution is strongly related to the statistical parameters of the fractional Brownian field, which dominate the global and local distribution of E in the field. As shown in Figure 2, a smaller H generates a rougher distribution of E, like a rough surface with spins (large E) and slots (small E), which are distributed in a short-range correlation; i.e., only elements with short distance may hold a similar modulus. As a result, the aggregation of elements with a similar modulus is infrequent in the field. The spins and slots serve as barriers and scatterers, occupying the propagation path of elastic waves, producing scattered coda waves and a distorted wavefield, which is highly different from the wavefield in homogeneous media. On the contrary, larger H generates a smooth surface with a concentrated region of a contiguous modulus as the moduli are distributed in a pattern of long-range correlation. The smoothly distributed E provides an unobstructed and gliding path for elastic wave transport, leaving few coda waves behind the anterior wave, and more energy can reach the right boundary. The wave generally keeps its planar shape similar to its incident appearance. However, the conspicuous concentration region causes a significantly asynchronous arrival phenomenon; i.e., the wavefront is tilted. The standard deviation of Young's modulus σ is a key factor influencing the magnitude of the modulus. Larger σ generates a wider range of E values, inducing a greater probability of any two elements in the field having larger differences in the modulus. Consequently, such higher heterogeneity is more visible to the wave, causing a clearer observation of anomalous wave transport, e.g., the generation of coda waves, rough profile of the wavefront, and asynchronous wave arrival.

3.2. Arrival Behavior of Elastic Wave

To eliminate the randomness caused by different random seeds in generating fractional Brownian fields, we generated 10 realizations for each combination of σ and H; i.e., in total, 200 fields were created. Figure 6 shows the temporal variation in the dimensionless roughness $\widetilde{R}$ of FFAW versus $\widetilde{t}$ for each case, where the red line in the subfigure is a result of the corresponding fractional Brownian field in Figure 2 and gray lines are the results of the other nine realizations. It can be clearly seen that the roughness of FFAW generally increases as the wave propagates through the random field across all realizations and parameter combinations, which is a consequence of the incremental effect of matrix heterogeneity on the wave transport due to the increasing propagation distance. When the field variability is small (σ = 1), the roughness evolves slowly and remains relatively low throughout the wave propagation, suggesting that minor heterogeneity has a limited effect on wavefront distortion. In addition, changes in H have minimal influence on roughness evolution, leading to similar roughness curves, which indicates that the spatial correlation of the modulus (H) plays a minor role in the wave arrival behavior when the field variability is low. As σ increases ($\sigma \geq 2$), the roughness evolution becomes significantly steeper with time, indicating a stronger scattering effect and more prominent wavefront distortion caused by larger modulus variability. In these cases with higher field variability, the modulus distribution (H) tends to amplify scattering effects, leading to a faster growth of roughness. We then plot the profile of FFAW at the breakthrough time $\widetilde{t}_b$ for each realization in Figure 7. When σ is small ($\leq$1), the width and roughness of the profile barely change with H while FFAW becomes wider and rougher with increasing H when σ exceeds 2. As for FFAWs with the same H, the shape changes more violently with larger σ.

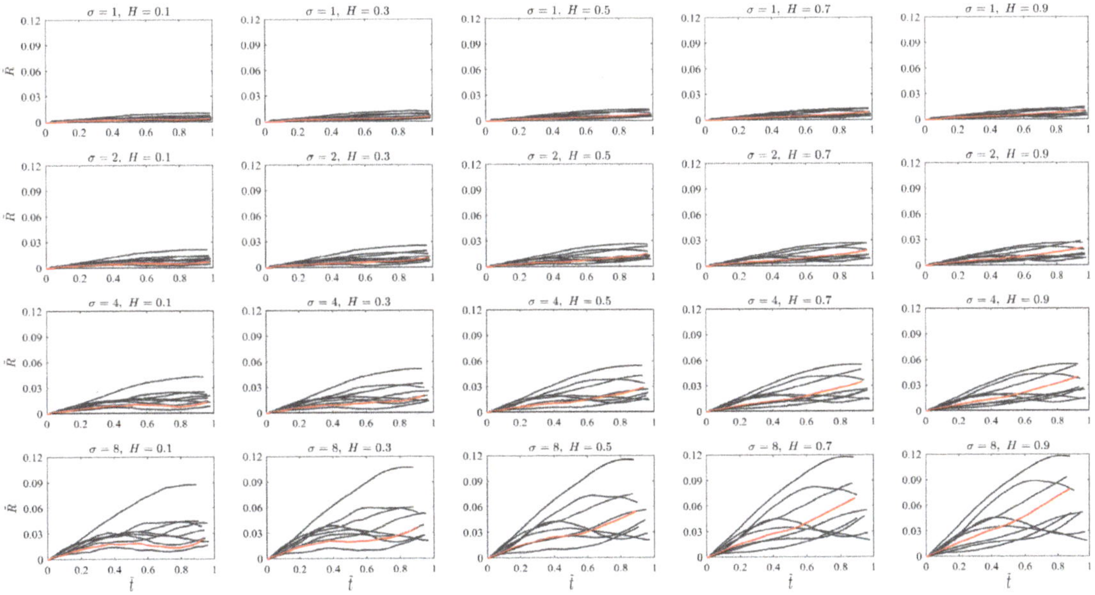

Figure 6. The temporal evolution of the dimensionless roughness $\widetilde{R}$ of FFAW versus the dimensionless time $\widetilde{t}$. The red line results from the fractional Brownian field in Figure 2, and gray lines result from the other 9 realizations.

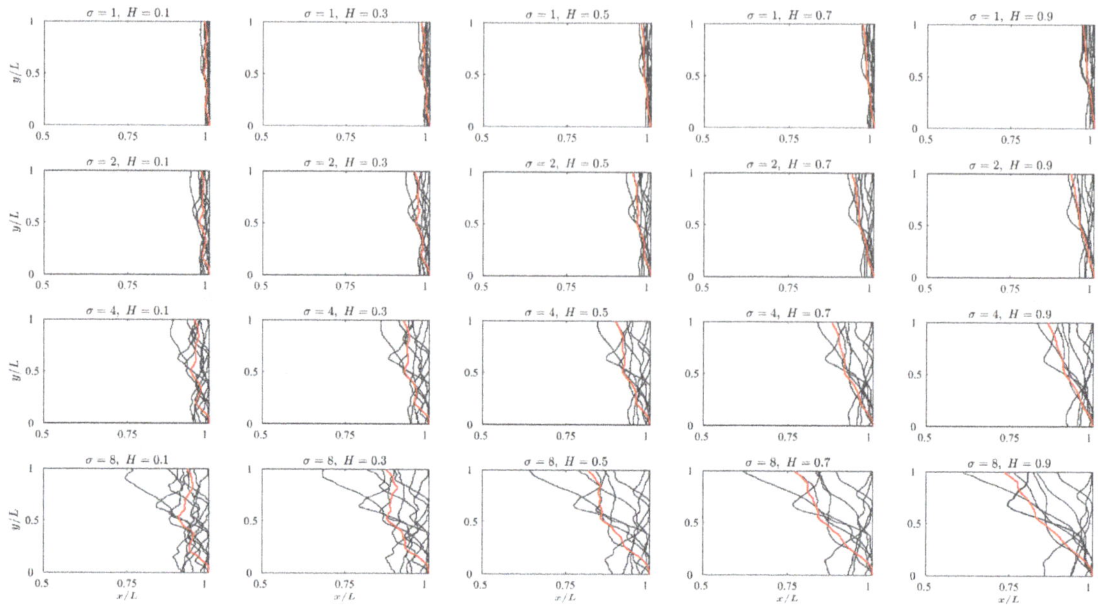

Figure 7. FFAW at the breakthrough time $\widetilde{t}_{\mathrm{b}}$. The red line results from the fractional Brownian field in Figure 2, and gray lines result from the other 9 realizations.

We further averaged the data over 10 realizations to show a universal tendency. Figure 8 gives the dimensionless roughness $\widetilde{R}$ of FFAW at $\widetilde{t}_{\mathrm{b}}$ versus H and σ. The roughness of FFAW shows a generally positive relationship with the Hurst exponent H when $\sigma \geq 2$, whereas the dependence in the cases in which $\sigma = 1$ is not well marked. In Figure 8b, $\widetilde{R}$ is linearly dependent on σ and the slope also increases. Such dependent manners are in accordance with the phenomenon demonstrated in Figures 6 and 7. We then determine the dimensionless breakthrough time $\widetilde{t}_{\mathrm{b}}$ and final arrival time $\widetilde{t}_{\mathrm{f}}$ versus H and σ, as plotted in Figure 9. $\widetilde{t}_{\mathrm{b}} < 1$ and $\widetilde{t}_{\mathrm{f}} > 1$ for all the cases because the randomly distributed Young's moduli provide interlaced channels that expedite or slow the wave speed during propagation (Figure 2), resulting in the arrival of some pioneering wave energy at the right boundary with some slowed energy still remaining in the study region. The opposite dependencies of $\widetilde{t}_{\mathrm{b}}$ and $\widetilde{t}_{\mathrm{f}}$ on σ (and H) also date from the transport patterns of wave energy in diverse random fields: higher modulus variability and concentricity produce more of a heterogeneous wavefield (Figure 4), a more fluctuant seismogram (Figure 5), and a rougher wavefront (Figures 4, 7 and 8), making the asynchronous arrival phenomenon more significant (Figures 4, 5 and 7). Thus, the difference derived from $\widetilde{t}_{\mathrm{f}} - \widetilde{t}_{\mathrm{b}}$ shows a positive relationship with these two heterogeneity indices, as shown in Figure 10. It should be noted that this dependence at $\sigma = 1$ is not evident, similar to Figures 8a and 9a, which is also consistent with the phenomena demonstrated in Figures 4 and 5.

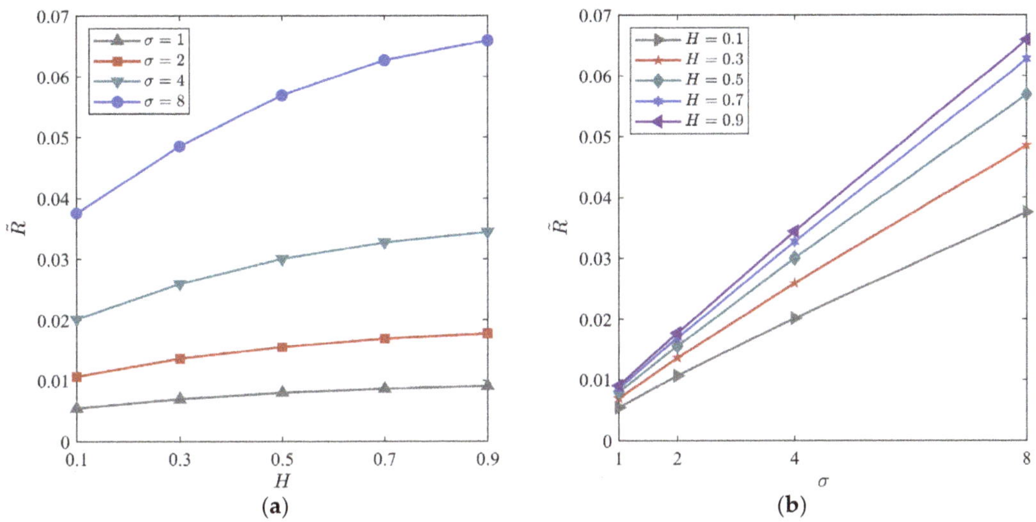

Figure 8. Dimensionless roughness $\widetilde{R}$ of FFAW at the breakthrough time $\widetilde{t}_b$ versus (**a**) H and (**b**) σ. Each marker is the averaged value over 10 realizations.

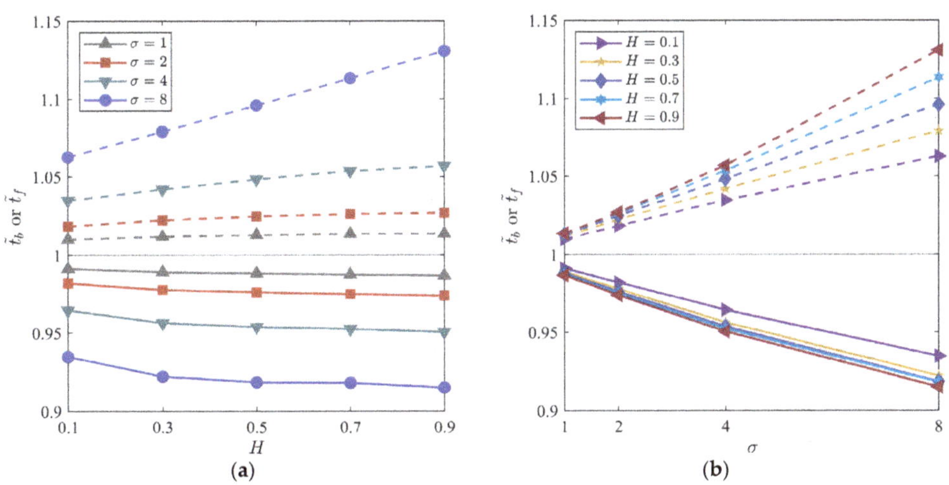

Figure 9. Breakthrough time $\widetilde{t}_b$ (solid lines) and final arrival time $\widetilde{t}_f$ (dashed lines) as a function of (**a**) H and (**b**) σ. Each marker is the averaged value over 10 realizations. The horizontal dotted line represents $\widetilde{t}_b = 1$, i.e., the breakthrough time of waves passing through a blank model without Young's modulus being randomly distributed.

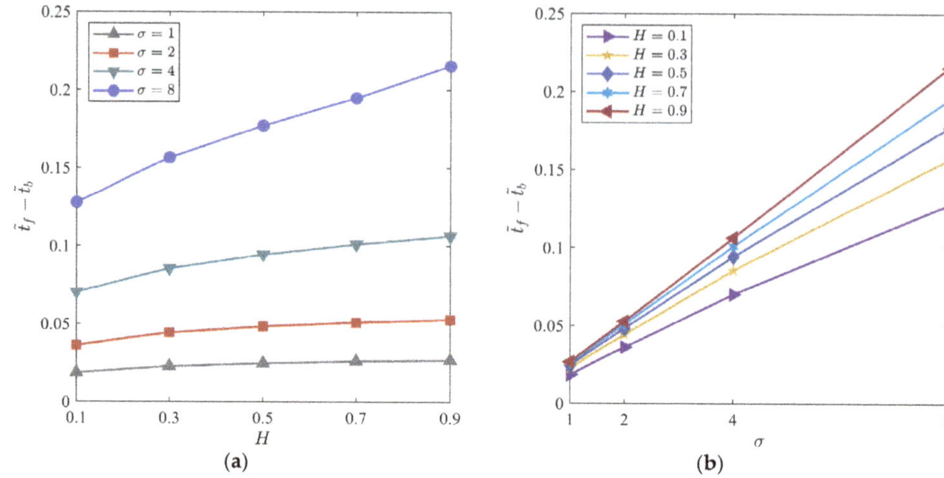

Figure 10. Difference in final arrival time $\tilde{t}_f$ minus breakthrough time $\tilde{t}_b$ as a function of (**a**) H and (**b**) σ. Each marker is the averaged value over 10 realizations.

3.3. *Decay of Elastic Wave Energy*

According to Equation (6), the relative spectral amplitudes as a function of the normalized traveling distance x/L from the source line are plotted in Figure 11. For a small σ, e.g., 1, the decay of amplitude is not significant and the slope of $\ln(A/A^0)$ versus x/L does not change when H changes. For a larger $\sigma \geq 2$, the inverse relationship between the wave amplitude and traveling distance becomes more obvious, showing that the matrix heterogeneity increasingly alters the wavefield as waves across the random field, leaving more energy scattered into the coda. This phenomenon agrees well with the wavefield evolution shown in Figure 4 and S1. We can also see that the decay of wave energy (denoted by the slope of the curve) exhibits an increasing relationship with σ, which is a result of the inducement of the higher variability of random fields with a larger σ. For the cases with the same σ, the decay of wave energy is negatively dependent on the traveling distance, showing that a smoother spatial distribution of Young's modulus is beneficial for energy transport and may produce less coda waves. We then calculate the inverse quality factor Q^{-1} based on the fitting lines of $\ln(A/A^0)$ versus x/L and plot its variation with H and σ in Figure 12. Here, we used the range of normalized distance x/L from 0.25 to 0.75 to derive the slope of $\ln(A/A^0)$ versus x/L because the wave transport observed in Figure 4 and S1 is generally in the propagation regime. It can be seen that Q^{-1} linearly decreases with increasing H showing that wave energies are more attenuated in random fields with a smaller H, and a smoother modulus distribution with larger H facilitates the transport of wave energies across the random media. The variability of Young's modulus σ seems to have no significant effect on the decreasing trend of Q^{-1} versus H. The inverse quality factor Q^{-1} is positively dependent on σ, while H does not change the variation trend. By comparing the variation patterns of Q^{-1} versus H and σ, the influence of matrix variability is more significant than that of the spatial distribution.

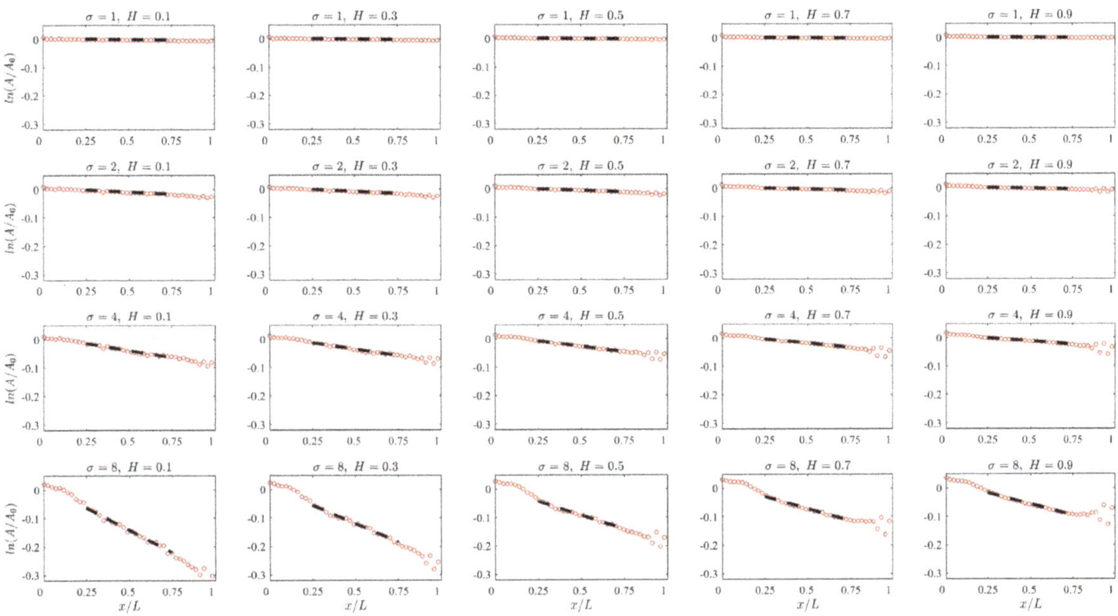

Figure 11. Relative spectral amplitude $\ln(A/A^0)$ varies with the normalized distance x/L from the source for different combinations of σ and H. Each scatter datum is calculated by averaging the mean values of data recorded by each column of 301 receivers equidistant from the source over 10 realizations. The black dashed line in each subfigure is the corresponding fitting line of data with normalized distance x/L from 0.25 to 0.75.

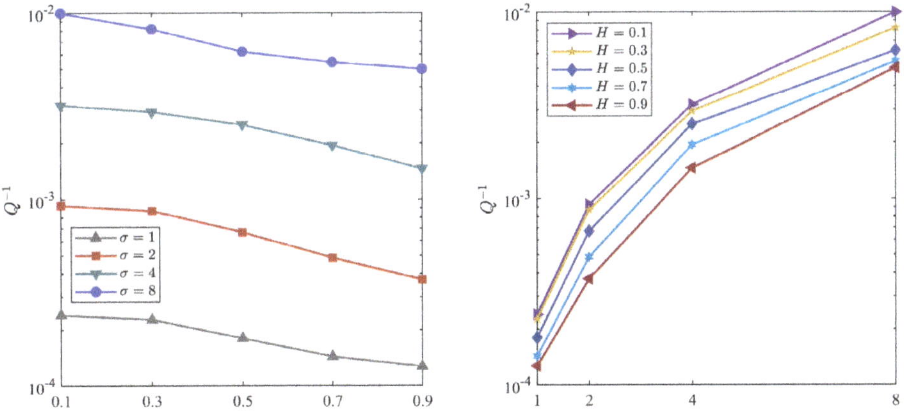

(a) (b)

Figure 12. Inverse quality factor Q^{-1} varies with (**a**) H and (**b**) σ. Each marker is the averaged value over 10 realizations.

The opposite dependencies of Q^{-1} on H and σ, i.e., negative and positive, respectively, give a straightforward revelation for the wavefield evolution shown in Figure 4 and S1. Higher σ induces stronger differences between element moduli, among which the elements with a smaller modulus slow down and scatter wave energies, intercepting their traveling paths at the right boundary. When H is small, the modulus distribution is rough and thus exacerbates the scattering effect. A smoother distribution with larger H provides more even channels for wave energies across the field in the pattern of gathering moduli with similar values in concentration clusters. The smoothness of connecting paths between clusters is also positively related to H. Such spatial characteristics of fractional Brownian fields generate fewer codas during waves passing through these smooth paths and clusters. Therefore, Q^{-1} increases and more wave energy decays faster in the fields with increasing σ and decreasing H.

4. Discussion

Our numerical results shed light on the effects of fractal heterogeneity on wavefield evolution, wavefront roughness, arrival behavior, and energy decay, aligning with and extending findings in the existing literature. The observed wavefront distortion and roughness at higher levels of Young's modulus variability (σ) and lower Hurst exponent (H) values underline the previous findings that such heterogeneity amplifies scattering and energy dispersion [4,6]. However, our study further quantifies this relationship by identifying specific parameter regimes in which wavefront irregularities and asynchronous arrivals become most pronounced, suggesting that fields with lower H and higher σ are particularly preferential to complex wave behavior.

The asynchronous arrival behavior observed in this study aligns with the demonstration of wave scattering in heterogeneous media causing delayed and spatially uneven arrivals [13] and acceleration in energy decay [22]. Our results further illustrate that, as H decreases, the heterogeneity becomes increasingly disruptive to wave coherence, producing asynchronous arrivals across the leading wave and emphasizing the significant influence of fractal correlation on wave behavior. This behavior has practical implications in geophysical applications, such as seismic imaging, where accurate arrival time measurements are crucial for subsurface mapping [34]. Specifically, understanding how fractal heterogeneity affects arrival behavior could inform strategies to account for scattering-induced delays, improving the precision of seismic models used in oil exploration [35] and fault detection [36].

In terms of energy decay, our findings indicate that higher σ values lead to increased attenuation due to scattering, consistent with Aki and Richards [14], who noted that random media often result in significant energy dissipation, similar to the effect of media porosity [37]. We expand on this by demonstrating a dual dependency on both σ and H: while high σ amplifies scattering and energy decay, higher H values reduce these effects by producing smoother modulus distributions, allowing more coherent wave transmission across the field. This relationship between heterogeneity parameters and energy decay is particularly relevant for rock mechanics and stability analyses, where accurate predictions of wave energy distribution are essential for assessing material integrity and failure risks in heterogeneous rock formations [38].

Our results also suggest potential applications in geological tomography. The sensitivity of wavefield evolution and energy decay to fractal characteristics could improve modeling of wave interactions in complex geological settings, aiding in the reconstruction of detailed subsurface images. Sahimi and Tajer [4] emphasized that fractal heterogeneity often dictates transport properties in natural materials; our findings build on this concept by showing how heterogeneity parameters (σ and H) jointly influence transport dynamics, especially wave attenuation patterns. This understanding is vital for accurately imaging subsurface structures in regions with heterogeneous geological compositions, such as fractured reservoirs or fault zones [39–41]. Numerical models that incorporate these insights can improve the accuracy of tomographic reconstructions, which rely on the correct interpretation of wave energy distribution and arrival times.

Furthermore, the dependence of wave behavior on fractal parameters in this study underscores the importance of characterizing geological heterogeneity for seismic tomography. The strong influence of σ and H on wavefield distortion and energy decay implies that highly heterogeneous regions with low H values could pose challenges in predicting seismic wave behavior, leading to amplified scattering and energy attenuation that could impact the accuracy of seismogram recognition. Future studies may benefit from exploring similar fractal heterogeneity effects in 3D models, as well as investigating additional rock mechanical properties, to develop more comprehensive frameworks for seismic analyses in complex media. As the Hurst exponent H is also an index manifesting the correlation of parameter distribution, another area of focus could be the influence of the correlation length of random fields [42] on wave transport.

5. Conclusions

To conclude, this study enhances the understanding of elastic wave propagation in heterogeneous fractal media by numerically modeling the effects of statistically heterogeneous parameters (Hurst exponent H and standard deviation σ) in a 2D fractional Brownian field. Our findings reveal that higher σ values, generating greater variability in Young's modulus, intensify wave scattering and cause the wavefront to become increasingly distorted. This effect is particularly notable when combined with lower H values, which produce rougher modulus distributions that act as barriers to coherent wave transmission. The asynchronous arrival phenomenon becomes more significant with increasing H and σ, as the time difference in leading waves and delayed ones is positively dependent on these parameters. We also found the strong dependence of energy decay on σ and H: higher σ leads to more scattered energy and faster decay, while higher H moderates this effect by providing smoother paths for wave propagation.

These insights contribute to the broader understanding of wave propagation in random, fractal-based media and offer a solid basis for potential implications for seismic analyses and geophysical exploration in heterogeneous geological media. Experimental verification or comparisons with analytical models are a valuable avenue for future study, e.g., laboratory-scale experiments or field-scale observations of wave transport in heterogeneous rocks or geological media. Another future work may extend this model to three-dimensional configuration and explore additional material parameters to capture a wider array of wave transport phenomena in naturally heterogeneous media in geological applications.

Supplementary Materials: The following supporting information can be downloaded at https://www.mdpi.com/article/10.3390/fractalfract8120750/s1, Animation S1: Wavefield evolution from $\tilde{t} = 0$ to 3.

Author Contributions: Conceptualization, Z.Z. and S.W.; methodology, S.W.; software, S.W.; validation, S.W.; formal analysis, S.W. and Z.Z.; investigation, S.W.; resources, S.W.; data curation, S.W.; writing—original draft preparation, S.W.; writing—review and editing, Z.Z.; visualization, S.W.; supervision, Z.Z.; project administration, S.W. and Z.Z.; funding acquisition, S.W. All authors have read and agreed to the published version of the manuscript.

Funding: This research was funded by the National Natural Science Foundation of China (grant number 52378407) and the China Postdoctoral Science Foundation (grant numbers 2023M732670 and 2024T170670). The first author would like to acknowledge funding from the Postdoctoral Fellowship Program of China Postdoctoral Science Foundation (grant number GZC20231948).

Data Availability Statement: Dataset available on request from the authors.

Conflicts of Interest: The authors declare no conflicts of interest.

References

1. Sato, H.; Fehler, M.C. *Seismic Wave Propagation and Scattering in the Heterogeneous Earth*, 2nd ed.; Springer: Berlin/Heidelberg, Germany, 2009; Volume 9783642230, ISBN 978-3-540-89622-7.
2. Aladejare, A.E.; Wang, Y. Evaluation of Rock Property Variability. *Georisk* **2017**, *11*, 22–41. [CrossRef]
3. Salah, M.K.; Alqudah, M.; El-Aal, A.K.A.; Barnes, C. Effects of Porosity and Composition on Seismic Wave Velocities and Elastic Moduli of Lower Cretaceous Rocks, Central Lebanon. *Acta Geophys.* **2018**, *66*, 867–894. [CrossRef]
4. Sahimi, M.; Tajer, S.E. Self-Affine Fractal Distributions of the Bulk Density, Elastic Moduli, and Seismic Wave Velocities of Rock. *Phys. Rev. E—Stat. Nonlinear, Soft Matter Phys.* **2005**, *71*, 046301. [CrossRef] [PubMed]

5. Özgen Karacan, C. Elastic and Shear Moduli of Coal Measure Rocks Derived from Basic Well Logs Using Fractal Statistics and Radial Basis Functions. *Int. J. Rock Mech. Min. Sci.* **2009**, *46*, 1281–1295. [CrossRef]
6. Varadan, V.K.; Ma, Y.; Varadan, V.V. Scattering and Attenuation of Elastic Waves in Random Media. *Pure Appl. Geophys.* **1989**, *131*, 577–603. [CrossRef]
7. Manolis, G.D.; Dineva, P.S.; Rangelov, T.V.; Wuttke, F. *Seismic Wave Propagation in Non-Homogeneous Elastic Media by Boundary Elements*; Springer International Publishing: Cham, Switzerland, 2016; ISBN 3319452053.
8. Shustak, M.; Landa, E. Time Reversal for Wave Refocusing and Scatterer Detection Using Machine Learning. *Geophysics* **2018**, *83*, T257–T263. [CrossRef]
9. Rawlinson, N.; Pozgay, S.; Fishwick, S. Seismic Tomography: A Window into Deep Earth. *Phys. Earth Planet. Inter.* **2010**, *178*, 101–135. [CrossRef]
10. Mandelbrot, B.B.; Van Ness, J.W. Fractional Brownian Motions, Fractional Noises and Applications. *SIAM Rev.* **1968**, *10*, 422–437. [CrossRef]
11. Mandelbrot, B.B. *The Fractal Geometry of Nature*; W. H. Freeman and Company: New York, NY, USA, 1982.
12. Tang, C.; Liu, H.; Lee, P.K.; Tsui, Y.; Tham, L. Numerical Studies of the Influence of Microstructure on Rock Failure in Uniaxial Compression—Part I: Effect of Heterogeneity. *Int. J. Rock Mech. Min. Sci.* **2000**, *37*, 555–569. [CrossRef]
13. Shapiro, S.A.; Kneib, G. Seismic Attenuation by Scattering: Theory and Numerical Results. *Geophys. J. Int.* **1993**, *114*, 373–391. [CrossRef]
14. Aki, K.; Richards, P.G. *Quantitative Seismology*; University Science Books: Sausalito, CA, USA, 2002; ISBN 9781891389634.
15. Allaei, S.M.V.; Sahimi, M. Shape of a Wave Front in a Heterogenous Medium. *Phys. Rev. Lett.* **2006**, *96*, 075507. [CrossRef] [PubMed]
16. Veres, I.; Ryzy, M.; Grabec, T. Finite-Element Modelling of Elastic Wave Propagation and Scattering within Heterogeneous Media. In Proceedings of the 2017 IEEE International Ultrasonics Symposium (IUS), Washington, DC, USA, 6–9 September 2017. [CrossRef]
17. Wang, S.F.; Zhang, Z.X.; Huang, X.; Lei, Q.H. Numerical Simulation of Elastic Wavefield Evolution in Heterogeneous Fractured Media Based on a Combined Displacement Discontinuity-Discrete Fracture Network Model. In Proceedings of the 58th US Rock Mechanics/Geomechanics Symposium, Golden, CO, USA, 23–26 June 2024. [CrossRef]
18. Frankel, A.; Clayton, R.W. Finite Difference Simulations of Seismic Scattering: Implications for the Propagation of Short-Period Seismic Waves in the Crust and Models of Crustal Heterogeneity. *J. Geophys. Res.* **1986**, *91*, 6465. [CrossRef]
19. Przybilla, J.; Korn, M.; Wegler, U. Radiative Transfer of Elastic Waves versus Finite Difference Simulations in Two-Dimensional Random Media. *J. Geophys. Res. Solid Earth* **2006**, *111*, B04305. [CrossRef]
20. Sahimi, M.; Rahimi Tabar, M.R.; Bahraminasab, A.; Sepehrinia, R.; Vaez Allaei, S.M. Propagation and Localization of Acoustic and Elastic Waves in Heterogeneous Materials: Renormalization Group Analysis and Numerical Simulations. *Acta Mech.* **2009**, *205*, 197–222. [CrossRef]
21. Garnier, J.; Sølna, K. On Effective Attenuation in Multiscale Composite Media. *Waves Random Complex Media* **2015**, *25*, 482–505. [CrossRef]
22. Hamzehpour, H.; Asgari, M.; Sahimi, M. Acoustic Wave Propagation in Heterogeneous Two-Dimensional Fractured Porous Media. *Phys. Rev. E* **2016**, *93*, 063305. [CrossRef]
23. Zhang, T.; Sens-Schönfelder, C.; Margerin, L. Sensitivity Kernels for Static and Dynamic Tomography of Scattering and Absorbing Media with Elastic Waves: A Probabilistic Approach. *Geophys. J. Int.* **2021**, *225*, 1824–1853. [CrossRef]
24. Shahbazi, F.; Bahraminasab, A.; Allaei, S.M.V.; Sahimi, M.; Tabar, M.R.R. Localization of Elastic Waves in Heterogeneous Media with Off-Diagonal Disorder and Long-Range Correlations. *Phys. Rev. Lett.* **2005**, *94*, 165505. [CrossRef]
25. Sheng, P. *Introduction to Wave Scattering Localization and Mesoscopic Phenomena*, 2nd ed.; Springer: Berlin/Heidelberg, Germany, 2006; ISBN 9783540291558.
26. Choi, W.; Skelton, E.; Lowe, M.J.S.; Craster, R. Unit Cell Finite Element Modelling for Ultrasonic Scattering from Periodic Surfaces. *AIP Conf. Proc.* **2013**, *1511*, 83–90. [CrossRef]
27. Pettit, J.R.; Walker, A.; Cawley, P.; Lowe, M.J.S. A Stiffness Reduction Method for Efficient Absorption of Waves at Boundaries for Use in Commercial Finite Element Codes. *Ultrasonics* **2014**, *54*, 1868–1879. [CrossRef]
28. COMSOL. *COMSOL Multiphysics Reference Manual, v5.5*; COMSOL: Stockholm, Sweden, 2019.
29. Wang, S.; Zhang, Z.; Huang, X.; Lei, Q. A Numerical Study of Elastic Wave Arrival Behavior in a Naturally Fractured Rock Based on a Combined Displacement Discontinuity-Discrete Fracture Network Model. *Rock Mech. Rock Eng.* **2023**, *56*, 2717–2736. [CrossRef] [PubMed]
30. Barabási, A.-L.; Stanley, H.E. *Fractal Concepts in Surface Growth*; Cambridge University Press: London, UK, 1995; ISBN 0521483085.
31. Barnsley, M.F.; Devaney, R.L.; Mandelbrot, B.B.; Peitgen, H.-O.; Saupe, D.; Voss, R.F. *The Science of Fractal Images*; Peitgen, H.-O., Saupe, D., Eds.; Springer: New York, NY, USA, 1988; ISBN 978-1-4612-8349-2.
32. Schmidt, V. *Stochastic Geometry, Spatial Statistics and Random Fields*; Schmidt, V., Ed.; Lecture Notes in Mathematics; Springer International Publishing: Cham, Switzerland, 2015; Volume 2120, ISBN 978-3-319-10063-0.
33. Torquato, S. *Random Heterogeneous Materials*; Interdisciplinary Applied Mathematics; Springer: New York, NY, USA, 2002; Volume 16, ISBN 978-1-4757-6357-7.

34. Bisrat, S.T.; DeShon, H.R.; Pesicek, J.; Thurber, C. High-Resolution 3-D P Wave Attenuation Structure of the New Madrid Seismic Zone Using Local Earthquake Tomography. *J. Geophys. Res. Solid Earth* **2014**, *119*, 409–424. [CrossRef]
35. Luo, C.; Li, X.; Huang, G. Application of Oil-Water Discrimination Technology in Fractured Reservoirs Using the Differences between Fast and Slow Shear-Waves. *J. Geophys. Eng.* **2017**, *14*, 723–738. [CrossRef]
36. Share, P.E.; Allam, A.A.; Ben-Zion, Y.; Lin, F.C.; Vernon, F.L. Structural Properties of the San Jacinto Fault Zone at Blackburn Saddle from Seismic Data of a Dense Linear Array. *Pure Appl. Geophys.* **2019**, *176*, 1169–1191. [CrossRef]
37. Saffari, P.R.; Thongchom, C.; Jearsiripongkul, T.; Saffari, P.R.; Keawsawasvong, S.; Kongwat, S. Porosity-Dependent Wave Propagation in Multi-Directional Functionally Graded Nanoplate with Nonlinear Temperature-Dependent Characteristics on Kerr-Type Substrate. *Int. J. Thermofluids* **2023**, *20*, 100408. [CrossRef]
38. Rafiei, H.; Derek, R.C.; Pedro, M.; Loren, V. Stability Analysis of Slopes with Spatially Variable Strength Properties. *Rock Mech. Rock Eng.* **2019**, *52*, 3791–3808. [CrossRef]
39. Li, Y.-G.; Vidale, J.E.; Aki, K.; Xu, F. Depth-Dependent Structure of the Landers Fault Zone from Trapped Waves Generated by Aftershocks. *J. Geophys. Res. Solid Earth* **2000**, *105*, 6237–6254. [CrossRef]
40. Yang, H. Recent Advances in Imaging Crustal Fault Zones: A Review. *Earthq. Sci.* **2015**, *28*, 151–162. [CrossRef]
41. Lei, Q.; Sornette, D. Transport and Localization of Elastic Waves in Two-Dimensional Fractured Media: Consequences on Scattering Attenuation. *J. Geophys. Res. Solid Earth* **2021**, *126*, e2020JB021178. [CrossRef]
42. Zeng, Y.; Lei, Q.; Wang, Z.; Ding, S.; Liu, K.; Huang, X.; Huang, Z.; Wang, X. Numerical Simulation of Fluid Injection-Induced Fault Slip in Heterogeneous Shale Formations. *Comput. Geotech.* **2021**, *134*, 104120. [CrossRef]

fractal and fractional

MDPI

Article

Penetration Grouting Mechanism of Bingham Fluid in Porous Media Based on Fractal Theory

Jiakun Gong [1,*], Chenxi Xu [1], Yuan Wang [2,*], Shan Gao [3], Haikuan Chen [4] and Lu Jia [4]

1 College of Mechanics and Engineering Science, Hohai University, Nanjing 211100, China
2 College of Water Conservancy and Hydropower Engineering, Hohai University, Nanjing 210098, China
3 Jiangsu Province South to North Water Diversion Project Construction Leading Group Office, Nanjing 210029, China
4 Center of Construction Management of Water Conservancy Projects, Huaian 223005, China
* Correspondence: jiakungong@hhu.edu.cn (J.G.); wangyuan@hhu.edu.cn (Y.W.)

Abstract: Penetration grouting is a significant grouting technique. The pore structure has important impacts on the infiltration mechanism of slurry in porous media. In this study, based on fractal theory, a theoretical penetration grouting model for Bingham fluid is established. An experimental apparatus for simulating the penetration process of Bingham fluid with a constant flow rate is developed. A series of penetration-grouting experiments are conducted to validate the theoretical model established in this study and analyze the impacts of the water–cement ratio and flow rate on the slurry injection pressure. The results show that the theoretical values of the slurry pressure along the penetration direction obtained from the penetration grouting model match the experimental values well. This indicates that the proposed model can better describe the process of slurry infiltration and provide valuable support for related grouting projects.

Keywords: fractal theory; Bingham fluid; porous media; penetration grouting

1. Introduction

Permeation grouting is one of the most important techniques used in grouting reinforcement engineering. It is widely applied in various fields, such as hydraulic engineering and highway construction, due to its limited disturbance to the grouted medium [1]. Extensive studies on the permeation-grouting theory have been conducted, including the slurry permeation and diffusion models for Newtonian fluids, Bingham fluids, and power-law fluids. Maag formulated the classical theoretical model for Newtonian fluid permeation grouting [1]. Yang et al. [2,3] investigated the permeation and diffusion mechanisms of Bingham fluids and power-law fluids in sand layers.

On the basis of these foundational studies, scholars have further extended the permeation grouting theory by considering varies factors such as time-dependent and spatial variations in slurry viscosity, seepage effects during the diffusion process, and the tortuosity effects in porous media. With respect to the time-dependent and spatial variability effects of slurry viscosity, Ruan [4] established time-dependent viscosity diffusion equations for cement-based and solution-based grouting materials. Yang et al. [5,6] investigated the permeation and diffusion models of Bingham fluids, considering time-dependent viscosity for spherical, cylindrical, and column-semi-spherical geometries. The cylindrical permeation grouting mechanisms of power-law cement slurries with time-varying rheological parameters was also explored [7]. Building on Maag's classical theory, Zhou [8] optimized Maag's formula by substituting the average viscosity with time-dependent viscosity. Additionally, Zhang et al. [9] analyzed the permeation and diffusion mechanisms of rapidly setting grouts, by considering the time–space variations in slurry viscosity. With respect to the seepage effect on the slurry diffusion process, Saada et al. [10–13] carried out laboratory experiments to examine the impact of density, consolidation stress, cement concentration, and grouting flow rate on the seepage effects of cement slurries. Feng et al. [14] proposed a theoretical model for calculating the diffusion distance of particulate slurries under laminar flow conditions. Wang et al. [15] discussed the impact of seepage and proposed an expression for the spatial distribution of viscosity-decay rate. Li et al. [16] analyzed the impact of slurry seepage on the porosity and permeability of sandy media during the cement slurry permeation process. With respect to tortuosity effects, Zhang et al. [17], Lu et al. [18], and Yang et al. [19] investigated the permeation and diffusion mechanisms for Newtonian fluids, Bingham fluids, and power-law fluids, through considering the diffusion pathways. Wang et al. [20] investigated the permeation grouting mechanism of viscous time-varying fluids, considering the diffusion path.

Fractal theory is a crucial tool for describing the pore characteristics of porous media, which makes it powerful in representing the influence of microscopic structural parameters of pores on slurry diffusion during the permeation grouting process. Yu et al. [21] provided a comprehensive discussion on the application of fractal theory in describing transport properties in porous media. Yun et al. [22] developed a model to calculate the initiation pressure gradient for Bingham fluid flow, considering the fractal characteristics of porous media and capillary pressure. Zhang et al. [23] derived a permeability calculation formula for power-law fluids, incorporating the fractal features of pore size distribution and the tortuous flow paths within the porous medium.

In summary, significant research on the permeation and diffusion mechanisms of Bingham fluids has been conducted. Fractal theory has also been utilized to characterize the tortuous flow paths of slurries. However, existing theoretical models for the permeation and diffusion of Bingham fluids still can be further enhanced to reflect the complex pore characteristics of porous media and their impact on slurry diffusion. Additionally, the parameters involved in fractal-based slurry permeation models are complex and difficult to determine from conventional geotechnical tests. This restricts the experimental validation of the theoretical models, thereby constraining their application.

This study aims to propose a theoretical model of permeation grouting for Bingham fluids based on fractal theory and validate the model by experiments. Pure cement slurry, which is widely used in grouting engineering practices, is adopted as the Bingham fluid [24] in the experiments. This work is organized as follows: Firstly, a theoretical model for the one-dimensional steady-flow permeation grouting of Bingham fluids is established based on fractal theory. Subsequently, slurry permeation and diffusion experiments are conducted with varying water–cement ratios and slurry flow rates. Finally, the theoretical and experimental results are compared to validate the applicability of the theoretical model. The spatial and temporal variations in grouting pressure under different water–cement ratios and grouting rates are also analyzed.

2. Theoretical Model of Permeation Grouting for Bingham Fluid Based on Fractal Theory

2.1. Assumptions

The assumptions of the permeation grouting theory model include the following:

1. The slurry is a homogeneous, incompressible Bingham fluid;
2. The viscosity variation of slurry is not considered;
3. The flow regime of slurry during diffusion is laminar, ignoring the effects of gravity and filtration;
4. The pore radius in the porous medium satisfies $r_{min}/r_{max} < 10^{-2}$.

2.2. The Diffusion Model of Bingham Fluid in a Circular Pipe

The rheological equation for Bingham fluid is as follows [25]:

$$\tau = \tau_0 + \mu_p \gamma \tag{1}$$

where τ is the shear stress, τ_0 is the yield stress, μ_p is the plastic viscosity, and γ is the shear rate ($\gamma = -dv/dr$, where v is the fluid velocity and r is the radius of the circular pipe).

In a complex morphology of pore structures in porous media, if the tortuous path of the slurry flow within the porous medium is not considered, it is often assumed that the flow channel can be represented as a long straight circular pipe. The flow of slurry in the porous medium is therefore considered as the superposition of the flow through all these channels [25]. The flow of the Bingham fluid in a single circular pipe is illustrated in Figure 1. The radius of the circular pipe is r_0. The radius of a small slurry element along the axis of the pipe is $r < r_0$, while the length is dl. The pressure at the two ends of the element is p and $p + dp$, respectively, with a shear stress τ acting opposite to the direction of the flow velocity.

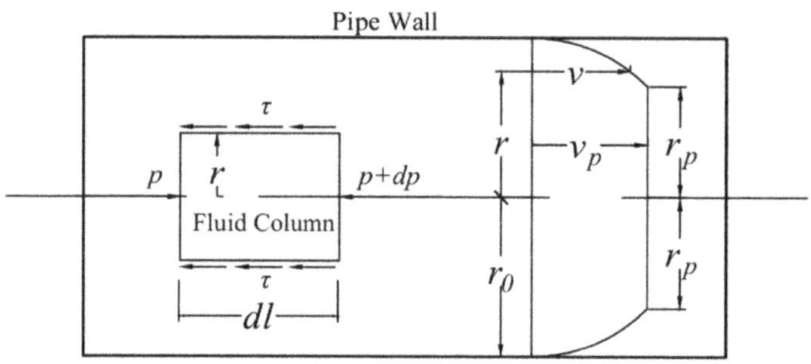

Figure 1. Schematic of Bingham fluid flow in a circular tube.

The force balance equation for the slurry element can be written as follows:

$$\pi r^2 dp + 2\pi r \tau dl = 0 \tag{2}$$

The shear stress distribution is as follows:

$$\tau = -\frac{r}{2}\frac{dp}{dl} \tag{3}$$

Around the axis of the pipe, the fluid experiences a lower shear stress. When $\tau \leq \tau_0$, there is no relative motion between fluid particles, leading to the existence of a central stagnant core region $0 \leq r \leq r_p$. In this region, fluid velocities are uniform and equal to v_p. For $r_p \leq r \leq r_0$, fluid particles are in a state of shear motion.

By substituting $r = r_p$ and $\tau = \tau_0$ into Equation (3), the following is obtained:

$$r_p = 2\tau_0 / \left(-\frac{dp}{dl}\right) \tag{4}$$

By substituting Equation (3) into Equation (1) and simplifying, the following is obtained:

$$\frac{dv}{dr} = \frac{1}{\mu_p}\left(\frac{r}{2}\frac{dp}{dl} + \tau_0\right) \tag{5}$$

By integrating Equation (5) and applying the boundary condition $r = r_0$, $v = 0$, the velocity distribution of the Bingham fluid in a circular pipe can be described as follows:

$$v = \frac{1}{\mu_p}\left[-\left(\frac{dp}{4dl}\right)\left(r_0^2 - r^2\right) - \tau_0(r_0 - r)\right] \quad (r_p \leq r \leq r_0) \tag{6}$$

and

$$v_p = \frac{1}{\mu_p}\left[-\left(\frac{dp}{4dl}\right)\left(r_0^2 - r_p^2\right) - \tau_0\left(r_0 - r_p\right)\right] \quad (0 \leq r \leq r_p) \tag{7}$$

In a single circular pipe, the flow rate q is the sum of the flow rates in the shear zone $(r_p \le r \le r_0)$ and the core region $(0 \le r \le r_p)$:

$$q = \pi r_p^2 v_p + \int_{r_p}^{r_0} 2\pi r v dr \tag{8}$$

Equations (4)–(7) are then incorporated into Equation (8) to obtain the total flow rate q for the Bingham fluid in a circular pipe.

$$q = \frac{\pi r_0^4}{8\mu_p}\left(-\frac{dp}{dl}\right)\left[1 - \frac{4}{3}\left(\frac{\frac{2\tau_0}{r_0}}{-\frac{dp}{dl}}\right) + \frac{1}{3}\left(\frac{\frac{2\tau_0}{r_0}}{-\frac{dp}{dl}}\right)^4\right] \tag{9}$$

In order to determine the condition that the flow rate q in the pipe is zero, Equation (9) is solved:

$$-\frac{dp}{dl} = \frac{2\tau_0}{r_0} = \delta \tag{10}$$

where δ represents the yield pressure gradient of the Bingham fluid.

Since the pressure gradient $-dp/dl$ is significantly greater than the yield pressure gradient δ during the grouting process, Equation (9) can be simplified as follows:

$$q = \frac{\pi r_0^4}{8\mu_p}\left(-\frac{dp}{dl}\right)\left[1 - \frac{4}{3}\left(\frac{\frac{2\tau_0}{r_0}}{-\frac{dp}{dl}}\right)\right] \tag{11}$$

The above equation describes the flow rate of the Bingham fluid in a single conduit, assuming that the fluid flow path in the porous medium can be approximated as a long and straight circular pipe.

2.3. *Fractal Theory-Based Diffusion Model of Bingham Fluid Penetration Grouting*

However, due to the complex pore structure of the porous medium, the actual seepage pathway of the Bingham fluid is tortuous, as illustrated in Figure 2.

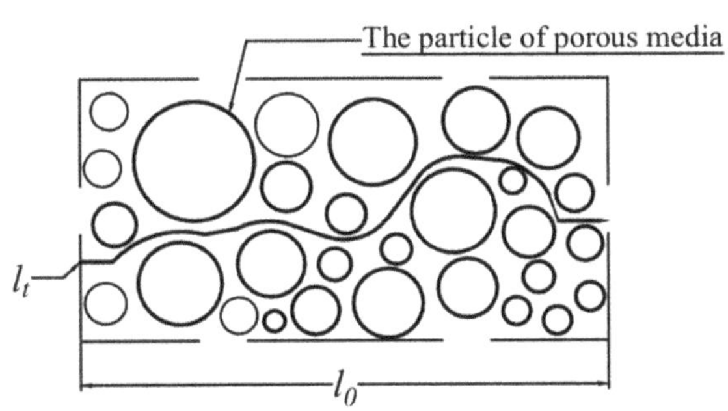

Figure 2. Fluid flow in porous media.

Porous media pores exhibit fractal characteristics [22], and the pore size distribution conforms to fractal scaling relationships:

$$N(\ge r_0) = \left(\frac{r_{max}}{r_0}\right)^{D_f} \tag{12}$$

where N is the number of pores; r_0 is the pore radius; r_{max} is the maximum pore radius; and D_f is the pore fractal dimension, $1 < D_f < 2$.

The length of curved pore channels l_t can be expressed in fractal theory [22]:

$$l_t = l_0^{D_T}(2r_0)^{1-D_T} \tag{13}$$

where l_t is the curved length of the channel; l_0 is the straight length of the channel; and D_T is the tortuosity fractal dimension, $1 < D_T < 2$. When $D_T = 1$, it indicates that the flow channel is straight.

By substituting l_t, which represents the straight pore channel in Equation (11), with the actual curved pore channel l represented by the pore fractal dimension, the following is obtained:

$$q = \frac{\pi r_0^4}{8\mu_p}\left(-\frac{dp}{dl_0}\right)\left[\frac{l_0^{1-D_T}(2r_0)^{D_T-1}}{D_T} - \frac{4}{3}\frac{\frac{2\tau_0}{r_0}}{\left(-\frac{dp}{dl_0}\right)}\right] \tag{14}$$

The above equation represents the flow rate of the Bingham fluid flowing through a pore with radius r_0. Therefore, the flow rate Q through the entire cross-section of the porous medium can be expressed as follows:

$$Q = -\int_{r_{min}}^{r_{max}} q dN \tag{15}$$

where Q is the total flow rate through the porous medium; r_{min} is the minimum pore radius; and r_{max} is the maximum pore radius.

Substituting Equation (14) into Equation (15) and integrating yields the following:

$$Q = \frac{\pi D_f}{8\mu_p}\left(-\frac{dp}{dl_0}\right)\left\{\frac{l_0^{1-D_T}2^{D_T-1}r_{max}^{D_T+3}}{D_T\left(D_T-D_f+3\right)}\left[1-\left(\frac{r_{min}}{r_{max}}\right)^{D_T-D_f+3}\right] - \frac{4r_{max}^3}{3\left(3-D_f\right)}\frac{2\tau_0}{\left(-\frac{dp}{dl_0}\right)}\left[1-\left(\frac{r_{min}}{r_{max}}\right)^{3-D_f}\right]\right\} \tag{16}$$

Generally, in porous media, $r_{min}/r_{max} < 10^{-2}$. It is considered that $1 < D_T < 2$ and $1 < D_f < 2$ [22]; therefore, $(r_{min}/r_{max})^{D_T-D_f+3} \ll 1$ and $(r_{min}/r_{max})^{3-D_f} \ll 1$.

Thus, neglecting higher-order terms, the expression for the flow rate of the Bingham fluid in porous media can be simplified as follows:

$$Q = \frac{\pi D_f r_{max}^{3+D_T}}{\mu_p l_0^{D_T-1}2^{4-D_T}D_T\left(D_T-D_f+3\right)}\left[\left(-\frac{dp}{dl_0}\right) - \frac{l_0^{D_T-1}\tau_0 D_T\left(D_T-D_f+3\right)}{3\left(3-D_f\right)2^{D_T-4}r_{max}^{D_T}}\right] \tag{17}$$

According to fractal theory, the cross-sectional area of a unit cell can be represented as follows [21]:

$$S = \frac{\pi D_f {r_{max}}^2(1-\varphi)}{\left(2-D_f\right)\varphi} \tag{18}$$

where S is the cross-sectional area of the unit cell and φ is the areal porosity.

The average flow velocity of Bingham fluid can then be expressed as follows:

$$\overline{v} = \frac{Q}{S} = \frac{r_{max}^{1+D_T}\left(2-D_f\right)\varphi}{\mu_p l_0^{D_T-1}2^{4-D_T}D_T\left(D_T-D_f+3\right)(1-\varphi)}\left[\left(-\frac{dp}{dl_0}\right) - \frac{l_0^{D_T-1}\tau_0 D_T\left(D_T-D_f+3\right)}{3\left(3-D_f\right)2^{D_T-4}r_{max}^{D_T}}\right] \tag{19}$$

where $\overline{v}$ is the average flow velocity on the cross-section of the unit cell.

In porous media, the Darcy seepage velocity V at any point in the unit cell satisfies the following relationship with the average flow velocity $\overline{v}$ [25]:

$$V = \phi\overline{v} \tag{20}$$

where V is the Darcy seepage velocity and ϕ is the volumetric porosity. The relationship between volumetric porosity ϕ and areal porosity φ can be expressed as follows [26]:

$$\phi = \Gamma\varphi \tag{21}$$

where Γ is the tortuosity, $\Gamma = 1$ indicates that the flow channel is straight, and the areal porosity and volumetric porosity are equal.

According to the law of mass conservation, the relationship between the grouting flow rate q, seepage velocity V, and the cross-sectional area A for fluid diffusion can be expressed as follows:

$$q = VA \tag{22}$$

where q is the grouting flow rate and A is the cross-sectional area for fluid diffusion.

By combining Equations (19)–(22) and simplifying, the following is obtained:

$$dp = -\frac{D_T\left(D_T - D_f + 3\right)}{l_0^{1-D_T} 2^{D_T-4} r_{max}^{D_T}}\left[\frac{\mu_p(\Gamma-\phi)q}{Ar_{max}\left(2-D_f\right)\phi^2} + \frac{\tau_0}{3\left(3-D_f\right)}\right] dl_0 \tag{23}$$

Therefore, the relationship between the grouting pressure p_c and the grout diffusion distance l_m can be expressed as follows:

$$p_c = \int_0^{l_m}\left\{-\frac{D_T\left(D_T - D_f + 3\right)}{l_0^{1-D_T} 2^{D_T-4} r_{max}^{D_T}}\left[\frac{\mu_p(\Gamma-\phi)q}{Ar_{max}\left(2-D_f\right)\phi^2} + \frac{\tau_0}{3\left(3-D_f\right)}\right]\right\} dl_0 + p_0 \tag{24}$$

In the case of $l_m = 0$, the grout diffusion distance is zero and the grout pressure $p_c = p_0$, where p_0 is the initial grouting pressure.

Under condition of a steady-state Bingham fluid (slurry) flow, the relationship between the grout pressure and fluid diffusion distance can be expressed as follows:

$$p_c = -\frac{\left(D_T - D_f + 3\right)}{2^{D_T-4} r_{max}^{D_T}}\left[\frac{\mu_p(\Gamma-\phi)q}{Ar_{max}\left(2-D_f\right)\phi^2} + \frac{\tau_0}{3\left(3-D_f\right)}\right] l_m^{D_T} + p_0 \tag{25}$$

2.4. Parameter Values

In the relationship between the grout pressure and diffusion distance represented by Equation (25), the slurry viscosity μ_p and yield stress τ_0 can be measured using a viscometer. The slurry flow rate q, fluid diffusion cross-sectional area A, grout diffusion distance l_m, and the volumetric porosity ϕ of the porous medium can be obtained through experiments.

The tortuosity Γ can be obtained from the volumetric porosity ϕ through the following expression [26]:

$$\Gamma = 1 - \frac{\phi}{2} + \frac{\sqrt{1-\phi}}{4} + \frac{\left(\phi + 1 + \sqrt{1-\phi}\right)\sqrt{9 - 5\phi - 8\sqrt{1-\phi}}}{8\phi} \tag{26}$$

The fractal dimension of porosity D_f can be represented by the maximum and minimum pore radii of the porous medium and the areal porosity φ according to the following relationship [21]:

$$D_f = 2 - \frac{ln\varphi}{ln\left(\frac{r_{min}}{r_{max}}\right)} \tag{27}$$

In Equation (27), the areal porosity φ can be derived from the measured volumetric porosity ϕ using Equation (21). The ratio of the minimum to maximum pore radii is related to the areal porosity φ according to the following relationship [21]:

$$\frac{r_{min}}{r_{max}} = \frac{\sqrt{2}}{d^{+}}\sqrt{\frac{1-\varphi}{1-0.342\varphi}} \tag{28}$$

where d^{+} is typically set to 24. Therefore, the fractal dimension of porosity D_f can be calculated from the volumetric porosity ϕ.

The maximum pore radius r_{max} is related to tortuosity Γ, the fractal dimension of porosity D_f, the areal porosity φ, and the permeability k according to the following relationship [26]:

$$r_{max} = \sqrt{\frac{8k\Gamma\left(4-D_f\right)(1-\varphi)}{\left(2-D_f\right)\varphi}} \tag{29}$$

where k is the permeability of the porous medium, which can be measured experimentally. Therefore, the maximum pore radius r_{max} can be calculated from the volumetric porosity ϕ and the permeability k.

The tortuosity fractal dimension D_T can be expressed in terms of the porosity fractal dimension D_f, the maximum and minimum pore radii, and the areal porosity φ as follows [21]:

$$D_T = 1 + ln\frac{1}{2}\left[1+\frac{1}{2}\sqrt{1-\varphi}+\sqrt{1-\varphi}\frac{\sqrt{\left(\frac{1}{\sqrt{1-\varphi}}-1\right)^2+\frac{1}{4}}}{1-\sqrt{1-\varphi}}\right]/ln\left(\frac{\frac{D_f-1}{D_f^{\frac{1}{2}}}\left[\frac{1-\varphi}{\varphi}\frac{\pi}{4\left(2-D_f\right)}\right]^{\frac{1}{2}}}{\frac{r_{min}}{r_{max}}}\right) \tag{30}$$

The tortuosity fractal dimension D_T can ultimately be transformed into an expression involving the volumetric porosity ϕ.

To summarize, the theoretical model for the permeation and grouting diffusion of the Bingham fluid based on fractal theory, developed in this study, relies on the following parameters: slurry viscosity μ_p and yield stress τ_0; volumetric porosity ϕ and permeability k of the porous medium; slurry flow rate q; fluid diffusion cross-sectional area A; and grout diffusion distance l_{m}. All these parameters can be determined from conventional geotechnical tests.

2.5. Scope of Application

Equation (25) is proposed based on the assumption of laminar flow; hence, it is not applicable for turbulent flow. The permeation and diffusion of the Bingham fluid in porous media can be determined based on its generalized Reynolds number Re. When $Re \leq 2000$, the slurry diffuses in a laminar flow regime. When $Re > 2000$, the slurry diffuses in a turbulent flow regime.

The Reynolds number Re can be determined using the following formula:

$$Re = \frac{\rho\overline{v}D}{\mu_p} \tag{31}$$

where ρ is the density of the Bingham fluid and D is the diameter of the pore channel in the porous medium.

3. Experiments of Permeation Grouting in Porous Media

To analyze the applicability and accuracy of the fractal-based theoretical model for Bingham fluid permeation grouting, a series of permeation grouting experiments are conducted. These experiments facilitate a comparative analysis between theoretical predictions and experimental results.

3.1. Experimental Apparatus

In order to study the diffusion patterns of the Bingham fluid with varying injection pressures and different fluid properties, a one-dimensional visual permeation grouting simulation apparatus is designed (Figure 3).

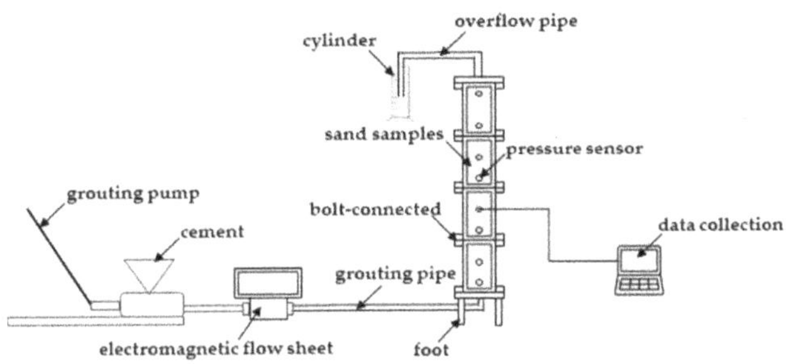

Figure 3. Schematic of experimental apparatus.

The grouting column model employs Polymethyl Methacrylate (PMMA), composed of four sections of identical dimensions, which enables the slurry flow to be visualized. Each cylindrical tube section measures Φ100 mm × 200 mm, with the total length of the apparatus reaching 80 cm. The bottom is equipped with adjustable supports. The tubes are filled with sand (the injected medium). The slurry is injected from the bottom and disperses upwards along the Polymethyl Methacrylate (PMMA). The inlet and outlet openings are covered with filter screens to prevent structural damage to the samples.

The slurry pressure measurements are conducted using a DM-YB1820 (Nanjing Danmo Electronic Technology Co., Ltd, Nanjing, China) static resistance strain gauge connected to infiltration pressure sensors. The pressure sensors are positioned at distances of 0, 10, 20, 30, 40, 50, 62.5, and 77.5 cm from the slurry inlet. The sensors located from 0 to 50 cm have a measurement range of 200 kPa, while those at 62.5 cm and 77.5 cm have a measurement range of 100 kPa, all with an accuracy of 0.3%.

3.2. Experimental Design

P.O 42.5 Portland cement is employed as the grouting material. During the experiment, the properties of the cement slurry are altered by adjusting the water–cement ratio (w/c ratio). The w/c ratios tested are 0.80, 1.00, and 1.25. The rheological equations for the cement slurry at different w/c ratios are shown in Table 1.

Table 1. Rheological equations of three cement grout.

Water–Cement Ratio	Rheological Equation
0.80	$\tau = 2.2078 + 0.0201\gamma$
1.00	$\tau = 0.8593 + 0.0169\gamma$
1.25	$\tau = 0.1136 + 0.0159\gamma$

The grouting medium used in the experiment is Chinese ISO standard sand with particle sizes ranging from 1 to 2 mm. The sand samples are initially in a dry state. Before each test, the sand samples are cleaned and dried. Based on the specified porosity, the required mass of sand samples is calculated and filled accordingly. The sand is filled in increments, with each filling step involving leveling the sand and assessing the filled height against the scale values on the model wall to ensure compliance. The permeability coefficient of the grouting medium is determined through the constant-head permeability tests [27]. The main parameters of the grouting medium are listed in Table 2.

Table 2. Main parameters of the injected medium.

Parameter	Bulk Density (-)	Particle Size Range (mm)	Volumetric Porosity (-)	Permeability (m^2)
value	2.65	1~2	0.394	2.54×10^{-8}

In this study, the permeation grouting experiments are conducted with the constant flow rate grouting method. The slurry flow rates are set at 1 L/min, 2 L/min, and 3 L/min, respectively. The experiments involve two primary variables: the water–cement ratio (W/C ratio) of the slurry and the slurry flow rate. To systematically investigate the effects of these variables on slurry dispersion, five distinct permeation grouting tests were designed (refer to Table 3).

Table 3. Design of test conditions.

Experimental Condition Number	Grouting Flow Rate (L/min)	Water–Cement Ratio (-)
1	1	1.00
2	2	1.00
3	3	1.00
4	2	0.80
5	2	1.25

In test conditions 1, 2, and 3, the water–cement ratio is held constant, in order to examine the impact of varying slurry flow rates on the dispersion characteristics of the slurry. Conversely, in test conditions 2, 4, and 5, the slurry flow rate is held constant, to predominantly analyze how variations in the water–cement ratio influence slurry dispersion. Throughout the experimental process, the slurry flow rate is meticulously monitored using an electromagnetic flowmeter. Simultaneously, the injection pressure and the pressure distribution along the grout diffusion path within the injected medium are continuously measured and recorded in real-time via pressure sensors. The duration of each test is precisely timed with a stopwatch.

3.3. Experimental Results and Analysis

3.3.1. Validation of Experimental Model Effectiveness

In the experimental setup, fluctuations in flow rate during slurry injection are observed. It is important to validate that the slurry flow rate can remain roughly constant, not strongly affected by the fluctuation in the slurry injection pump. As illustrated in Figure 4, the injection time and the volume of injected slurry exhibit a linear relationship, suggesting a constant flow rate of slurry. The slurry flow rate is approximately 3.18 L/min with an error within 10% (condition 3 in Table 3), indicating that the experimental conditions are achieved as designed.

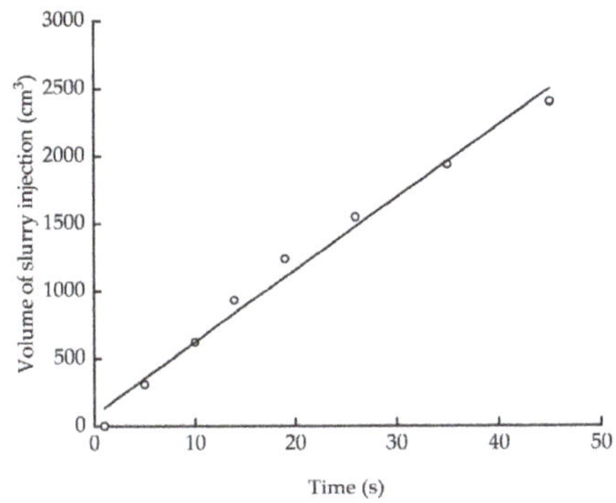

Figure 4. Changing flow rate with time under condition 3.

3.3.2. Impact of Grouting Parameters on Grouting Pressure

Figures 5 and 6 present the grouting pressure (slurry injection pressure). The results show that the grouting pressure increases nonlinearly over time. As the grouting process continues, both the grouting pressure and its rate of increase rise progressively. This increase in grouting pressure is primarily due to the flow resistance of slurry in the injected medium. Over time, the resistance builds up, leading to a corresponding increase in grouting pressure.

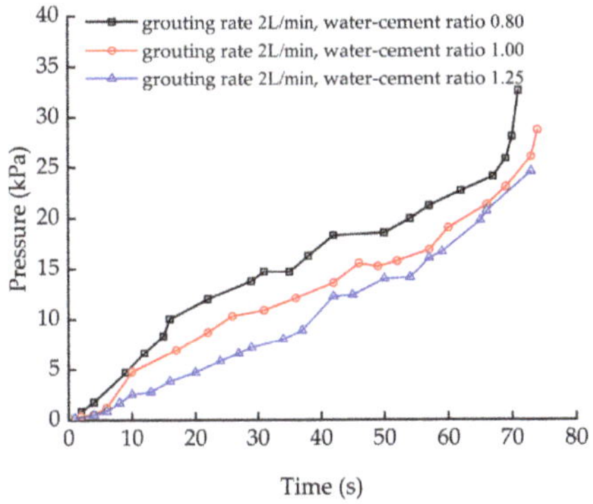

Figure 5. Impact of water–cement ratio on grouting pressure.

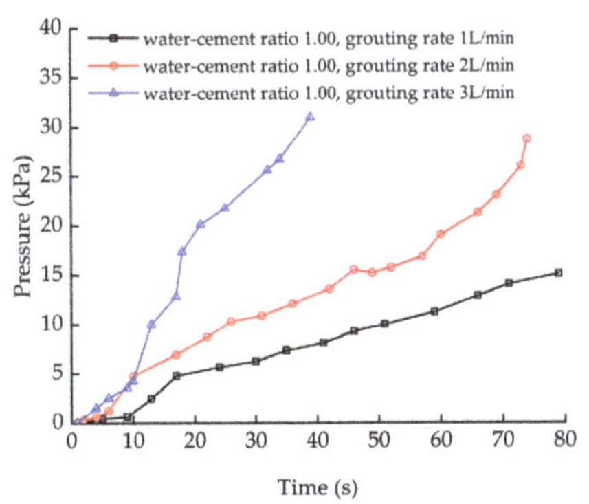

Figure 6. Impact of slurry injection rate on grouting pressure.

Figure 5 illustrates the effect of the water–cement ratio on the grouting pressure associated with slurry permeation and diffusion. Across different water–cement ratios, the grouting pressure consistently increases with grouting time (or the volume of injected slurry). This means that as the duration of grouting increases, the slurry injection pressure gradually rises. However, at the same grouting time (or volume), an increase in the water–cement ratio results in a decrease in the slurry injection pressure. The primary reason for this trend is that a higher water–cement ratio reduces the plastic viscosity and yield stress of the slurry. Consequently, the resistance that the slurry encounters during its diffusion process is lower with a higher water–cement ratio, in turn, leading to a decrease in the grouting pressure.

Figure 6 shows the impact of the slurry injection rate on the grouting pressure associated with slurry permeation and diffusion. Across different slurry injection rates, the grouting pressure consistently increases with the duration of the grouting process (or the volume of injected slurry). This indicates that as the grouting time extends, the slurry injection pressure progressively rises. As the slurry injection rate increases, the distance that the slurry permeates and diffuses within the same period is greater. This results in a higher flow resistance that the slurry must overcome, leading to an increase in the slurry injection pressure.

4. Comparison of Theoretical and Experimental Results

In this section, we conduct a comparative analysis between the results of permeation grouting experiments under different conditions and the theoretical calculations from the fractal theory-based diffusion model. This comparison aims to validate the fractal theory-based model of Bingham fluid permeation and diffusion proposed in this study. In addition to the model proposed in this study, the analysis also includes other two theoretical models established in earlier studies: the model for Bingham fluid permeation and diffusion without considering the permeation and diffusion paths, and the model for Bingham fluid permeation and diffusion considering the permeation and diffusion paths. A brief overview of these models is provided as follows.

4.1. The Theoretical Model of Permeation and Diffusion for Bingham Fluid

In the theoretical model of permeation and diffusion for Bingham fluid without considering the permeation and diffusion paths, the pressure spatiotemporal distribution is expressed as follows [28]:

$$p_c = -\left(\frac{\mu_p q}{Ak} + \frac{2\tau_0}{3}\sqrt{\frac{2\phi}{k}}\right) l_m + p_0 \tag{32}$$

In the theoretical model considering the permeation and diffusion paths, the pressure spatiotemporal distribution is expressed as follows [18]:

$$p_c = -\left(\frac{\mu_p q}{Ak} l_m + \frac{\tau_0 \phi}{6k\eta} l_m^2\right) + p_0 \tag{33}$$

where η represents the length ratio of the porous channels, which is related to the areal porosity φ by the following equation:

$$\eta = 6ln\left(\frac{\sqrt{2}}{24\varphi}\sqrt{\frac{1-\varphi}{1-0.342\varphi}}\right)\sqrt{\frac{2(1-0.342\varphi)\pi}{\left[2ln\left(\frac{\sqrt{2}}{24}\sqrt{\frac{1-\varphi}{1-0.342\varphi}}\right) - ln\varphi\right]\varphi ln\varphi}} \tag{34}$$

4.2. Comparative Analysis of Experimental Results and Theoretical Calculations

The calculation parameters for theoretical models corresponding to the permeation grouting experiments are listed in Table 4. These parameters are substituted into Equations (25), (32), and (33), respectively, for each experimental condition, yielding the relationship curve of slurry pressure with diffusion distance under grouting pressure p_0 (Figure 7).

Table 4. Calculation parameters.

Experimental Condition Number	q (L/min)	p_0 (kPa)	t (s)	A (m^2)
1	1	23.84	130	
2	2	28.73	80	
3	3	31.04	46	0.00785
4	2	32.60	80	
5	2	24.60	80	

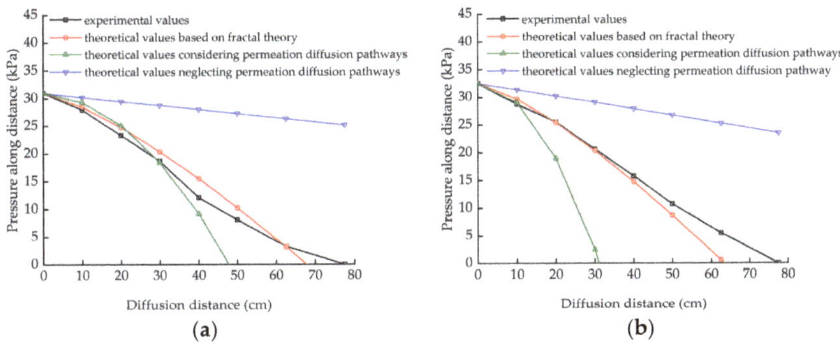

Figure 7. Comparison of experimental and theoretical results: (**a**) condition 3 and (**b**) condition 4.

As shown in Figure 7, taking experimental conditions 3 and 4 as examples, the theoretical model that neglecting permeation diffusion pathways exhibit a linear decay with an increasing slurry diffusion distance. This significantly overestimates the slurry pressure obtained from experiments. On the other hand, the model that considers both the slurry permeation path and the fractal theory-based model can capture the general changing trend over the diffusion distance. The fractal theory-based model can better predict the slurry diffusion distance under certain grouting pressure, and better describe the slurry pressure over distance.

In the experiments conducted in this section, the slurry injection stops at the point at which the slurry is just about to flow out of the overflow pipe. Therefore, the slurry diffusion distance in all tests is set to 80 cm. Under different experimental conditions, the calculated slurry diffusion distances from the model neglecting permeation pathways are 3.5 to 6.3 times larger than the experimental values, showing significant discrepancies. The reason is that the approach assumes slurry diffusion occurs uniformly in a circular pipe, disregarding the influence of pore structures, thereby causing notable deviations from the experimental results.

In contrast, considering permeation diffusion pathways reduces the calculated slurry diffusion distances to 0.4 to 1.4 times of the experimental values, but it overlooks the effects of porous medium structural parameters on slurry diffusion.

On the other hand, the slurry diffusion distances calculated based on fractal theory range from 0.8 to 1.1 times of the experimental values (Figure 8), showing the greatest agreement. Overall, the proposed Bingham fluid permeation diffusion model based on fractal theory comprehensively characterizes slurry diffusion paths and pore-structure features. Its theoretical pressure calculations closely match experimental values.

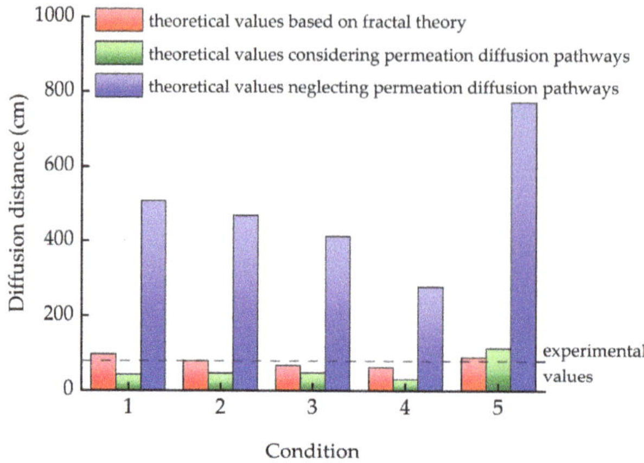

Figure 8. Diffusion distance at initial grouting pressure under different conditions.

4.3. Error Analysis of Theoretical Results

Figures 9 and 10 show the deviation in slurry pressure between theoretical (fractal theory-based diffusion model) and experiment results under various conditions. As presented in the figures, the deviation is positively correlated with the slurry diffusion distance. As the slurry diffusion distance increases, more pores become clogged by the slurry, resulting in faster pressure attenuation observed in experiments, thereby increasing the deviation from theoretical values.

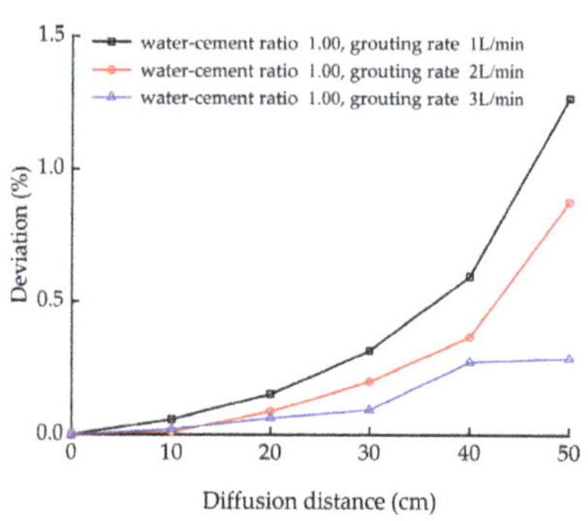

Figure 9. Impact of grouting rate on deviation.

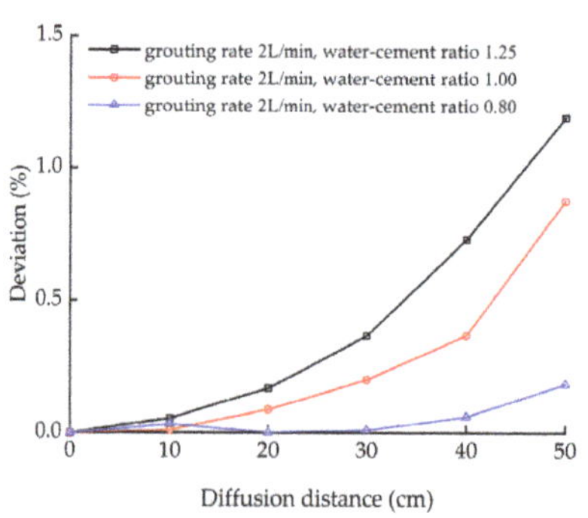

Figure 10. Impact of water–cement ratio on deviation.

As presented in Figure 9, the higher the grouting rate (slurry injection rate), the smaller the deviation between the theoretical and experimental values. One possible reason is that at a lower grouting rate, the slurry flows for a longer time through the porous medium, causing a greater difference in the viscosity of the slurry along the diffusion distance. There is a significant difference in slurry viscosity between the vicinity of the injection point and the distant areas. However, the theoretical model based on fractal theory proposed in this study does not account for the temporal variability in slurry viscosity.

Figure 10 reveals that under experimental condition 4 (slurry water–cement ratio of 0.8), the theoretical values of pressure based on the fractal-theory model closely match the experimental results. As the slurry water–cement ratio increases, however, the deviation between theoretical and experimental values becomes larger. One possible reason is the filtration effect when slurry flows through a porous medium. In this process, some particles in the slurry are retained in pores, which reduces flow resistance at greater diffusion distances, resulting in lower slurry pressure. Moreover, higher water–cement ratios enhance the filtration effect [14]. The experimental observations in this study also confirm significant filtration effects (Figure 11).

Figure 11. Image of the bottom and top of the sample after slurry penetration.

Figure 11 depicts a comparison between the bottom and top sections after the experiment at condition 5 (slurry water–cement ratio of 1.25). The bottom of the sample (inlet) appears darker and denser, indicating a higher slurry concentration, whereas the top of the sample (outlet) appears lighter and more diluted, illustrating noticeable slurry filtration effects. However, the fractal theory-based Bingham fluid permeation diffusion theoretical model does not consider the filtration effects of the slurry. The theoretical calculations of the slurry pressure at higher water–cement ratios tend to be larger than the experimental values, with discrepancies increasing as the water–cement ratio rises.

5. Conclusions

A theoretical penetration grouting model for Bingham fluid based on fractal theory was proposed in this study and validated by pure-cement slurry diffusion experiments. The primary conclusions are listed as follows:

(1) In addressing the influence of pore structure on the diffusion of Bingham fluid slurry, this study establishes a theoretical model based on fractal theory. The parameters involved in the proposed model can be determined from conventional geotechnical tests.

(2) A one-dimensional permeation grouting simulation apparatus was developed. The experimental setup allows for the comprehensive analysis of the spatiotemporal pressure variations of the pure-cement slurry under different experimental conditions.

(3) The theoretical model for Bingham fluid permeation grouting based on fractal theory proposed in this study aligns closely with the experimental results. It comprehensively considers the pore structure and the tortuosity effect on the pure-cement slurry

permeation and diffusion. This model provides a significant reference for the design of grouting operations and can contribute valuable insights to related construction practices.

Author Contributions: J.G.: formal analysis, methodology, and writing—original draft. C.X.: data curation and writing—original draft. Y.W.: supervision and writing—review and editing. S.G.: investigation. H.C.: conceptualization and project administration. L.J.: resources. All authors have read and agreed to the published version of the manuscript.

Funding: This research was funded by the National Natural Science Foundation of China, grant number U2240210, 52279098; and the Fundamental Research Funds for the Central Universities, grant number B230201021.

Data Availability Statement: Data is contained within the article.

Conflicts of Interest: The authors declare no conflicts of interest.

References

1. Kuang, J. *Theory and Project Example of Grout in Geotechnical Engineering*; Science Press: Beijing, China, 2001.
2. Yang, X.; Wang, X.; Lei, J. Study on grouting diffusion radius of Bingham fluids. *J. Hydraul. Eng.* **2004**, *35*, 75–79.
3. Yang, X.; Lei, J.; Xia, L.; Wang, X. Study on grouting diffusion radius of exponential fluids. *Rock Soil Mech.* **2005**, *26*, 1803–1806.
4. Ruan, W. Research on diffusion of grouting and basic properties of grouts. *Chin. J. Geotech. Eng.* **2005**, *27*, 69–73.
5. Yang, Z.; Hou, K.; Guo, T.; Ma, Q. Study on penetration grouting mechanism based on Bingham fluid of time–dependent behavior. *J. Sichuan Univ. Eng. Sci. Ed.* **2011**, *43*, 67–72.
6. Yang, Z.; Hou, K.; Guo, T.; Ma, Q. Study of column–hemispherical penetration grouting mechanism based on Bingham fluid of time–dependent behavior of viscosity. *Rock Soil Mech.* **2011**, *32*, 2698–2703.
7. Yang, Z.; Niu, X.; Hou, K.; Zhou, Z.; Liang, W.; Guo, Y.; Lu, Y.; Yang, B.; Cheng, Y. Columnar diffusion of cement grout with time dependent rheological parameters. *Chin. J. Rock Mech. Eng.* **2015**, *34*, 1415–1425.
8. Zhou, J.; Lu, J.; Zhang, Y.; Wang, Z. Calculation formula of permeation grouting considering slurry viscosity variation. *J. Mater. Sci. Eng.* **2019**, *37*, 758–762.
9. Zhang, L.; Zhang, Q.; Liu, R.; Li, S.; Wang, H.; Li, W.; Zhang, S.; Zhu, G. Penetration grouting mechanism of quick setting slurry considering spatiotemporal variation of viscosity. *Rock Soil Mech.* **2017**, *38*, 443–452.
10. Saada, Z.; Canou, J.; Dormieux, L.; Dupla, J.C. Evaluation of elementary filtration properties of a cement grout injected in a sand. *Can. Geotech. J.* **2006**, *43*, 1273–1289. [CrossRef]
11. Chupin, O.; Saiyouri, N.; Hicher, P.Y. The effects of filtration on the injection of cement–based grouts in sand columns. *Transp. Porous Media* **2008**, *72*, 227–240. [CrossRef]
12. Saada, Z.; Canou, J.; Dormieux, L.; Dupla, J.C.; Maghous, S. Modeling of cement suspension flow in granular porous media. *Int. J. Numer. Anal. Methods Geomech.* **2005**, *29*, 691–711. [CrossRef]
13. Dupla, J.C.; Canou, J.; Gouvenot, D. An advanced experimental set–up for studying a mono directional grout injection process. *Ground Improv.* **2004**, *8*, 91–99. [CrossRef]
14. Feng, X.; Xia, C.; Wang, F.; Zhang, B. Variation of granular slurry diffusion distance in sand medium. *J. Shandong Univ. Eng. Sci.* **2020**, *50*, 20–25.
15. Wang, Q.; Zhu, Y.; Li, W.; Cui, P.; Pan, Y. Study on the mechanism of column permeation grouting of Bingham fluid considering the spatial attenuation of viscosity. *Chin. J. Rock Mech. Eng.* **2022**, *41*, 1647–1658.
16. Li, S.; Feng, X.; Liu, R.; Zhang, L.; Han, W.; Zheng, Z. Diffusion of grouting cement in sandy soil considering filtration effect. *Rock Soil Mech.* **2017**, *38*, 925–933.
17. Zhang, Q.; Wang, H.; Liu, R.; Li, S.; Zhang, L.; Zhu, G.; Zhang, L. Infiltration grouting mechanism of porous media considering diffusion paths of grout. *Chin. J. Geotech. Eng.* **2018**, *40*, 918–924.
18. Lu, Q.; Yang, Z.; Yang, Z.; Yu, R.; Zhu, Y.; Yang, Y.; Zhang, B.; Wang, R.; Fang, Y.; Yu, D.; et al. Penetration grouting mechanism of Bingham fluid considering diffusion paths. *Rock Soil Mech.* **2022**, *43*, 385–394.
19. Yang, Z.; Lu, J.; Wang, Y.; Zhang, Z.; Yang, Y.; Zhu, Y.; Zhang, J.; Guo, Y.; Chen, X. Column penetration grouting mechanism for power–law fluids considering tortuosity effect of porous media. *Chin. J. Rock Mech. Eng.* **2021**, *40*, 410–418.
20. Wang, H.; Dong, L.; Zhang, Q.; Li, Z.; Zhang, P. Permeation grouting mechanism of viscous time–varying fluid considering diffusion path. *Front. Earth Sci.* **2023**, *10*, 1100196. [CrossRef]
21. Yu, B.; Li, J. Some fractal characters of porous media. *Fractals* **2001**, *9*, 365–372. [CrossRef]
22. Yun, M.; Yu, B.; Cai, J. A fractal model for the starting pressure gradient for Bingham fluids in porous media. *Int. J. Heat Mass Transf.* **2007**, *51*, 1402–1408. [CrossRef]
23. Zhang, B.; Yu, B.; Wang, H.; Yun, M. A fractal analysis of permeability for power–law fluids in porous media. *Fractals* **2006**, *14*, 171–177. [CrossRef]

24. Yang, Z.; Chen, M.; Ding, Y.; Yang, Y.; Zhu, Y.; Guo, Y.; Wang, R.; Zhang, B.; Fang, Y.; Yu, D.; et al. Influence of coupling effects of time and water–to–cement ratio on rheological properties of Bingham cement grouts. *Adv. Mater. Sci. Eng.* **2021**, *2021*, 5781753. [CrossRef]
25. Kong, X. *Advanced Mechanics of Fluid in Porous Media*; University of Science and Technology of China Press: Beijing, China, 1999.
26. Hou, X.; Mo, H.; Zhao, W.; Huang, Y. Infiltration grouting mechanism in porous media based on fractal theory. *J. Yangtze River Sci. Res. Inst.* **2024**, *41*, 106–113.
27. *GB/T50123–2019*; Standard of Geotechnical Test Method. China Planning Press: Beijing, China, 2019.
28. Zhang, L. Study on Penetration and Reinforcement Mechanism of Grouting in Sand Layer Disclosed by Subway Tunnel and Its Application. Doctor's Thesis, Shandong University, Jinan, China, 2017.

fractal and fractional

MDPI

Article

Fractal Analysis of Particle Size and Morphology in Single-Particle Breakage Based on 3D Images

Ruidong Li [1], Xiang Gao [1], Shao-Heng He [1,*], Dongheng Ru [2] and Zhi Ding [3]

[1] Department of Civil and Environmental Engineering, The Hong Kong Polytechnic University, Hong Kong, China; 22072096r@connect.polyu.hk (R.L.); xiang.gao@connect.polyu.hk (X.G.)
[2] School of Aerospace Engineering and Applied Mechanics, Tongji University, Shanghai 200092, China; rudongheng@tongji.edu.cn
[3] Department of Civil Engineering, Hangzhou City University, Hangzhou 310015, China; dingz@zucc.edu.cn
* Correspondence: shaoheng.he@polyu.edu.hk

Abstract: The accurate modeling of single-particle breakage based on three-dimensional (3D) images is crucial for understanding the particle-level mechanics of granular materials. This study aims to propose a systematic framework incorporating single-particle breakage experiments and numerical simulations based on a novel 3D particle reconstruction technique for fractal analysis of particle size and morphology in single-particle breakage. First, the vision foundation model is used to generate accurate particles from 3D images. The numerical approach is validated by simulating the single-particle breakage test with multiple Fujian sand particles. Then, the breakage processes of reconstructed sand particles under axial compression are numerically modeled. The relationship between 3D fractal dimensions and particle size, particle crushing strength, and morphology is meticulously investigated. Furthermore, the implications of these relationships on the particle breakage processes are thoroughly discussed, shedding light on the underlying mechanisms that govern particle breakage. The framework offers an effective way to investigate the breakage behavior of single sand particles, which will enhance understanding of the mechanism of the whole particle breakage process.

Keywords: fractal dimension; 3D μCT images; vision foundation models; particle breakage; numerical simulation

1. Introduction

Particle breakage is an extremely important phenomenon in nature, industrial processes, and geotechnical engineering. For example, particle breakage extensively exists in soils and exerts a significant influence on soil behavior by changing soil microstructure [1,2]. The interplay between particle breakage and soil response is critical, dictating the need for a nuanced understanding to ensure the resilience and performance of geotechnical structures. Furthermore, optimizing the percentage of particle breakage during ball milling can effectively increase economic benefits. Therefore, the ability to predict particle breakage in nature or industrial processes has sparked great interest in soil/rock mechanics but has also been a persistent challenge. The mechanism of particle breakage is still poorly understood. Complicated interactions among the contributions of the mineral composition, stress condition, loading rate, particle morphology, and particle distribution will affect or even lead to particle breakage [3,4]. The most efficient way to investigate particle breakage is to conduct single-particle breakage loading tests [5,6]. Single-particle breakage loading tests have been demonstrated to be effective in analyzing energy utilization in comminution processes and material mechanical responses under different stresses, as well as the effect of particle size, morphology, and physical properties on the breakage characteristics [7–9].

Apart from experiments, numerical simulations such as the discrete element method (DEM) have been widely used in investing particle breakage [10,11]. Compared to experi-

mental investigations, numerical simulations are more flexible, supporting the visualization of microscopic characteristics, e.g., the interior fractures. Generally, there are two prevalent approaches utilized in the DEM for breakage simulations. The first one is based on the premise that parent particles are composed of distinct, bonded, and completely resolved child particles from the beginning of the simulation [12,13]. The alternative way, called particle replacement, refers to ways where particles are replaced with child particles, such as spheres and superquadrics. Once the specific failure threshold is reached, those child particles will occupy the geometric space of the parent particle to mimic particle breakage [14,15]. The evolution of grading, anisotropy, energy dissipation, and plastic change during single-particle breakage have been investigated by the DEM widely [15,16]. The effect of particle morphology on single-particle breakage is also well summarized and validated [17,18]. However, most existing studies focus on particles with controlled shapes instead of realistic morphology. The accurate and reliable modeling of the breakage process based on single-particle breakage experiments and three-dimensional (3D) μCT images remains scarce.

In recent years, fractal geometry techniques have found widespread applications in many fields, including medicine, material science, geography, etc. Relatively, there have been a few cases of the application of fractal geometry in geotechnical engineering, especially for single-particle breakage. Research has indicated that the fragmentation of granular materials can be described by a fractal dimension [19]. Arasan et al. analyzed the relationship between the fractal dimension and shape properties of particles but only in two-dimensional (2D) fractals [20]. The investigation of 3D fractal dimensions for single-particle breakage is still unclear, especially for scenarios that combine experiments and numerical simulations. Furthermore, there is currently no comprehensive research examining the effect of particle size and morphology on 3D fractal dimensions.

To this end, this study explores the relationship between fractal dimension and particle size and morphology in single-particle breakage from in experiments to numerical simulations based on 3D images. A novel 3D particle reconstruction technique is proposed and used to obtain the particle morphologies. Furthermore, a series of single-particle loading tests are conducted to investigate the mechanical behaviors of Fujian sand particles selected from the sample for reconstruction. Subsequently, the DEM model is established and calibrated based on experimental results. Finally, a series of numerical simulations are conducted to explore and analyze the relationship between fractal dimensions and particle size and morphology.

2. Materials and Methods

2.1. Materials and Equipment

The Tiniusolsen 50ST multi-parameter testing machine was used for conducting the single-particle crushing tests, as shown in Figure 1a. A variety of load cells are available for differing test materials and methods, providing precise applied load measurements. A compression loading cell was employed to crush the particle that was placed between two load plates, and the loading measuring capacity of the compression loading cell was 5 kN, with a resolution of 0.01 N. During the tests, the displacement ratio of 0.1 mm/min was used until the final fracture of the testing particle occurred, and the corresponding force and displacement were measured for further analysis [21].

The tested material was Fujian sand, known as the Chinese standard sand, which is widely employed in soil mechanics research. It is characterized by a predominant quartz composition that imparts high intrinsic strength to the material. Furthermore, the stable distribution in particle size, shape, and inter-flaw of this material ensures that experimental results are reproducible. In this study, 30 single-particle crushing tests were conducted for the statistical analysis of the fragmentation properties, and the origin-tested sand particles were randomly selected from the following sample used for CT scanning and reconstruction, as shown in Figure 1b. The particle size range used in the crushing test was from 1.2 mm to 2.0 mm, and the average particle size was 1.57 mm.

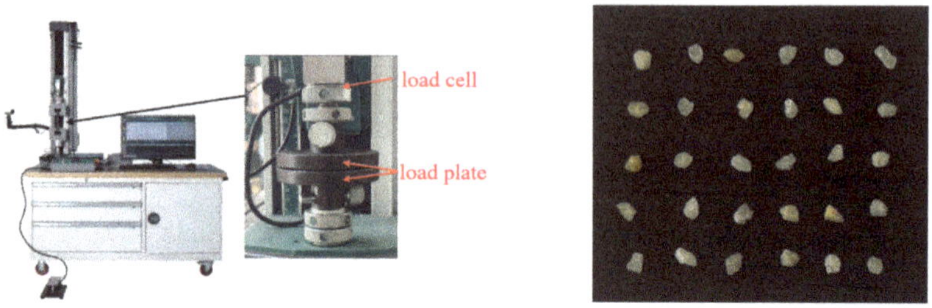

(**a**) (**b**)

Figure 1. Illustrations of experiment: (**a**) loading apparatus; (**b**) tested sand particles.

2.2. Particle Crushing Strength

Tensile failure is considered the primary cause of particle breakage, and a characteristic particle crushing strength derived from the tensile stress defined by Jaeger was employed to analyze the experiment results [22]:

$$\sigma = \frac{F}{d^2} \tag{1}$$

where d is the particle diameter and F is the force applied on the tested particle, measured by the testing machine. In this study, the catastrophic splitting fracture is focused, and the corresponding peak force F_f is employed in the particle strength analysis:

$$\sigma_f = \frac{F_f}{d^2} \tag{2}$$

where σ_f is the particle crushing strength. Therefore, the probability of the survival of a particle at a given stress level σ_c is defined as P_s:

$$P_s = \frac{N\big|\sigma_f \geq \sigma_c}{N_t} \tag{3}$$

where N_t is the total number of particles; $N \mid \sigma_f \geq \sigma_c$ represents the number of particles whose strength σ_f is greater than the given stress σ_c. P_s can be estimated by the Weibull distribution [23]:

$$P_s = \exp[-(\frac{\sigma_f}{\sigma_0})^m] \tag{4}$$

where m is the Weibull modulus and σ_0 is the characteristic stress of the particles, at such a stress level that 37% of the tested particles survived.

2.3. 3D Particle Reconstruction

The Nikon XT H225ST 2X μCT system was used to capture the microstructure of the Fujian sand particles using 3D μCT images. This system consists of an X-ray beam tub, a high-voltage generator, a sample manipulator, and a flat panel detector. The generated X-ray passes through the sample, casting the shadow image on the detector to form one projection. For high-quality μCT images, 3000 projections were taken in this study for each scan as the sample rotated 360°. The dimension sizes of the used 3D μCT images were 1400 × 1400 × 90 (width × height × depth) voxels, including 318 separate Fujian sand particles, as shown in Figure 2a. To reconstruct the 3D Fujian sand particles from the μCT images accurately, the first vision foundation model, the Segment Anything Model (SAM), was used [24]. The main components of the SAM are an image encoder, a prompt encoder, and a mask decoder, as shown in Figure 2b.

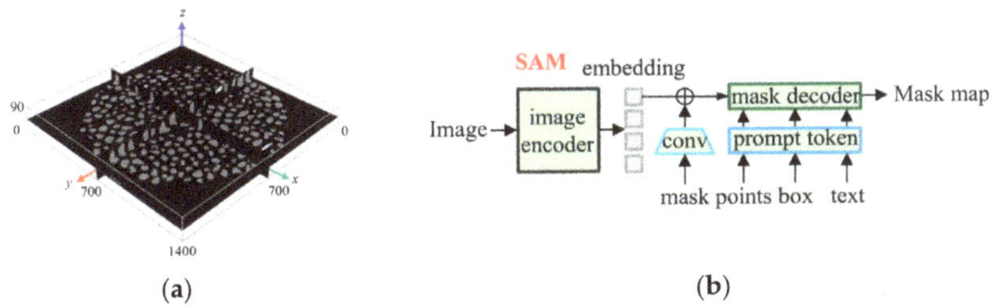

Figure 2. Illustration of (**a**) 3D µCT images; (**b**) main components of SAM.

In this study, the $n \times n$ regular points were chosen as input prompts to identify all particles in the image, as shown in Figure 3a. n denotes the number of points in a row or column, taken as 32 for balancing efficiency and accuracy. All µCT slices along the z-axis were transformed into a 2D mask map by the SAM, as shown in Figure 3b. These 2D mask maps were stitched along the z-axis to form a 3D mask map. Finally, the marching cube algorithm [25] was used to convert all masks into 3D particles, as shown in Figure 3c.

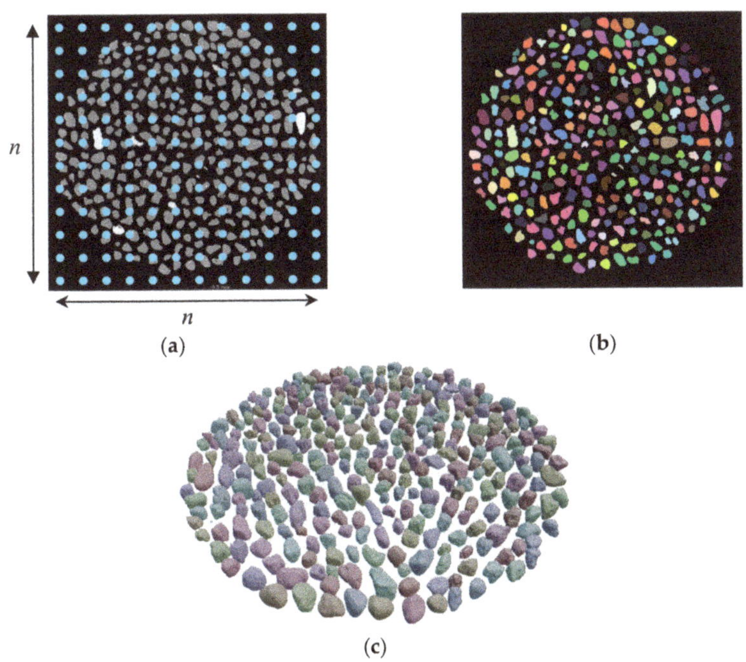

Figure 3. Illustration of (**a**) input prompts; (**b**) 2D mask map; (**c**) 3D reconstructed particles.

2.4. Numerical Simulation

The DEM software particle flow code (PFC3D 7.0) [26] was utilized to conduct numerical single-particle crushing simulations, leveraging the reconstructed 3D particle geometries. The common bonded methods employed in the simulation of particle breakage are the bonded particle method (BPM) and the bonded block method (BBM). In the BPM, the basic cell is depicted as a rigid spherical ball, whereas the BBM employs rigid polyhedral blocks to represent the basic cell. Consequently, unexpected voids are inevitably formed between bonded spheres with only slight overlaps. On the other hand, polyhedral blocks can be modeled to ensure that adjacent faces align perfectly or slightly overlap, thereby minimizing the number of voids. Considering the limited inter-flaws and inter-voids in Fujian sand, the simulation based on the BBM is more useful than that based on the BPM. The numerical geometry model was established based on the high-accuracy 3D reconstructed geometry in Section 2.3. The geometry used in the numerical simulation was also randomly selected from the sample for reconstruction, and the size range was the same as the tested particles. In addition, some operations on the movement and rotation of the geometry were necessary to ensure that the position and orientation during loading were consistent with the experiment. Firstly, the centroid of the geometry was translated to the origin to ensure it was centered on the loading plate. Subsequently, the geometry was rotated such that its principal axes aligned sequentially with the x, y, and z axes in descending order of their magnitudes, thereby standardizing the orientation.

A refined geometry particle modeling method for the BBM proposed by Fang et al. [27] was adopted in this study. Firstly, a specified number of balls were generated within the imported geometric shape, and then the radius of the balls gradually increased until they filled the geometric space. Furthermore, an equal number of Voronoi cells (rblocks in PFC) were generated based on the balls. Subsequently, geometry was employed again to cut the assembly of the Voronoi cells to obtain the precise numerical model. The effect of the number of cells (N_c) on the peak force decreased as the N_c increased; therefore, the N_c = 500 was adopted in this study, with limited and tolerable discrepancies of peak force and strength [28].

After the DEM geometry model was established, the linear parallel bond model, a built-in model in PFC, was applied between the blocks with the specific gap limitation, as shown in Figure 4a. The detailed behavior is described in PFC documentation [26]. This contact model provides two types of interface behavior: bonded behavior and frictional behavior. When the activated contact between the rblocks is in the bonded state, the interaction behavior is controlled by the microparameters of the parallel bond group; otherwise, the microparameters of the linear group dominate the interface behavior. If the normal and shear stress of the bonded contact exceeds the corresponding tensile and cohesive strength, the bonds of contact break and the bonded contact degenerates into a friction contact.

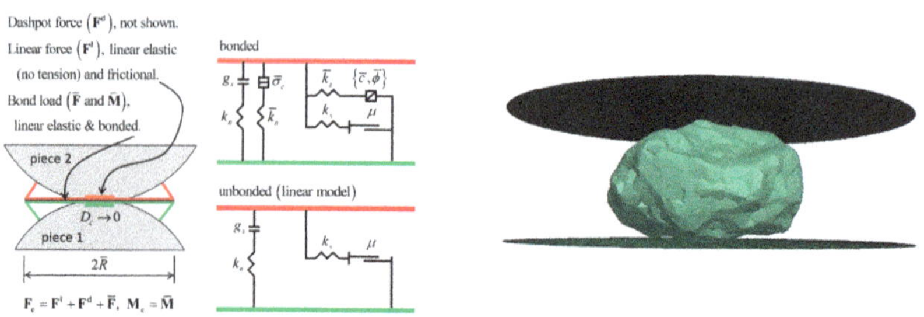

(a) (b)

Figure 4. Illustration of (**a**) linear parallel bond model; (**b**) numerical model of the single-particle crushing test.

Two circular walls were generated at the top and bottom of the particle model, serving as the loading platens, as illustrated in Figure 4b. Considering the realistic velocity in the experiment cannot be employed in the numerical simulation due to the limited calculational efficiency, a series of numerical tests with different loading rates ranging from 0.01 m/s to 0.5 m/s were conducted to investigate the dynamic effect, as shown in Figure 5. It was observed that the peak force decreased as the loading velocity decreased and gradually stabilized. The variance in the peak force between the loading velocities of 0.4 m/s and 0.1 m/s was a mere 0.2%, which was deemed inconsequential. Therefore, a constant velocity of 0.04 m/s was employed in the following numerical model, at which the dynamic effect decreased. Moreover, to avoid dynamic stress shock in the internal rblocks, the

velocity of the loading platens was gradually increased to the specified speed at the start stage.

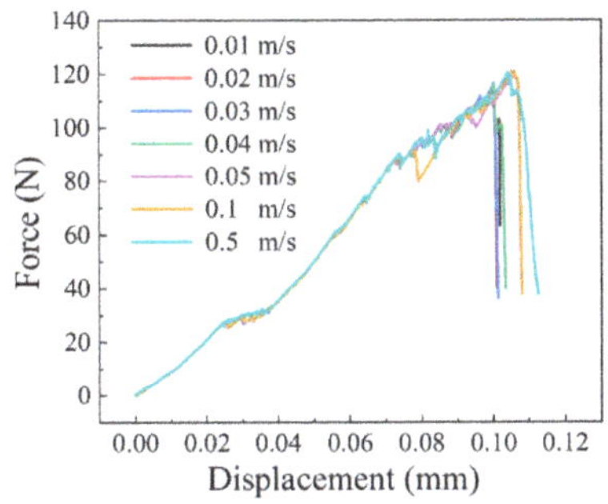

Figure 5. Force–displacement curves with different loading rates.

2.5. Shape Indexes and Fractal Dimensions

Particle morphology is generally evaluated by using shape indexes. In this study, three common shape indexes were used: the elongation index (*EI*), flatness index (*FI*), and sphericity index (*SI*). The *EI* and *FI* were used to quantify the form of the particles and are defined as follows [29]:

$$EI = \frac{b}{a} \tag{5}$$

$$FI = \frac{c}{b} \tag{6}$$

where a, b, and c are the major, intermediate, and minor principal dimensions of a particle, respectively. The *SI* is defined as the surface area of a sphere, enclosing the same volume V as the 3D object. This index is commonly used for the measuring of compactness and can be derived from the classical isoperimetric inequality and calculated as follows:

$$SI = \frac{\sqrt[3]{36\pi V^3}}{A} \tag{7}$$

where V and A denote the volume and surface area of the particle, respectively. In this study, the equivalent volume diameter d of a particle was used to quantify the particle size:

$$d = \sqrt[3]{\frac{6V}{\pi}} \tag{8}$$

The 3D particles or fragments were first converted to voxels (3D images). Then, the 3D fractal dimension F_D of the particles was calculated by using the box-counting method [30], expressed as follows:

$$F_D = \lim_{r \to 0} \frac{\log(N_r)}{\log(1/r)} \tag{9}$$

where r is the cube size, which ranges from the smallest unit in the voxel scale to 1/2 of the entire volume, and N_r is the total number of cubes containing the object of interest in the r scale. More details about the computation of the 3D fractal dimensions of particles can be found in [31]. The coefficient of determination (R^2) and mean squared error (MSE) were adopted to evaluate the fitting quality:

$$R^2 = 1 - \frac{SS_{\text{res}}}{SS_{\text{tot}}} \tag{10}$$

$$\text{MSE} = \frac{1}{n}\sum_{i=1}^{n} \left(Y_i - \hat{Y}_i\right)^2 \tag{11}$$

where SS_{res} and SS_{tot} are the sum of the squares of the residuals and the total sum of the squares, respectively, n denotes the total number, and Y_i and $\hat{Y}_i$ are the observed and predicted values, respectively.

3. Analysis of Experimental and Numerical Simulations

3.1. Experimental Results

Figure 6 presents typical force–displacement curves alongside the corresponding photographs of the representative tested particles before and after undergoing the crushing test. Due to the fine engineering properties of Fujian sand, most of the particles exhibit a smooth (result 1 in Figure 6) or slight sawtooth (result 2 in Figure 6) force–displacement curve preceding a sudden force decrease as the final catastrophic fracture. The tested particles tended to break into two to four primary fragments, occasionally accompanied by a scattering of minor fragments and some fine powder, as shown in Figure 7. This behavior aligned with the identification of the two fracture mechanisms reported by Nakata et al. [21]. During compression, localized damage at the contact points or edges led to the initial sawtooth pattern observed in the force–displacement curves and resulted in the formation of minor fragments. Subsequently, catastrophic failure occurred, causing the particle to split into two or three major fragments. These two fracture patterns are defined as asperity fracture and splitting fracture, respectively. The strength and the probability of the survival of each particle are depicted in Figure 8a and Figure 8b, respectively. The characteristic stress σ_0 in the test was 50.65 MPa, consistent with previous studies [21].

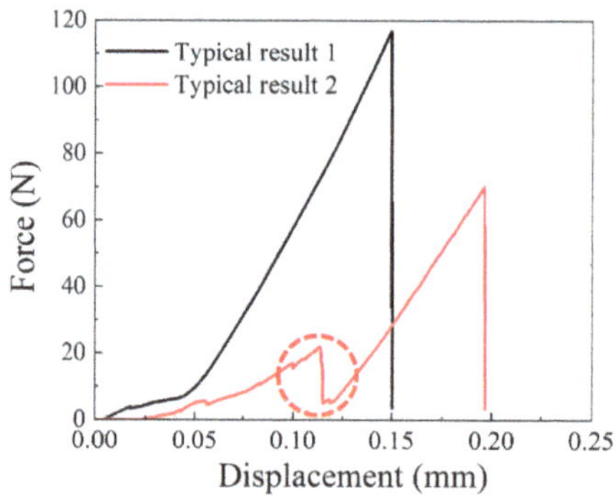

Figure 6. Typical force–displacement curves. The slight sawtooth is marked by the dotted circle.

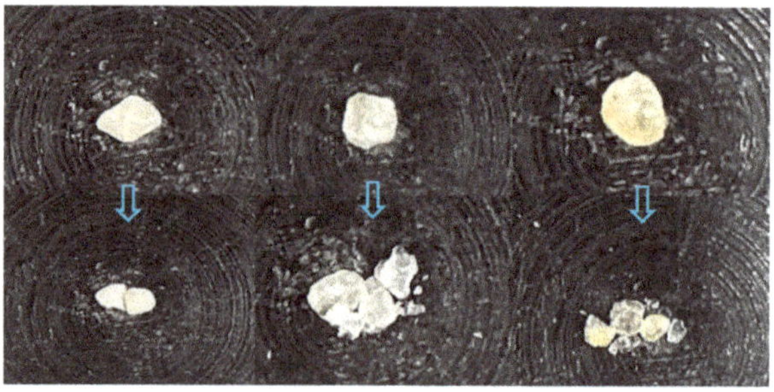

Figure 7. Representative fragmentations of sand particles.

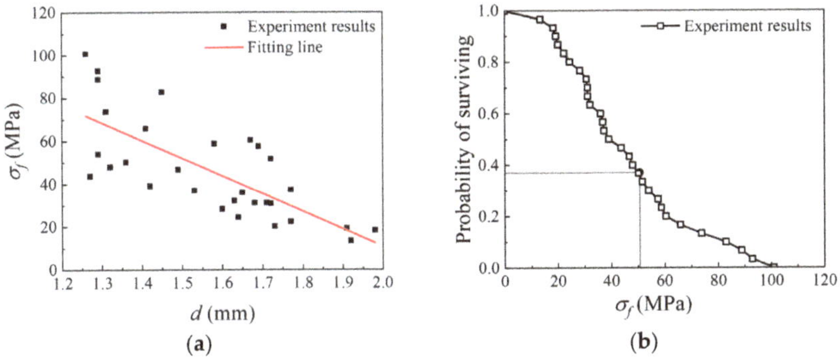

Figure 8. Experimental results: (**a**) single-particle crushing strength; (**b**) probability of survival.

3.2. Numerical Calibration

The microparameters of the contact model were calibrated by matching the characteristic strength σ_0 of the particles calculated from the results of the single-particle crushing test. After finely adjusting the microparameters of the linear parallel bond model, the distribution of the probability of surviving was similar to the experimental results as shown in Figure 9a, and the characteristic stress in the simulation was 51.38 MPa, well in agreement with the 50.65 MPa in the experiment. Furthermore, the typical force–displacement curves of the simulation results are depicted in Figure 9b, capturing the different fracture patterns illustrated in Figure 6. This correlation serves as a compelling validation of the numerical model's applicability and its capacity to accurately emulate single-particle breakage behaviors. The calibrated microparameters used in the DEM are listed in Table 1.

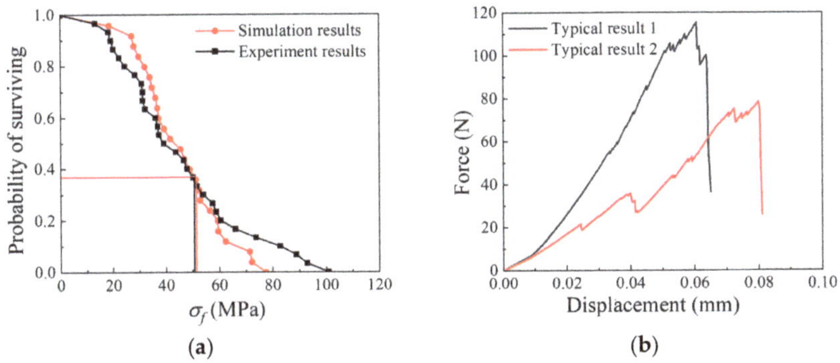

Figure 9. Calibration of numerical simulations: (**a**) distribution of the probability of surviving; (**b**) typical force–displacement curves.

It can be observed from Figure 9a that discrepancies exist between the experimental and simulated curves, particularly at their extremities, which can be attributed to two principal factors. Firstly, in the experiment, the radius of the particles was manually measured using an electronic micrometer caliper, whereas in the simulation, the radius of the particles was represented by the distance between two loading platens. Overall, the measurement radius in the simulation was smaller than that in the experiment. Secondly, the simulation did not account for flaws and incomplete homogeneity within the particles. As a result, the variation in strength observed in the simulations was less pronounced than that in the experiments. This disparity is reflected in the Weibull modulus (m) values, with an experimentally determined value of m = 2.21 compared to a value of m = 3.15 obtained from the simulations.

Table 1. Microparameters used in DEM simulation.

Microscale Parameters	Value
Particle density (kg/m^3)	2650
Damping	0.5
Frictional coefficient	0.5
Bond effective modulus (GPa)	1.9
Bond normal to shear stiffness ratio	2.5
Bond tensile strength (MPa)	29
Bond cohesive strength (MPa)	28.5
Rblock effective modulus (GPa)	1.9
Rblock normal to shear stiffness ratio	2.5
Wall elastic modulus (GPa)	1.9
Wall normal to shear stiffness ratio	1

Regarding the fragmentations produced by the catastrophic failure of the particle, as shown in Figure 10, the numerical model yielded a primary fragment count ranging from 2 to 4 pieces, thereby demonstrating a close concurrence with the experimental observations. Furthermore, within the contact zone, the isolated block marked in navy blue represents the abrasion and asperity fracture of the particle.

Figure 10. Numerical results of particles before and after breakage. Each color represents a particle.

4. Relations Between Fractal Dimensions, Particle Size, Crushing Strength, and Morphology

The application of fractal theory to investigate the relationship between 3D fractal dimensions, particle size, particle crushing strength, and morphology holds significant practical implications. Understanding these relationships can lead to improved material performance in various industries. Furthermore, by establishing a quantitative link between fractal dimensions and mechanical properties, engineers can design materials with tailored characteristics, optimizing them for specific applications.

4.1. Relations Between Fractal Dimensions and Particle Size

The statistical results of the particle size *d* before and after the breakage are presented in Figure 11a and Figure 11b, respectively. It is clear that particle sizes ranged from 1.2 mm to 1.8 mm before the single-particle breakage. The number of particles with a size between 1.4 mm and 1.5 mm was the highest. The dominant particle size ranged from 0.2 mm to 0.4 mm after the breakage, consistent with the experimental results. Small-size fragments were mostly observed in single-particle breakage fragments. Notably, the fine fragments or powder-like particles within the size range of 0 mm to 0.2 mm were excluded from the statistical analysis, as they were not encompassed within the defined size range for the blocks under consideration. Therefore, the count in this range was zero.

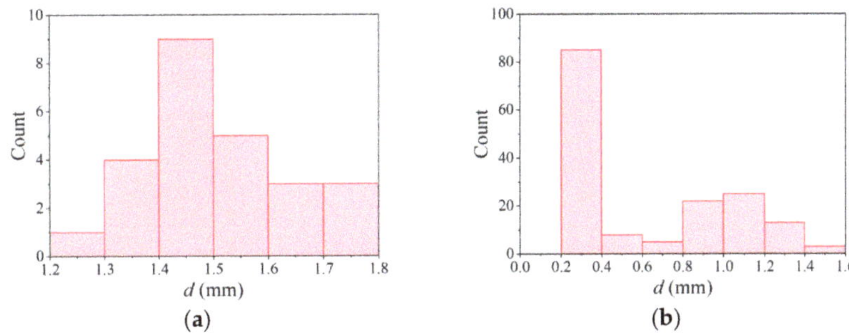

Figure 11. Statistical results of particle size d: (**a**) before breakage; (**b**) after breakage.

The correlation between the 3D fractal dimension of the particles and the particle size before and after the particle breakage are presented in Figure 12a and Figure 12b, respectively. It can be seen that the particle size decreased while the fractal dimension increased. The particle size before the breakage and the 3D fractal dimension can be fitted by a straight line with R^2 = 0.625. The unsatisfactory R^2 value may be attributed to the insufficient number of simulations. The fitting results could be improved by incorporating a greater number of simulations. Moreover, the accuracy of the reconstructed 3D models also plays a significant role. Inaccurate surface details in the 3D model can introduce noise, which in turn may lower the R^2 value. Interestingly, the relation between the particle size after the breakage and the 3D fractal dimension can be interpreted by an exponential function with a high R^2 = 0.931. These results suggest that the fractal dimension after particle breakage is significantly affected by the particle size. The sizes of many small fragments are similar, but the individual fractal dimensions exhibit significant variation. A possible explanation is that smaller particle fragments are more likely to be irregular in shape, resulting in varying fractal dimensions. The difference in the relation between the particle size distribution and 3D fractal dimension before and after breakage can be used as an indicator for identifying the breakage degree.

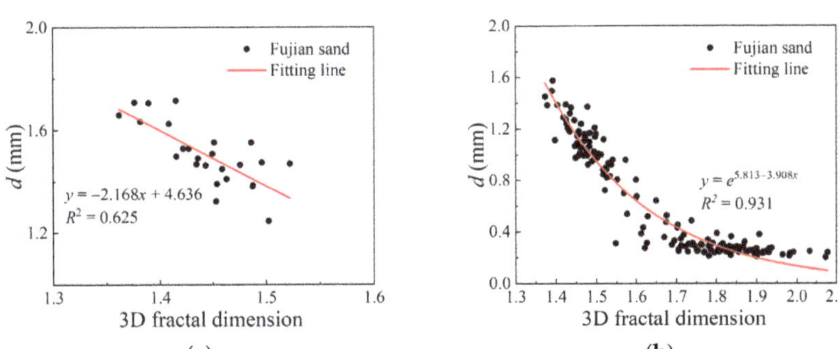

Figure 12. Relations between 3D fractal dimension and particle size: (**a**) before breakage; (**b**) after breakage.

The relation between the particle crushing strength σ_f and 3D fractal dimension is presented in Figure 13. The result indicates that the particle crushing strength σ_f increased with the 3D fractal dimension. These scatter points can be fitted by a straight line with MSE = 73.091. It can be concluded that the particles with a higher 3D fractal dimension resulted in higher particle crushing strength. This provides an auxiliary to predict the particle crushing strength based on the 3D fractal dimension. However, the data used in this study are insufficient to obtain an accurate relationship between the 3D fractal dimension and particle crushing strength. Nevertheless, a more complicated model could be established if more experimental data are provided, guaranteeing a better understanding of the relationship between the 3D fractal dimension and particle crushing strength. More importantly, in practice, the crushing strength of particles can be quickly estimated by measuring their 3D fractal dimension, and the macroscopic mechanical properties of materials can be thereby inferred. This can save a lot of experimental time and costs and optimize material selection and design processes.

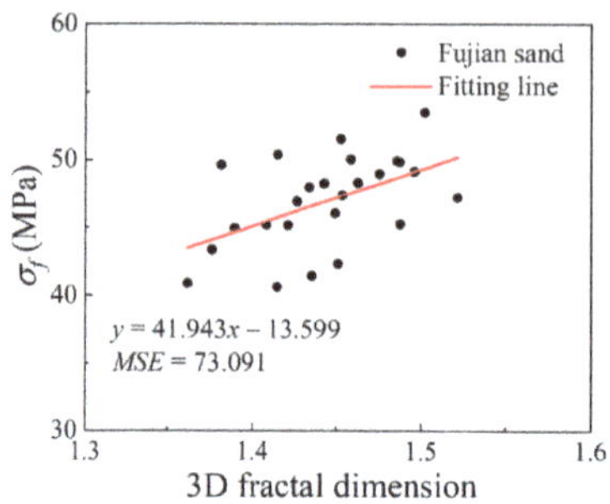

Figure 13. Relations between 3D fractal dimension and particle crushing strength.

4.2. Relations Between Fractal Dimensions and Morphology

The correlation between the 3D fractal dimension of the particles before the breakage and particle shape indexes (*EI*, *FI*, and *SI*) of granular materials are shown in Figure 14a, Figure 14b, and Figure 14c, respectively. It can be observed that the *EI* and *SI* increased as the 3D fractal dimension increased. However, the shape index *FI* decreased as the 3D fractal dimension increased. This is an expected result since higher *FI* values indicate higher particle shape irregularities, and it is well known that increasing particle irregularities increase fractal dimension [32]. The *MSE* value of the fitting line for the *SI* is the lowest one, suggesting that the *SI* is more appropriate to be used for examining the 3D fractal dimension of intact particles indirectly.

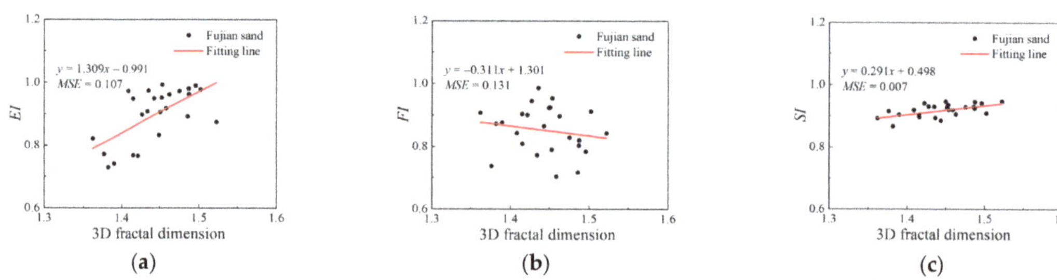

Figure 14. Relations between 3D fractal dimension and particle shape indexes before breakage: (**a**) *EI*; (**b**) *FI*; (**c**) *SI*.

Similarly, the correlation between the 3D fractal dimension of the particles after the breakage and particle shape indexes (*EI*, *FI*, and *SI*) of granular materials are shown in Figure 15a, Figure 15b, and Figure 15c, respectively. Conversely, the shape indexes *EI* and *SI* decreased as the 3D fractal dimension increased, while the *FI* increased as the 3D fractal dimension increased. The relation between the 3D fractal dimension and shape indexes is completely different before and after breakage. This result may be related to the change in the particle size distribution, where the irregular small fragments dominate the fragmented particles. The quick increase in the small irregular fragments alters the relationship between the 3D fractal dimension and shape indexes. The relationship between the 3D fractal dimension and shape indexes is extremely different before and after breakage.

This result suggests that the 3D fractal dimension can also be a good indicator for describing particle morphology, thus predicting the mechanical behaviors of particles.

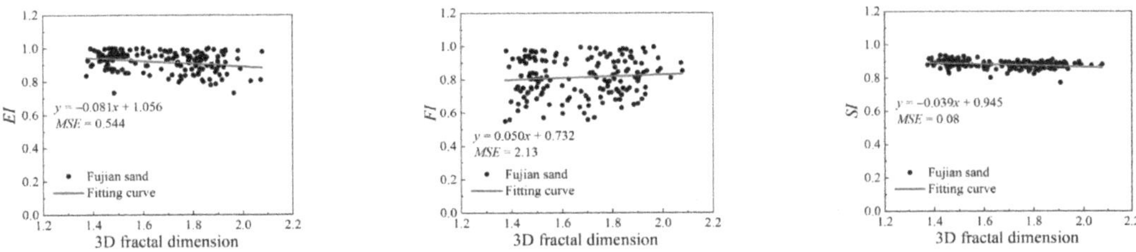

(a) (b) (c)

Figure 15. Relations between 3D fractal dimension and particle shape indexes after breakage: (**a**) *EI*; (**b**) *FI*; (**c**) *SI*.

5. Conclusions

A systematic framework for investigating single-particle breakage from experiments to numerical simulations based on three-dimensional (3D) images was developed in this study. A novel 3D particle reconstruction method utilizing the vision foundation model was proposed and demonstrated to be effective. This approach provides a novel alternative for reconstructing particles from 3D images. The reconstructed 3D accurate particles were calibrated and modeled by the discrete element method (DEM) with the bonded block method (BBM), based on one-to-one mapping single-particle loading tests. The results showed that the calibrated numerical simulations were sufficient to mimic the mechanical behavior of the single-particle breakage. The numerical simulations for the single-particle loading tests were performed using calibrated parameters. Moreover, fractal theory was used to investigate the relationship between the 3D fractal dimensions and particle size, particle crushing strength, and morphology. The relationship between the 3D fractal dimension and shape indexes changed significantly before and after the breakage. The results indicate that the particle size decreased as the fractal dimension increased. The trend of the particles before the crushing can be fitted with a straight line, while the trend of the particles after the crushing can be fitted with an exponential function. The particles with a higher 3D fractal dimension had a higher particle crushing strength. After the particle breakage, the *EI* and *FI* exhibited a slight decrease as the 3D fractal dimension increased. Conversely, the *SI* decreased as the 3D fractal dimension increased. The 3D fractal dimension is a good indicator for describing particle morphology and particle crushing strength. A more comprehensive conclusion can be found if more experimental data are available.

The proposed framework achieved a robust and accurate simulation for single-particle breakage, providing an auxiliary approach for investigating particle breakage from experiments to numerical simulations based on fractal theory. However, particle morphology could be accurately captured but did not consider the scenarios in which particles are connected. The influence of multi-point contact on single-particle breakage was also not revealed and will be investigated in future work. Moreover, a model comparing particle shape, crushing strength, 3D fractal dimension, and macroscopic mechanical properties can be developed by the physical-informed neural network (PINN) to obtain more accurate numerical simulation tools and provide reliable prediction results for engineering practice.

Author Contributions: Conceptualization, R.L. and X.G.; methodology, S.-H.H.; software, X.G.; experiment, D.R.; formal analysis, R.L.; data curation, S.-H.H.; writing—original draft preparation, R.L. and X.G.; writing—review and editing, S.-H.H.; supervision, S.-H.H. and Z.D.; project administration, S.-H.H.; funding acquisition, S.-H.H. and Z.D. All authors have read and agreed to the published version of the manuscript.

Funding: This research was financially supported by the National Natural Science Foundation of China under Grant No. 52178400.

Data Availability Statement: Data will be available upon reasonable request.

Conflicts of Interest: The authors declare no conflicts of interest.

References

1. Yu, F. Particle breakage in granular soils: A review. *Part. Sci. Technol.* **2021**, *39*, 91–100. [CrossRef]
2. Wang, P.; Xu, C.; Yin, Z.-Y.; Song, S.; Xu, C.; Dai, S. A DEM-based Generic Modeling Framework for Hydrate-Bearing Sediments. *Comput. Geotech.* **2024**, *171*, 106287. [CrossRef]
3. Tavares, L.M. Review and further validation of a practical single-particle breakage model. *KONA Powder Part. J.* **2022**, *39*, 62–83. [CrossRef]
4. Tavares, L.M. Chapter 1 Breakage of Single Particles: Quasi-Static. In *Handbook of Powder Technology*; Elsevier: Amsterdam, The Netherlands, 2007; pp. 3–68. [CrossRef]
5. Tavares, L.M. Energy absorbed in breakage of single particles in drop weight testing. *Miner. Eng.* **1999**, *12*, 43–50. [CrossRef]
6. Salman, A.D.; Biggs, C.A.; Fu, J.; Angyal, I.; Szabo, M.; Hounslow, M.J. An experimental investigation of particle fragmentation using single particle impact studies. *Powder Technol.* **2002**, *128*, 36–46. [CrossRef]
7. Rumpf, H. Physical aspects of comminution and new formulation of a law of comminution. *Powder Technol.* **1973**, *7*, 145–159. [CrossRef]
8. Saeidi, F.; Yahyaei, M.; Powell, M.; Tavares, L.M. Investigating the effect of applied strain rate in a single breakage event. *Miner. Eng.* **2017**, *100*, 211–222. [CrossRef]
9. Zhang, T.; Yang, W.; Zhang, C.; Hu, C. Particle Breakage Effect on Compression Behavior of Realistic Granular Assembly. *Int. J. Geomech.* **2021**, *21*, 04021105. [CrossRef]
10. Kh, A.B.; Mirghasemi, A.A.; Mohammadi, S. Numerical simulation of particle breakage of angular particles using combined DEM and FEM. *Powder Technol.* **2011**, *205*, 15–29.
11. Zhang, C.; Ren, Z.; Hao, D.; Zhang, T. Numerical Simulation of Particle Size Influence on the Breakage Mechanism of Broken Coal. *Arab. J. Sci. Eng.* **2020**, *45*, 9171–9185. [CrossRef]
12. Li, H.; McDowell, G.R.; Lowndes, I. Discrete-element modelling of rock breakage using dense random packing agglomerates. *Géotechnique Lett.* **2013**, *3*, 98–102. [CrossRef]
13. Potyondy, D.O.; Cundall, P.A. A bonded-particle model for rock. *Int. J. Rock Mech. Min. Sci.* **2004**, *41*, 1329–1364. [CrossRef]
14. Barrios, G.K.P.; Jiménez-Herrera, N.; Tavares, L.M. Simulation of particle bed breakage by slow compression and impact using a DEM particle replacement model. *Adv. Powder Technol.* **2020**, *31*, 2749–2758. [CrossRef]
15. Delaney, G.W.; Morrison, R.D.; Sinnott, M.D.; Cummins, S.; Cleary, P.W. DEM modelling of non-spherical particle breakage and flow in an industrial scale cone crusher. *Miner. Eng.* **2015**, *74*, 112–122. [CrossRef]
16. Washino, K.; Chan, E.L.; Nishida, Y.; Tsuji, T. Coarse grained DEM simulation of non-spherical and poly-dispersed particles using Scaled-Up Particle (SUP) model. *Powder Technol.* **2023**, *426*, 118676. [CrossRef]
17. Pan, J.; Pinzón, G.; Wang, R.; Andò, E.; Viggiani, G.; Zhang, J.-M. Lessons learned from matching 3D DEM and experiments at macro, meso and fabric scales for triaxial compression tests on lentils. *J. Mech. Phys. Solids* **2024**, *183*, 105494. [CrossRef]
18. Scholtès, L.; Donzé, F.-V. A DEM model for soft and hard rocks: Role of grain interlocking on strength. *J. Mech. Phys. Solids* **2013**, *61*, 352–369. [CrossRef]
19. Brown, G.J.; Miles, N.J.; Jones, T.F. A fractal description of the progeny of single impact single particle breakage. *Miner. Eng.* **1996**, *9*, 715–726. [CrossRef]
20. Arasan, S.; Akbulut, S.; Hasiloglu, A.S. The relationship between the fractal dimension and shape properties of particles. *KSCE J. Civ. Eng.* **2011**, *15*, 1219–1225. [CrossRef]
21. Nakata, A.F.L.; Hyde, M.; Hyodo, H.; Murata. A probabilistic approach to sand particle crushing in the triaxial test. *Géotechnique* **1999**, *49*, 567–583. [CrossRef]
22. Jaeger, J.C. Failure of rocks under tensile conditions. *Int. J. Rock Mech. Min. Sci. Geomech. Abstr.* **1967**, *4*, 219–227. [CrossRef]
23. McDowell, G.R. Statistics of soil particle strength. *Géotechnique* **2001**, *51*, 897–900. [CrossRef]
24. Kirillov, A.; Mintun, E.; Ravi, N.; Mao, H.; Rolland, C.; Gustafson, L.; Xiao, T.; Whitehead, S.; Berg, A.C.; Lo, W.-Y.; et al. Segment Anything. *arXiv* **2023**. [CrossRef]
25. Lorensen, W.E.; Cline, H.E. Marching cubes: A high resolution 3D surface construction algorithm. *SIGGRAPH Comput. Graph.* **1987**, *21*, 163–169. [CrossRef]
26. Itasca, User's Manual for pfc3d. 2019. Available online: https://docs.itascacg.com/pfc600/contents.html (accessed on 19 January 2024).
27. Fang, C.; Gong, J.; Nie, Z.; Li, B.; Li, X. DEM study on the microscale and macroscale shear behaviours of granular materials with breakable and irregularly shaped particles. *Comput. Geotech.* **2021**, *137*, 104271. [CrossRef]
28. Fang, C.; Nie, Z.; Gong, J.; Li, B.; Hu, W.; Mohammed, A. Discrete element simulation of effects of multicontact loading on single particle crushing. *Particuology* **2022**, *69*, 49–60. [CrossRef]

29. Zingg, T. Beitrag zur Schotteranalyse. Ph.D. Thesis, ETH Zurich, Zurich, Switzerland, 1935. Available online: https://www.research-collection.ethz.ch/bitstream/handle/20.500.11850/135183/1/eth-21472-01.pdf (accessed on 19 January 2024).
30. So, G.-B.; So, H.-R.; Jin, G.-G. Enhancement of the box-counting algorithm for fractal dimension estimation. *Pattern Recognit. Lett.* **2017**, *98*, 53–58. [CrossRef]
31. Li, J.; Du, Q.; Sun, C. An improved box-counting method for image fractal dimension estimation. *Pattern Recognit.* **2009**, *42*, 2460–2469. [CrossRef]
32. Kolay, E.; Kayabali, K. Investigation of the effect of aggregate shape and surface roughness on the slake durability index using the fractal dimension approach. *Eng. Geol.* **2006**, *86*, 271–284. [CrossRef]

fractal and fractional

MDPI

Article

Research on Slope Early Warning and Displacement Prediction Based on Multifractal Characterization

Xiaofei Sun [1], Ying Su [1], Chengtao Yang [2,*], Junzhe Tan [2] and Dunwen Liu [2,*]

[1] Road & Bridge International Co., Ltd., Beijing 101100, China; 225511019@csu.edu.cn (X.S.); 225512135@csu.edu.cn (Y.S.)

[2] School of Resources and Safety Engineering, Central South University, Changsha 410083, China; 235511012@csu.edu.cn

* Correspondence: 225512138@csu.edu.cn (C.Y.); dunwen@csu.edu.cn (D.L.)

Abstract: The occurrence of landslide hazards significantly induces changes in slope surface displacement. This study conducts an in-depth analysis of the multifractal characteristics and displacement prediction of highway slope surface displacement sequences. Utilizing automated monitoring devices, data are collected to analyze the deformation patterns of the slope surface layer. Specifically, the multifractal detrended fluctuation analysis (MF-DFA) method is employed to examine the multifractal features of the monitoring data for slope surface displacement. Additionally, the Mann–Kendall (M-K) method is combined to construct the α indicator and $f(\alpha)$ indicator criteria, which provide early warnings for slope stability. Furthermore, the long short-term memory (LSTM) model is optimized using the particle swarm optimization (PSO) algorithm to enhance the prediction of slope surface displacement. The results indicate that the slope displacement monitoring data exhibit a distinct fractal sequence characterized by $h(q)$, with values decreasing as the fluctuation function q decreases. Through this study, the slope landslide warning classification has been determined to be Level III. Moreover, the PSO-LSTM model demonstrates superior prediction accuracy and stability in slope displacement forecasting, achieving a root mean square error (RMSE) of 0.72 and a coefficient of determination (R^2) of 91%. Finally, a joint response synthesis of the slope landslide warning levels and slope displacement predictions resulted in conclusions. Subsequent surface displacements of the slope are likely to stabilize, indicating the need for routine monitoring and inspection of the site.

Keywords: tunnel deformation; multiple fractal theory; deformation warning; PSO-LSTM

1. Introduction

With the continuous development of transportation infrastructure in China, a significant number of highways, railroads, and other projects have been established. Due to the terrain, many highway and railroad projects must traverse mountainous areas, making slope engineering crucial in the construction of high-speed infrastructure in these regions [1,2]. Slopes are common structures in geotechnical engineering, and their stability is influenced by numerous factors characterized by randomness, ambiguity, and uncertainty [3–7]. In recent years, the growth of highway transportation, mining, and related industries in China has heightened the demand for slope engineering, leading to an increase in slope failure incidents. This trend poses safety risks to the public and generates substantial economic losses for the country [8–14]. Consequently, accurately determining the slope warning level is a critical issue in geotechnical engineering.

During the service life of slope engineering, monitoring and controlling surface displacement is a key technological challenge. As such, the evaluation of slope surface displacement stability is particularly important. Current methods for slope stability analysis include numerical simulation techniques [15–20], cluster analysis [21–27], image recognition [28–31], and machine learning approaches [32–37]. However, due to the complexity of the geological environment and various deformation-inducing factors. This leads to the time-series curve of slope surface displacement showing obvious volatility and strong non-linear characteristics. Thus, multifractal detrended fluctuation analysis (MF-DFA) provides an effective tool for analyzing tunnel deformation.

Analyzing the deformation patterns in slope monitoring data offers technical support and theoretical guidance for early warning systems, which is vital for enhancing slope safety. For instance, Yuanfeng Dong et al. [38] investigated the Chongqing Tongnan orchard landslide, analyzing various monitoring results to grade the landslide warning levels accurately and issue critical slip warnings, thereby preventing property damage and casualties. Similarly, Xiaopeng Deng [39] utilized cusp catastrophe theory to evaluate the stability of the Bazimen landslide in the Three Gorges Reservoir area, achieving a comprehensive assessment of landslide warning through limit displacement criteria and V/S analysis. Shuang Zhou et al. [40] employed an intensity reduction finite difference calculation approach, combined with monitoring data, but did not achieve effective slope deformation warnings. While previous studies have explored early warning mechanisms for slopes, they often overlooked the investigation of multifractal characteristics. In contrast, Heng Lei et al. [41] and Haoyu Mao et al. [42] applied multifractal theory to slope deformation warnings, successfully revealing warning signals through displacement data and micro-seismic signal monitoring, demonstrating the potential of this theory in early warning applications. Thus, further exploration of multifractal characteristics, based on slope surface displacement monitoring data and evaluation of early warning classifications, remains a valuable avenue for research.

Accurate prediction of slope deformation is also crucial for disaster prevention and warning systems. Machine learning methods have been increasingly employed for slope deformation prediction. Long short-term memory (LSTM) networks are proven to be particularly effective due to their ability to capture long-term dependencies and process temporal information [43]. Jiangbo Xu et al. [44] developed an LSTM model for slope displacement prediction, leveraging the maximum mutual information coefficient and the XGBoost algorithm, ultimately concluding that the model demonstrates high reliability. Haiping Xiao et al. [45] proposed a slope deformation prediction model integrating genetic algorithms and LSTM, achieving high accuracy and stability. Moreover, swarm intelligence optimization algorithms, such as particle swarm optimization (PSO), can enhance PSO-LSTM performance, leading to recent applications of the LSTM model in slope deformation prediction. Therefore, further application of the PSO-LSTM model for predicting slope displacements is warranted.

The study of disaster early warning is of paramount importance for disaster prevention and mitigation. With advancements in science and technology, monitoring and early warning systems have emerged as crucial tools for the proactive prevention of geological disasters. Currently, the LSTM models [46–50] and MF-DFA methods [51–55] have been widely applied in numerous studies focused on early warning for geological disasters. By utilizing a real-time monitoring and warning model for landslides, we can achieve dynamic tracking and timely alerts. Based on the grading results of landslide disaster early warnings, rapid decision-making and appropriate emergency measures can be implemented. This approach effectively reduces casualties and property losses associated with landslide disasters.

Building on these findings, research into the early warning predictions of slope landslides is essential. This paper utilizes monitoring data of slope surface displacement to first conduct a MF-DFA and Mann–Kendall (M-K) analysis, focusing on the multifractal features and early warning classifications of slope landslides. Subsequently, PSO-LSTM is employed to predict slope surface displacement data, forecasting displacement for the next 112 h. Finally, the results from the multifractal analysis and displacement predictions are combined to comprehensively evaluate the warning level of slope deformation, providing essential theoretical guidance for practical slope management.

2. Materials and Methods

2.1. Projiect Overview and Monitoring Data

This study focuses on a slope located within the Keqiao to Zhuji Expressway Project, specifically in the bidding section TJ03. This contract section spans Diankou Town and Yaojiang Town in Zhuji City, Shaoxing District. The slope begins in the middle of Keqiao, extends southwest through the southern part of Diankou, and terminates in the northern region of Zhuji City. The topography of the area is characterized by a low elevation at both ends and a higher elevation in the middle, resulting in a varied landscape.

The slope is part of a hilly terrain, where the section under investigation is an open quarry. Excavation is clearly defined, revealing medium weathering bedrock with exposed characteristics. At the summit of the slope, there exists a thin layer of residual slope deposits containing gravel and powdery clay. The underlying bedrock consists of Aurignacian Zhitang group tuff sandstone, which exhibits significant joint fissures and a broken rock structure.

The slope extends from K20 + 415 to K20 + 510, measuring a total length of 111 m, with a maximum height of 31.3 m. On the left flank of the K20 + 460 section, the slope is constructed in three tiers, each with a gradient of 1:2, where the third tier is excavated directly to the top, achieving a maximum excavation height of 30 m. Conversely, the right side of the slope is excavated with a gradient of 1:1.5, reaching a maximum height of 4 m. The overall slope configuration is step-like, with each tier height measuring 10 m and a 2-m-wide crumbling platform situated between each tier. Following the completion of the excavation, the slope configuration is illustrated in Figure 1a.

(a)

(b)

Figure 1. Site plan of the slope: (**a**) slopes, (**b**) monitoring devices.

To ensure the safety and stability of the slope post-excavation, an automatic monitoring system has been installed. This system monitors surface displacements of the slope at a frequency of once every 8 h to collect data, as depicted in Figure 1b.

To accurately assess the stability of the slope, continuous monitoring of surface displacement is conducted. Initially, the monitoring area and the locations of appropriate monitoring points are established. The safety conditions of the slope are evaluated based on the collected monitoring data. In this study, a total of 440 sets of slope monitoring data, collected from 29 January 2024 to 24 June 2024, are analyzed. The data from the monitoring points are compiled to generate the slope displacement–time diagram, as illustrated in Figure 2.

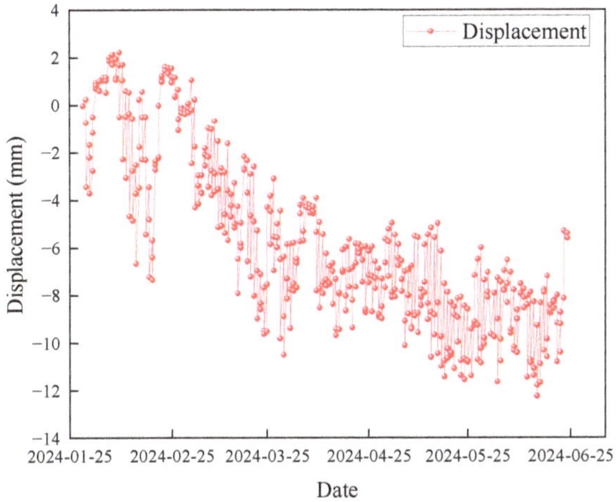

Figure 2. Displacement–time diagram of the surface layer of the slope.

2.2. MF-DFA

In the contexts of stochastic processes, chaos theory, and time series analysis, detrended fluctuation analysis (DFA) serves as a method for calculating the α (or Hurst exponent) to assess the statistical self-similarity of a signal. However, traditional DFA computes only second-order statistical moments and assumes that the underlying process follows a normal distribution. In contrast, the MF-DFA evaluates all q-order statistical moments $h(q)$, providing a more comprehensive characterization of nonlinear data and non-smooth signals compared to traditional DFA. The calculation flow for MF-DFA is illustrated in Figure 3.

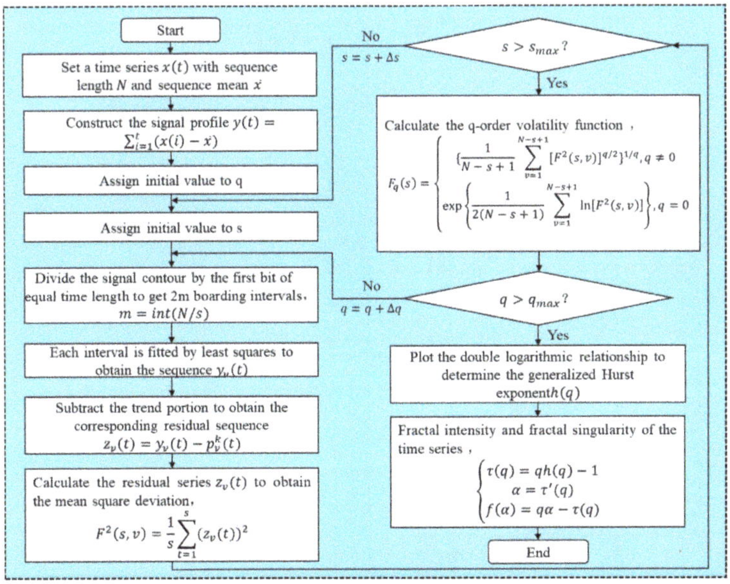

Figure 3. Flowchart of MF-DFA calculation.

2.3. M-K Test Method

The M-K test method is a non-parametric method. The specific calculation steps are as follows:

Step 1: There is a sample size of $\{x_1, x_2, \ldots\ldots, x_n\}$ of time series. For all the $k(j \leq n$ and $k \neq j)$, the distributions of x_k and x_j are different, and the difference function $Sgn(x_j - x_k)$ is computed:

$$Sgn(x_j - x_k) = \begin{cases} +1, & (x_j - x_k) > 0 \\ 0, & (x_j - x_k) = 0 \\ -1, & (x_j - x_k) < 0 \end{cases} \tag{1}$$

Step 2: Calculate the test statistic S:

$$S = \sum_{k=1}^{n-1} \sum_{j=k+1}^{n} Sgn(x_j - x_k) \tag{2}$$

Step 3: S is normally distributed with mean 0. Calculate the variance $Var(S)$:

$$Var(s) = \frac{n(n-1)(2n+5)}{18} \tag{3}$$

Step 4: Calculate the standard normal statistical variable Z:

$$Z = \begin{cases} \frac{S-1}{\sqrt{Var(s)}}, & S > 0 \\ 0, & S = 0 \\ \frac{S+1}{\sqrt{Var(s)}}, & S < 0 \end{cases} \tag{4}$$

Step 5: The trend characteristics of the evaluation object can be determined by the magnitude of Z. The Z_α value represents the critical threshold at the specified significance level α. In this study, a significance test is conducted at a 99% confidence level, resulting in $\alpha = 0.01$ and $Z_{0.01} = 2.32$.

If $Z \geq Z_\alpha$, it indicates that the indicator criterion has a tendency to increase.

If $-Z_\alpha < Z < Z_\alpha$, it indicates that the indicator criterion has a smooth trend.

If $Z \leq -Z_\alpha$, it indicates that the indicator criterion has a decreasing trend.

2.4. PSO-LSTM Prediction Modeling

2.4.1. LSTM

The primary advantage of LSTM models over other common machine learning algorithms lies in their unique "gate" structure. This structure allows the LSTM model to evaluate information based on the "memory" of the network. Information is selectively retained or discarded by multiplying by 1 or 0. The unitary state effectively solves the gradient vanishing problem associated with short-term memory by retaining sequence-relevant information throughout the sequence.

The gate structure employs a sigmoid activation function that compresses values between 0 and 1, which facilitates the updating of retained information while discarding less relevant data.

The LSTM model features three main gates:

1. Forgetting Gate: This gate determines whether information should be discarded or retained. It processes relevant information through the sigmoid function, producing an output between 0 and 1, where values closer to 0 indicate less importance and greater likelihood of being discarded, while values closer to 1 signify critical information.
2. Input Gate: This gate updates the cell state. After processing by both sigmoid and hyperbolic tangent (*tanh*) functions, a final output value closer to 0 indicates less importance, whereas a value closer to 1 indicates significant information.

3. Output Gate: Similar to the input gate, the output gate determines the value of the next hidden state in the cell structure. The processed value from this gate is used to decide the information the hidden state should carry, which is then passed along with the new cell state.

$$f_t = \sigma\left(w_f[h_{t-1}, x_t] + b_f\right) \quad (5)$$

$$i_t = \sigma(w_i[h_{t-1}, x_t] + b_i) \quad (6)$$

$$\check{C}_t = tanh(w_c[h_{t-1}, x_t] + b_c) \quad (7)$$

$$C_t = f_t C_{t-1} + i_t \check{C}_t \quad (8)$$

$$o_t = \sigma(w_o[h_{t-1}, x_t] + b_o) \quad (9)$$

$$h_t = O_t tanh C_t \quad (10)$$

$$tanhx = \left(e^x - e^{-x}\right) / \left(e^x + e^{-x}\right) \quad (11)$$

where x represents the input vector, h denotes the output vector, and f, i, and o refer to the forgetting gate, input gate, and output gate, respectively. The variable t indicates the time step. The activation functions σ and $tanh$ are nonlinear. C represents the cell state, while w signifies the trainable weight matrix, and b refers to the bias matrix.

2.4.2. Improvement of PSO Algorithm

The PSO algorithm was developed by researchers inspired by the foraging behavior of birds. In this algorithm, each particle simulates the foraging behavior of a bird, with each particle in the swarm maintaining its own established direction while searching for the optimal value. Additionally, each particle communicates its current value and position to the swarm. Consequently, each particle adjusts its search direction based on its own experience as well as the position of the optimal value recorded by the swarm. The flow of the PSO algorithm is illustrated in Figure 4.

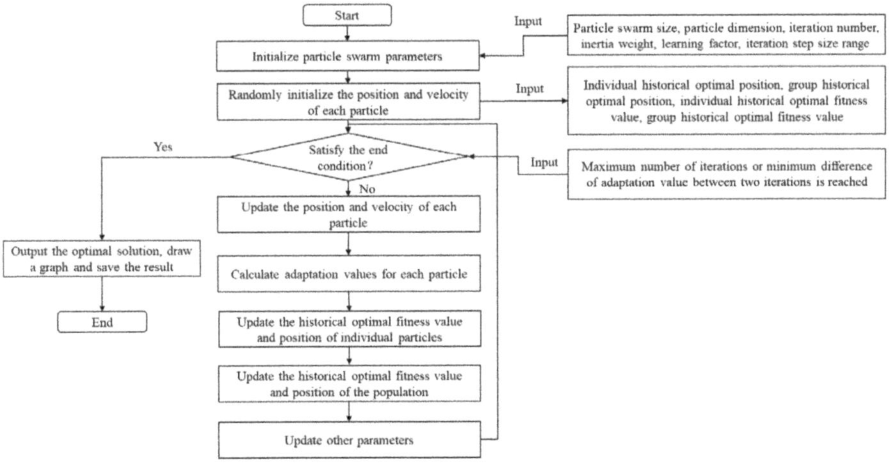

Figure 4. Flowchart of PSO algorithm.

The mathematical expression for the base element of the PSO algorithm is

$$V_{id}^{k+1} = WV_{id}^{k} + C_1 r_1 \left(p_{id,pbest}^{k} - x_{id}^{k}\right) + C_2 r_2 \left(p_{id,gbest}^{k} - x_{id}^{k}\right) \quad (12)$$

$$x_{id}^{k+1} = x_{id}^{k} + V_{id}^{k+1} \quad (13)$$

where V_{id}^{k+1} represents the d-dimensional component of the velocity of particle i at the $k+1$ iteration; W denotes the inertia weight; and V_{id}^{k} is the d-dimensional component of the particle's velocity at the k iteration. C_1 and C_2 are the acceleration coefficients, while r_1 and r_2 are random numbers uniformly distributed between 0 and 1. $p_{id,pbest}^{k}$ indicates the d-dimensional component of particle i's historical optimal position at the k iteration, and x_{id}^{k} represents the position of the particle at the k iteration. $p_{id,gbest}^{k}$ denotes the d-dimensional component of the optimal position recorded by the particle swarm throughout its history at the k iteration. The term $C_1 r_1$ represents the learning weight of the particle's own experience, whereas $C_2 r_2$ represents the learning weight based on the experiences of the population. Finally, x_{id}^{k+1} is the d-dimensional component of the particle's position at the $k+1$ iteration.

2.4.3. PSO-LSTM

The slope surface displacement data are input into the network structure of the LSTM model. Initially, the parameters of the LSTM model are established, followed by the initialization of particle swarm parameters. The positions and velocities of the particles are generated randomly, and the fitness values are computed. Subsequently, the individual and collective velocities and positions within the particle swarm are updated. After each update, the fitness values are recalculated to assess whether the maximum number of iterations has been reached. Upon reaching the maximum number of iterations, the optimal parameters are identified. The PSO-LSTM model is then constructed, and the data is both trained and tested, ultimately yielding the output results and evaluation metrics.

The coefficient of determination (R^2), mean absolute error (MAE), and root mean square error (RMSE) were selected as evaluation criteria for the model. Their respective expressions are as follows:

$$\begin{cases} R^2 = 1 - \frac{\sum_{i=1}^{n}[y_c(i)-y_0(i)]^2}{\sum_{i=1}^{n}[y_0(i)-\overline{y_c}]^2} \\ MAE = \frac{1}{n}\sum_{i=1}^{n}|y_c(i) - y_0(i)| \\ RMSE = \sqrt{\frac{1}{n}\sum_{i=1}^{n}[y_c(i) - y_0(i)]^2} \end{cases} \quad (14)$$

where n represents the number of predicted outcomes. $y_o(i)$ represents the true outcome. $y_c(i)$ represents the predicted outcome. $\overline{y_o}$ represents the mean of the true values.

3. Results and Discussion

3.1. Multifractal Characterization of Slope Surface Displacement

Before calculating the multifractal characteristics of the monitoring data, it is essential to segment the data into four groups, each containing 110 data points. The groups are defined as follows: The first group spans from 29 January 2024 to 6 March 2024. The second group covers the period from 6 March 2024 to 11 April 2024. The third group extends from 12 April 2024 to 18 May 2024. The fourth group ranges from 18 May 2024 to 24 June 2024.

To perform multifractal analysis on each group of surface displacement monitoring sequences of the slope, we employed a sliding time window optimization method using the MF-DFA. In this analysis, the fluctuation order q is varied over the range of $[-10, 10]$, while the scale s is set within the range of [10, 100]. The sliding window step is defined as 1. The resulting double logarithmic scatter plot of $logF_q(s) - logs$ is illustrated in Figure 5.

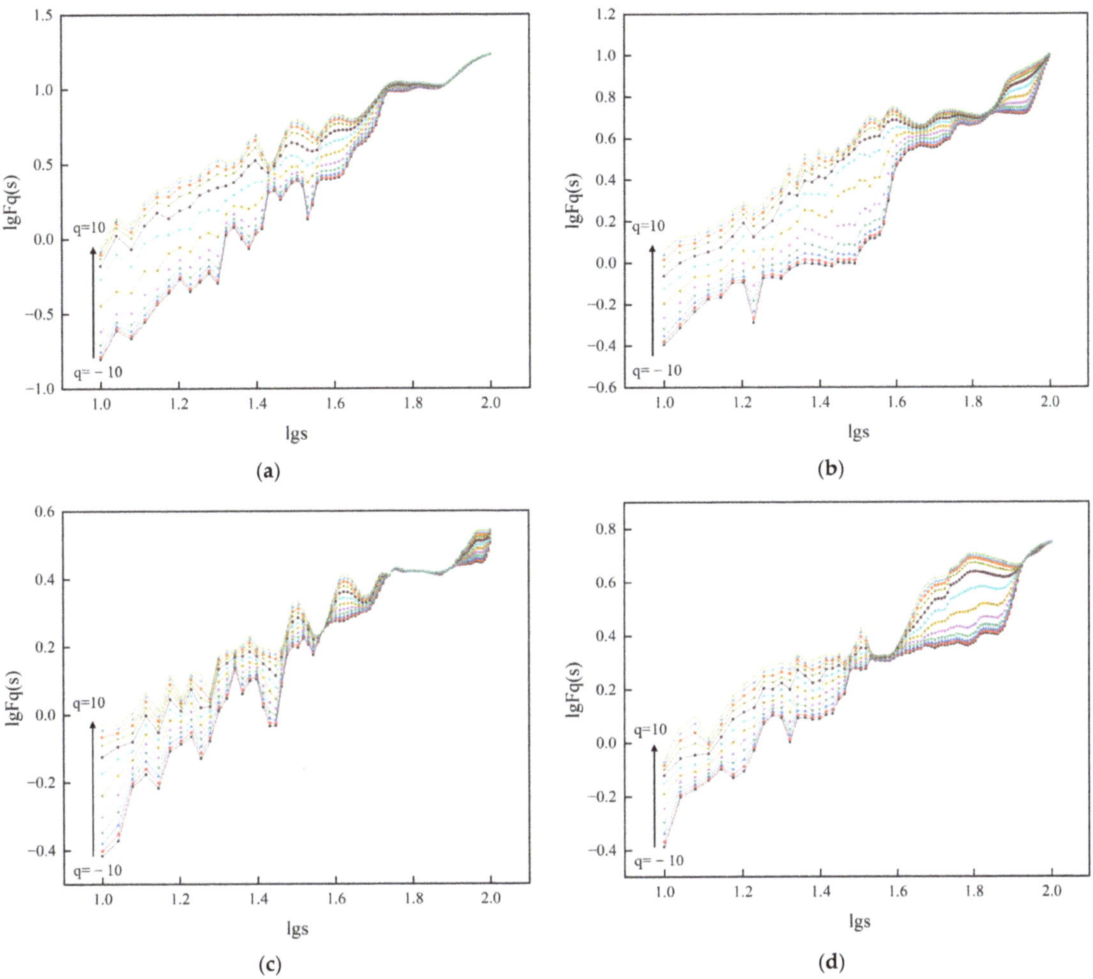

Figure 5. q-order fluctuation function $logF_q(s) - logs$ trend plot of double logarithmic fit: (**a**) Group I, (**b**) Group II, (**c**) Group III, (**d**) Group IV.

Using the aforementioned parameters, the four data series were analyzed through multifractal analysis with Matlab (R2018b) software. The generalized Hurst exponent and Renyi exponent—specifically, the scalar function—for each group of displacement sequences were computed across varying values of $\tau(q)$. The changes in the indices corresponding to the measurement point sequences are presented in Figure 6. Furthermore, the multifractal spectra for each group of displacement sequences are displayed in Figure 7.

As illustrated in Figure 6a, the generalized Hurst exponent of the surface displacement data series for the slope exhibits a nonlinear decreasing trend as q varies within the range of $[-10, 10]$. This trend indicates that the surface displacement data at the monitoring sites is characterized by multifractal properties. Notably, for different fluctuation orders q, the generalized Hurst index curves of the second, third, and fourth groups are positioned at lower fluctuations compared to the first group, suggesting a weaker multifractal feature in these groups. However, the $h(q)$ values for each group's displacement series are all greater than 0.5. Indicating that the displacement sequences possess strong memory and long-range correlations from the overall structure to local components.

Additionally, Figure 6b demonstrates a good consistency in the scale functions of the surface displacement sequences across all groups, with the central part of the scaling function curve exhibiting a convex shape that satisfies the relationship $\tau(0) = -1$. This finding further confirms that each group's unique surface layer sequences exhibit multifractal characteristics.

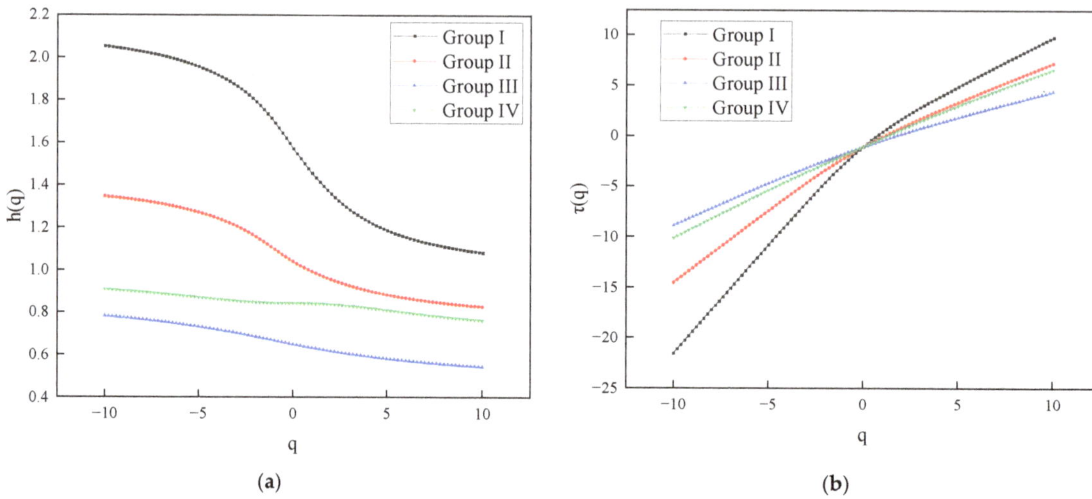

Figure 6. Variation of each index of displacement series: (**a**) generalized Hurst index, (**b**) scale function τ, q.

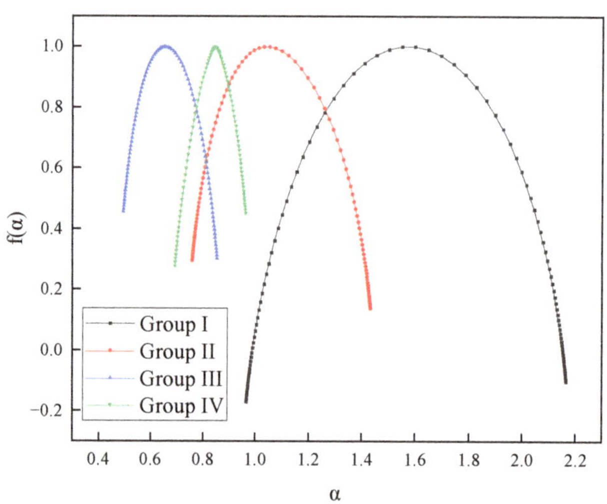

Figure 7. Multiple fractal spectra of surface displacements for each group of side slopes.

As shown in Figure 7, the multifractal spectra for each group of slope surface displacement sequences exhibit a typical single-peak convex distribution, resembling a quadratic function curve. The local scales of these spectra vary, indicating the diversity of local variations across different time points. The singularity intensity α of most displacement sequences is distributed along both sides of the graph, reflecting the uneven distribution of the fractal structure within each data series. This further underscores the multifractal properties of the slope surface displacement sequences. Moreover, the multifractal spectral curves are generally symmetrical, suggesting a stable overall developmental state.

Utilizing Equation (8), we calculated the multifractal characteristic statistics for each group of surface displacement sequences. The results of these calculations are presented in Table 1.

Table 1. Multifractal characterization statistics for surface displacements for each group of slopes.

Eigenvalue (Math.)	Group I	Group II	Group III	Group IV
$\Delta\alpha$	1.19744	0.67596	0.36077	0.2713
$\Delta f(\alpha)$	0.06878	−0.1566	−0.1539	0.17519

Table 1 reveals a comparison of the widths of the multifractal spectra for each group of displacement sequences, denoted as $\Delta\alpha$. The multifractal spectral width of the first group $\Delta\alpha$ is significantly greater than that of the other three groups, indicating that the multifractal intensity of the surface layer displacement sequences in the first group is higher, and the displacement fluctuations are more complex.

Furthermore, when comparing the proportions of large and small fluctuations ($\Delta f(\alpha)$) in the displacement series across each group, the fourth group shows a slightly higher $\Delta f(\alpha)$ compared to the other three groups. This suggests that the displacement sequences for the fourth group exhibit a greater prevalence of small fluctuations.

These findings indicate that the calculated multifractal eigenvalues of the surface displacement sequences are more consistent with actual monitoring results. Thus, the eigenvalues can be utilized for the study of landslide early warning grading in slopes.

3.2. Early Warning Grading Study of landslides on Slopes

3.2.1. Criteria for Classifying the Warning Level of Landslides on Slopes

Building upon the results of existing research on landslide warning levels, we utilized the surface displacement monitoring data for each group to construct warning criteria based on the $\Delta\alpha$ and $\Delta f(\alpha)$ parameters. This framework enables the classification of landslide warning levels into three categories: Level I, Level II, and Level III.

1. Level I warnings indicate that slope deformation is trending in an extremely unfavorable direction, posing a significant risk of damage and serving as a precursor to imminent disaster. In this scenario, it is recommended to implement necessary disaster prevention and management measures, including evacuation and relocation, to mitigate potential losses.
2. Level II warnings indicate that deformation is moving in an unfavorable direction, presenting a general risk of damage.
3. Level III warnings suggest that deformation is trending towards stabilization.

The specific criteria for these warning levels are detailed in Table 2.

Table 2. Criteria for classifying the warning level of landslides on slopes.

Warning Level	$\Delta\alpha$ Indicator Criterion	$\Delta f(\alpha)$ Indicator Criterion	Treatment Measures
I	Decreasing trend	Increasing trend	Suspend construction and carry out necessary disaster prevention and management or relocation to avoid disaster damage.
II	Increasing trend	Decreasing trend	Enhance the frequency of monitoring and patrols and make disaster preparedness plans.
III	Steady trend	Steady trend	Normal monitoring and patrolling.

3.2.2. Slope Landslide Warning Classification

Through the multiple fractal analysis and computations described in Section 2.2, we obtained the necessary $\Delta\alpha$ and $\Delta f(\alpha)$ parameter sets, as illustrated in Figure 8.

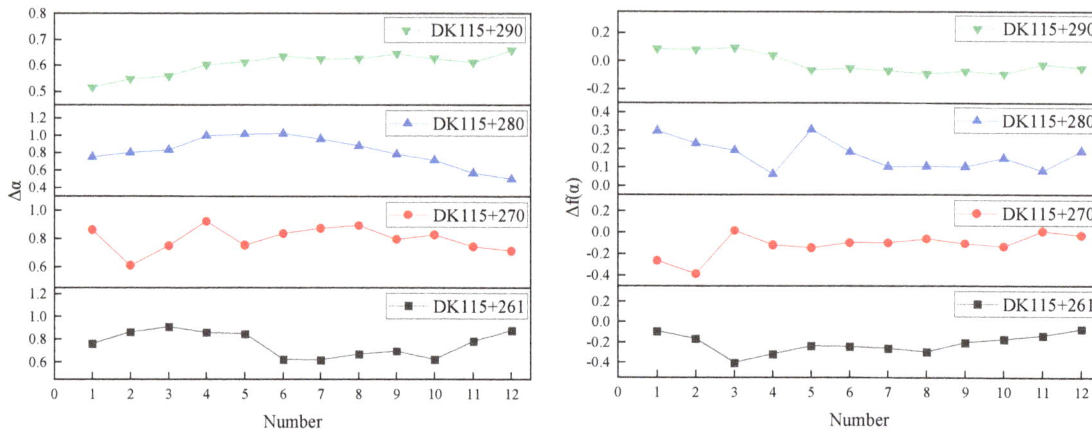

(**a**) (**b**)

Figure 8. Values of slope landslide warning parameters: (**a**) $\Delta\alpha$ parameter values, (**b**) $\Delta f(\alpha)$ parameter values.

The trends of the two discriminant indicators were assessed using the M-K test to establish the early warning grading for tunnel displacement. The results are analyzed as follows:

From Table 3, the analysis of the $\Delta\alpha$ indicator criterion yields a calculated $Z = -2.0412$. This value falls within the range of $-Z_{0.01} < Z < Z_{0.01}$, indicating a stable trend, which corresponds to warning Level III.

Table 3. Results of landslide early warning analysis.

Indicators	Z-Value	Growing Trend	Warning Level	Integrated Early Warning
$\Delta\alpha$	−2.0412	steady trend	III	III
$\Delta f(\alpha)$	0.4082	steady trend	III	

For the $\Delta f(\alpha)$ indicator, the calculated $Z = 0.4082$, which is positioned at the level of $-Z_{0.01} < Z < Z_{0.01}$. This analysis also confirms a warning level of III.

Considering both indicators and applying the principle of unfavorability, the final warning level is determined to be III. This suggests that the slope surface displacement and deformation are trending towards stabilization, and normal monitoring and inspections should continue.

3.3. Prediction of Slope Surface Displacements

3.3.1. Optimization of Model Parameters

Matlab (R2018b) software was utilized to develop a program aimed at optimizing the parameters of the LSTM prediction model using the PSO algorithm. The parameters for the PSO algorithm were configured as follows: The number of search particles was set to 4, the maximum number of iterations was set to 300, and the number of optimization parameters was limited to 4. By systematically adjusting the range of parameter values, we determined the optimal parameter range, which is as follows: learning rate range: $(1 \times 10^{-3}, 1 \times 10^{-2})$; number of neurons in the hidden layer range: (10, 30); regularization coefficients range: (1×10^{-4}, 1×10^{-1}); and iteration number range: (100, 200). Each parameter of the PSO-LSTM prediction model was initialized accordingly. The optimal parameter values obtained through the PSO optimization procedure were 0.010, 19.227, 0.100, and 183.581, respectively.

3.3.2. Model Predictions

The hyperparameters obtained from the PSO optimization were utilized as inputs for the LSTM model, which was subsequently trained on the input data. Through multiple iterations of training, the model's prediction accuracy was improved to an acceptable level. The 440 slope surface displacement data presented in Figure 4 were chosen for prediction, with 70% of the data designated as the training set and the remaining 30% allocated as the test set. The PSO-LSTM model underwent both training and testing, with the results of the slope surface displacement prediction illustrated in Figure 9.

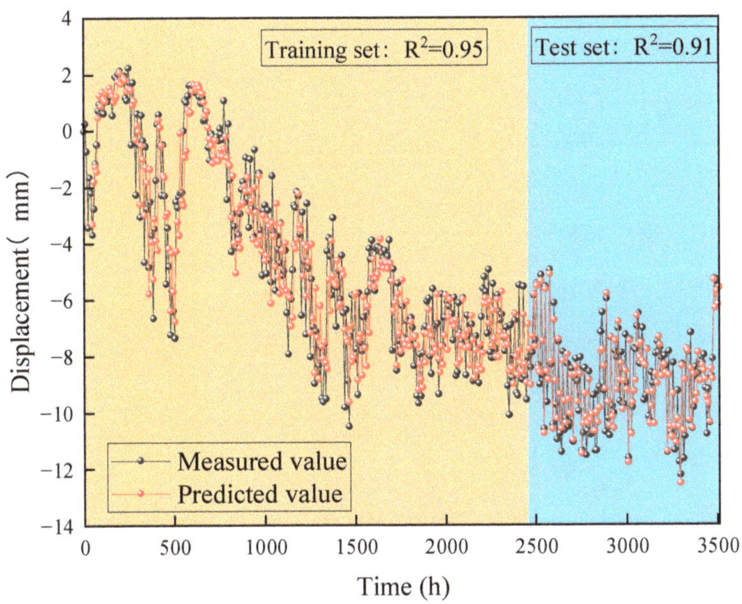

Figure 9. Predicted results of slope surface displacements.

As illustrated in Figure 9, the prediction results of the PSO-LSTM model for slope surface displacement closely align with the actual monitoring results. For the training set, the model achieved an R^2 value of 0.95, a MAE of 0.84, and a RMSE of 1.10. For the test set, the corresponding values were $R^2 = 0.91$, MAE = 0.55, and RMSE = 0.72. The prediction curves indicate that the overall trend of the predicted slope surface displacements mirrors that of the actual monitoring data, demonstrating the effectiveness of the PSO-LSTM model in predicting slope surface displacement.

To further assess the discrepancy between the PSO-LSTM model's predictions and the true values, the calculated prediction errors for slope surface displacements are presented in Figure 10.

As shown in Figure 10, a comparison of the sample error calculations from the PSO-LSTM model for predicting the slope surface displacement indicates that the maximum errors for the training set and the test set are 4.47 mm and −2.02 mm, respectively. Overall, approximately 70.7% of the samples exhibited prediction errors within ±1 mm, while about 99.3% of the samples demonstrated errors within ±3 mm. Notably, the maximum prediction error for the model's test set is significantly smaller than that observed for the training set, suggesting that the PSO-LSTM model yields lower sample errors in predicting slope surface displacement. This finding further underscores the superior performance of the PSO-LSTM prediction model.

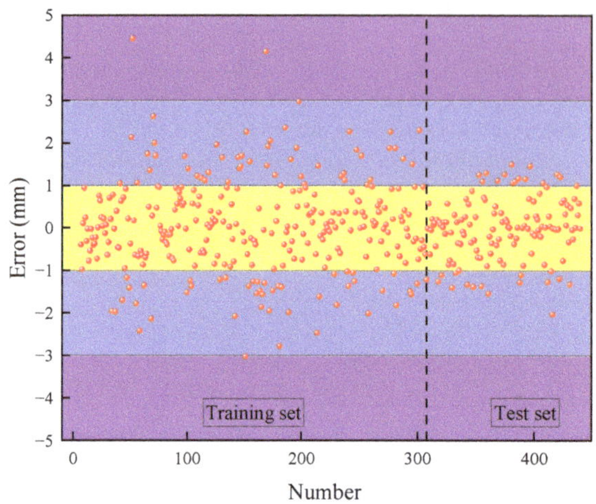

Figure 10. Error map of slope displacement prediction.

To extend the assessment of slope displacement, the previously constructed PSO-LSTM prediction model was employed to forecast displacement results for the subsequent 112 h, as illustrated in Figure 11.

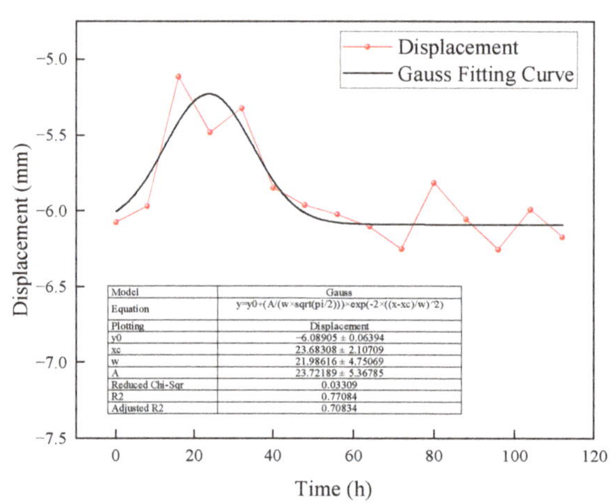

Figure 11. Predicted and fitted slope displacements.

To better visualize the stabilization trend of the data, we applied a fitting method to the predictions. Among the various fitting techniques available, the Gaussian fitting method is particularly effective in capturing the overall trend of the data. Given that the predicted dataset exhibits a single-peaked distribution, the Gaussian fitting method is suitable for modeling the prediction results.

As illustrated in Figure 11, analysis of the prediction results reveals that the slope surface displacements initially increase, followed by a decrease, and ultimately fluctuate around a stable value. Gaussian fitting indicates that the slope surface displacement eventually stabilizes, further validating the accuracy of the slope Level III warning results derived from multiple fractal eigenvalue analyses.

In summary, the results from the slope landslide warning levels and the predictions of slope surface displacements demonstrate a coherent response, indicating that the slope surface displacement is trending toward stability. Continuous monitoring and inspection are necessary to ensure the safety and stability of the slope. The slope site inspection is shown in Figure 12.

Figure 12. Slope site walk-through map.

4. Conclusions

Analyzing and predicting the deformation patterns of slope surface displacements can provide essential technical support and theoretical guidance for early warning systems related to slope safety. This paper focuses on a slope in Zhuji, Zhejiang Province, combining on-site monitoring and measurement with multiple fractal analysis and a PSO-LSTM model to conduct an in-depth study of slope warning levels and surface displacement predictions. The main conclusions are as follows:

1. The application of the MF-DFA method reveals that the slope surface displacements exhibit multiple fractal characteristics, indicating a stable developmental trend toward stabilization.
2. The PSO-LSTM prediction model was employed to forecast the deformation trends of slope surface displacements. The results for the test set yielded $R^2 = 0.91$, MAE = 0.55, and RMSE = 0.72. The prediction errors associated with the PSO-LSTM model were minimal, demonstrating that the model effectively meets the requirements for slope surface displacement prediction.
3. Synthesis of results from the analysis of multifractal characteristics and deformation predictions indicates that the current warning level for the slope is III, with subsequent deformations trending toward stabilization. Continued routine monitoring and inspections are recommended.
4. The slopes analyzed in this study were characterized by a homogenous rock body and limited monitoring point locations. In future studies, a comprehensive fractal characterization of surface displacement monitoring results across multiple slopes with varying rock properties will be conducted. Additionally, numerical modeling of these slopes will be performed to further validate the accuracy of the proposed method. Additionally, incorporating more influencing factors related to slope deformation could further enhance the predictive accuracy of the PSO-LSTM model.

Author Contributions: Conceptualization, C.Y. and J.T.; methodology, C.Y.; software, C.Y. and Y.S.; validation, Y.S. and X.S.; formal analysis, D.L.; investigation, X.S. and J.T.; resources, D.L.; data curation, X.S. and Y.S.; writing—original draft preparation, C.Y.; writing—review and editing, J.T. and D.L.; visualization, X.S. and Y.S.; supervision, X.S. and Y.S.; project administration, D.L.; funding acquisition, D.L. All authors have read and agreed to the published version of the manuscript.

Funding: This research received no external funding.

Data Availability Statement: The data that support the findings of this study are available upon request from the authors.

Acknowledgments: The authors would like to thank Road and Bridge International Co., Ltd. for their assistance with conducting the monitoring of data.

Conflicts of Interest: Authors Xiaofei Sun and Ying Su were employed by Road & Bridge International Co., Ltd. The remaining authors declare that the research was conducted in the absence of any commercial or financial relationships that could be construed as a potential conflict of interest.

References

1. Cao, S.; Ye, H.; Zhan, Y. Cliff roads: An ecological conservation technique for road construction in mountainous regions of China. *Landsc. Urban Plan* **2010**, *94*, 228–233. [CrossRef]
2. Park, H.J.; West, T.R.; Woo, I. Probabilistic analysis of rock slope stability and random properties of discontinuity parameters, Interstate Highway 40, Western North Carolina, USA. *Eng. Geol.* **2005**, *79*, 230–250. [CrossRef]
3. Niu, P.; Zhou, A.; Huang, H. Assessing model of highway slope stability based on optimized SVM. *China Geol.* **2020**, *3*, 339–344. [CrossRef]
4. Lin, H.; Chang, S.; Wu, J.; Juang, C.H. Neural network-based model for assessing failure potential of highway slopes in the Alishan, Taiwan Area: Pre- and post-earthquake investigation. *Eng. Geol.* **2009**, *104*, 280–289. [CrossRef]
5. Shinoda, M.; Miyata, Y.; Kurokawa, U.; Kondo, K. Regional landslide susceptibility following the 2016 Kumamoto earthquake using back-calculated geomaterial strength parameters. *Landslides* **2019**, *16*, 1497–1516. [CrossRef]
6. Song, J.; Rodriguez-Marek, A.; Feng, T.; Ji, J. A generalized seismic sliding model of slopes with multiple slip surfaces. *Earthq. Eng. Struct. D* **2021**, *50*, 2595–2612. [CrossRef]
7. Chen, L.; Zhao, C.; Li, B.; He, K.; Ren, C.; Liu, X.; Liu, D. Deformation monitoring and failure mode research of mining-induced Jianshanying landslide in karst mountain area, China with ALOS/PALSAR-2 images. *Landslides* **2021**, *18*, 2739–2750. [CrossRef]
8. Kanungo, D.P.; Arora, M.K.; Sarkar, S.; Gupta, R.P. A comparative study of conventional, ANN black box, fuzzy and combined neural and fuzzy weighting procedures for landslide susceptibility zonation in Darjeeling Himalayas. *Eng. Geol.* **2006**, *85*, 347–366. [CrossRef]
9. Kundu, S.; Saha, A.K.; Sharma, D.C.; Pant, C.C. Remote Sensing and GIS Based Landslide Susceptibility Assessment using Binary Logistic Regression Model: A Case Study in the Ganeshganga Watershed, Himalayas. *J. Indian Soc. Remote* **2013**, *41*, 697–709. [CrossRef]
10. Wubalem, A.; Meten, M. Landslide susceptibility mapping using information value and logistic regression models in Goncha Siso Eneses area, northwestern Ethiopia. *SN Appl. Sci.* **2020**, *2*, 807. [CrossRef]
11. Ayalew, L.; Yamagishi, H. Slope failures in the Blue Nile basin, as seen from landscape evolution perspective. *Geomorphology* **2004**, *57*, 95–116. [CrossRef]
12. Temesgen, B.; Mohammed, M.U.; Korme, T. Natural hazard assessment using GIS and remote sensing methods, with particular reference to the landslides in the Wondogenet area, Ethiopia. *Phys. Chem. Earth Part C—Sol. Terrestial Planet. Sci.* **2001**, *26*, 665–675. [CrossRef]
13. Singh, T.N.; Gulati, A.; Dontha, L.; Bhardwaj, V. Evaluating cut slope failure by numerical analysis—A case study. *Nat. Hazards* **2008**, *47*, 263–279. [CrossRef]
14. Ramakrishnan, D.; Singh, T.N.; Verma, A.K.; Gulati, A.; Tiwari, K.C. Soft computing and GIS for landslide susceptibility assessment in Tawaghat area, Kumaon Himalaya, India. *Nat. Hazards* **2013**, *65*, 315–330. [CrossRef]
15. Cheng, Y.M.; Lansivaara, T.; Wei, W.B. Two-dimensional slope stability analysis by limit equilibrium and strength reduction methods: Reply. *Comput. Geotech.* **2008**, *35*, 309–311. [CrossRef]
16. Liu, S.Y.; Shao, L.T.; Li, H.J. Slope stability analysis using the limit equilibrium method and two finite element methods. *Comput. Geotech.* **2015**, *63*, 291–298. [CrossRef]
17. Morales-Esteban, A.; Luis De Justo, J.; Reyes, J.; Miguel Azanon, J.; Durand, P.; Martinez-Alvarez, F. Stability analysis of a slope subject to real accelerograms by finite elements. Application to San Pedro cliff at the Alhambra in Granada. *Soil Dyn. Earthq. Eng.* **2015**, *69*, 28–45. [CrossRef]
18. Stianson, J.R.; Chan, D.; Fredlund, D.G. Role of admissibility criteria in limit equilibrium slope stability methods based on finite element stresses. *Comput. Geotech.* **2015**, *66*, 113–125. [CrossRef]
19. Moawwez, M.A.; Wang, J.; Hussain, M.A. Development of empirical correlations for limit equilibrium methods of slope stability analysis. *Arab. J. Geosci.* **2021**, *14*, 2020. [CrossRef]

20. Tesfaye, M.; Regassa, B.; Garo, T. Rock slope stability modeling using kinematic and limit equilibrium methods along Woliso to Wonchi lake road, central Ethiopia. *Model. Earth Syst. Environ.* **2024**, *10*, 331–347. [CrossRef]
21. Deparis, J.; Garambois, S.; Hantz, D. On the potential of Ground Penetrating Radar to help rock fall hazard assessment: A case study of a limestone slab, Gorges de la Bourne (French Alps). *Eng. Geol.* **2007**, *94*, 89–102. [CrossRef]
22. Medeiros Santos, A.E.; Lana, M.S.; Cabral, I.E.; Pereira, T.M.; Naghadehi, M.Z.; Santos Da Silva, D.D.F.; Dos Santos, T.B. Evaluation of Rock Slope Stability Conditions Through Discriminant Analysis. *Geotech. Geol. Eng.* **2019**, *37*, 775–802. [CrossRef]
23. Sarkar, S.; Kanungo, D.P.; Kumar, S. Rock Mass Classification and Slope Stability Assessment of Road Cut Slopes in Garhwal Himalaya, India. *Geotech. Geol. Eng.* **2012**, *30*, 827–840. [CrossRef]
24. Vishal, V.; Siddique, T.; Purohit, R.; Phophliya, M.K.; Pradhan, S.P. Hazard assessment in rockfall-prone Himalayan slopes along National Highway-58, India: Rating and simulation. *Nat. Hazards* **2017**, *85*, 487–503. [CrossRef]
25. Verma, A.K.; Kumar, N.; Sardana, S.; Singh, T.N. Rockfall Analysis and Optimized Design of Rockfall Barrier Along a Strategic Road near Solang Valley, Himachal Pradesh, India. *Indian Geotech. J.* **2018**, *48*, 686–699. [CrossRef]
26. Sardana, S.; Verma, A.K.; Verma, R.; Singh, T.N. Rock slope stability along road cut of Kulikawn to Saikhamakawn of Aizawl, Mizoram, India. *Nat. Hazards* **2019**, *99*, 753–767. [CrossRef]
27. Sardana, S.; Verma, A.K.; Singh, A.; Laldinpuia. Comparative analysis of rockmass characterization techniques for the stability prediction of road cut slopes along NH-44A, Mizoram, India. *Bull. Eng. Geol. Environ.* **2019**, *78*, 5977–5989. [CrossRef]
28. Cignetti, M.; Godone, D.; Wrzesniak, A.; Giordan, D. Structure from Motion Multisource Application for Landslide Characterization and Monitoring: The Champlas du Col Case Study, Sestriere, North-Western Italy. *Sensors* **2019**, *19*, 2364. [CrossRef]
29. Do, T.M.T.; Artieres, T. Learning mixture models with support vector machines for sequence classification and segmentation. *Pattern Recogn.* **2009**, *42*, 3224–3230. [CrossRef]
30. Yang, Y.; Song, S.; Yue, F.; He, W.; Shao, W.; Zhao, K.; Nie, W. Superpixel-based automatic image recognition for landslide deformation areas. *Eng. Geol.* **2019**, *259*, 105166. [CrossRef]
31. Li, Q.; Song, D.; Yuan, C.; Nie, W. An image recognition method for the deformation area of open-pit rock slopes under variable rainfall. *Measurement* **2022**, *188*, 110544. [CrossRef]
32. Wang, G.; Zhao, B.; Wu, B.; Zhang, C.; Liu, W. Intelligent prediction of slope stability based on visual exploratory data analysis of 77 in situ cases. *Int. J. Min. Sci. Technol.* **2023**, *33*, 47–59. [CrossRef]
33. Lin, S.; Zheng, H.; Han, B.; Li, Y.; Han, C.; Li, W. Comparative performance of eight ensemble learning approaches for the development of models of slope stability prediction. *Acta Geotech.* **2022**, *17*, 1477–1502. [CrossRef]
34. Zhang, W.; Li, H.; Han, L.; Chen, L.; Wang, L. Slope stability prediction using ensemble learning techniques: A case study in Yunyang County, Chongqing, China. *J. Rock Mech. Geotech.* **2022**, *14*, 1089–1099. [CrossRef]
35. Qi, C.; Tang, X. Slope stability prediction using integrated metaheuristic and machine learning approaches: A comparative study. *Comput. Ind. Eng.* **2018**, *118*, 112–122. [CrossRef]
36. Kardani, N.; Zhou, A.; Nazem, M.; Shen, S. Improved prediction of slope stability using a hybrid stacking ensemble method based on finite element analysis and field data. *J. Rock Mech. Geotech.* **2021**, *13*, 188–201. [CrossRef]
37. Khanh, P.; Kim, D.; Park, S.; Choi, H. Ensemble learning-based classification models for slope stability analysis. *Catena* **2021**, *196*, 104886. [CrossRef]
38. Dong, Y.; Lan, Y.; Li, B.; Li, W. Study on deformation characteristics and early warning analysis of rainfall-type shallow soil landslides—Tongnan Orchard in Chongqing as an example. *Yangtze River* **2020**, *51*, 97–101. [CrossRef]
39. Deng, X. Landslide deformation stability evaluation and early warning analysis based on information decomposition. *Yangtze River* **2021**, *52*, 101–107. [CrossRef]
40. Zhou, Z.; Shen, J.; Shu, J.; Duan, W.; Yang, R.; Li, Y. Study of prediction and early warning of slope deformation of accumulation layer in front of a hydropower station dam in Southwest China. *J. Chengdu Univ. Technol. (Sci. Technol. Ed.)* **2020**, *47*, 481–491.
41. Lei, H.; Zhou, X.; Wang, Y. Research on landslide early warning and prediction based on combined response of multifractal characteristics and sub item prediction. *J. Geod. Geodyn.* **2022**, *42*, 885–891. [CrossRef]
42. Mao, H.; Zhang, M.; Li, B.; Xu, N. Stability Analysis of the Left Bank Slope of Baihetan Hydropower Station Based on the MF-DFA Method. *Adv. Civ. Eng.* **2020**, *2020*, 8898318. [CrossRef]
43. Ding, L.; Li, C.; Lei, Z.; Zhang, C.; Wei, L.; Guo, Z.; Li, Y.; Fan, X.; Qi, D.; Wang, J. Spatiotemporal evolution of deformation and LSTM prediction model over the slope of the deep excavation section at the head of the South-North Water Transfer Middle Route Canal. *Heliyon* **2024**, *10*, e26301. [CrossRef]
44. Xu, J.; Hou, X.; Wu, X.; Liu, Y.; Sun, G. Research on slope displacement prediction based on MIC-XGBoost-LSTM model. *China J. Highw. Transp.* **2024**, 1–12. Available online: https://kns.cnki.net/KCMS/detail/detail.aspx?dbcode=CAPJ&dbname=CAPJLAST&filename=ZGGL20240221002&uniplatform=OVERSEA&v=1DM6sN0rIQ2AlZhs-Qi66TChuhveI3aSVq8otiIQOGTZAw56q-C9LbQiyTY2598- (accessed on 26 February 2024).
45. Xiao, H.; Wang, S.; Chen, L.; Fan, Y.; Wan, J. An optimization network model for slope deformation prediction based on GA and LSTM fusion and tis application. *J. Geod. Geodyn.* **2024**, *44*, 491–496. [CrossRef]
46. Yang, S.; Jin, A.; Nie, W.; Liu, C.; Li, Y. Research on SSA-LSTM-Based Slope Monitoring and Early Warning Model. *Sustainability* **2022**, *14*, 10246. [CrossRef]
47. Yang, X.; Fan, X.; Wang, K.; Zhou, Z. Research on landslide susceptibility prediction model based on LSTM-RF-MDBN. *Environ. Sci. Pollut. R* **2024**, *31*, 1543–1561. [CrossRef]

48. Zhang, X.; Zhu, C.; He, M.; Dong, M.; Zhang, G.; Zhang, F. Failure Mechanism and Long Short-Term Memory Neural Network Model for Landslide Risk Prediction. *Remote Sens.* **2022**, *14*, 166. [CrossRef]
49. Lin, Z.; Sun, X.; Ji, Y. Landslide Displacement Prediction Model Using Time Series Analysis Method and Modified LSTM Model. *Electronics* **2022**, *11*, 1519. [CrossRef]
50. Khalili, M.A.; Guerriero, L.; Pouralizadeh, M.; Calcaterra, D.; Di Martire, D. Monitoring and prediction of landslide-related deformation based on the GCN-LSTM algorithm and SAR imagery. *Nat. Hazards* **2023**, *119*, 39–68. [CrossRef]
51. Xiong, Y. Correlation analysis between multifractal characteristics of regional geomorphology and development of geological disasters. *Earth Sci. Res. J.* **2021**, *25*, 49–55. [CrossRef]
52. Sun, B.; Ren, F.; Liu, D. Research on the failure precursors of layered slate based on multifractal characteristics of acoustic emission. *Rock Soil Mech.* **2022**, *43*, 749–760. [CrossRef]
53. Yang, C.; Huang, R.; Liu, D.; Qiu, W.; Zhang, R.; Tang, Y. Analysis and Warning Prediction of Tunnel Deformation Based on Multifractal Theory. *Fractal Fract.* **2024**, *8*, 108. [CrossRef]
54. Niu, Y.; Liu, P.; Zhang, C.; Hu, Y.; Wang, J. Mechanical properties and dynamic multifractal characteristics of shale under anisotropic stress using AE technology. *Geoenergy Sci. Eng.* **2023**, *226*, 211748. [CrossRef]
55. Telesca, L.; Chamoli, A.; Lovallo, M.; Stabile, T.A. Investigating the Tsunamigenic Potential of Earthquakes from Analysis of the Informational and Multifractal Properties of Seismograms. *Pure Appl. Geophys.* **2015**, *172*, 1933–1943. [CrossRef]

fractal and fractional

MDPI

Article

Strength and Fractal Characteristics of Artificial Frozen–Thawed Sandy Soft Soil

Bowen Kong [1,2], Yuntian Yan [1], Huan He [1], Jing Yu [2], Baoping Zou [1,3,*] and Qizhi Chen [1,3,*]

1 School of Civil Engineering and Architecture, Zhejiang University of Science and Technology, Hangzhou 310023, China; kbw@zust.edu.cn (B.K.); yanyuntian1013@163.com (Y.Y.); hehuan465@163.com (H.H.)
2 Department of Civil Engineering, Hangzhou City University, Hangzhou 310015, China; 15397142671@163.com
3 Zhejiang-Singapore Joint Laboratory for Urban Renewal and Future City, Hangzhou 310023, China
* Correspondence: zoubp@zust.edu.cn (B.Z.); chenqizhi@zust.edu.cn (Q.C.); Tel.: +86-187-5808-4001 (B.Z.); +86-189-6877-8088 (Q.C.)

Abstract: In regions with sandy soft soil strata, the subway foundation commonly undergoes freeze–thaw cycles during construction. This study focuses on analyzing the microstructural and fractal characteristics of frozen–thawed sandy soft soil to improve our understanding of its strength behavior and stability. Pore size distribution curves before and after freeze–thaw cycles were examined using nuclear magnetic resonance technology. Additionally, fractal theory was applied to illustrate the soil's fractal properties. The strength properties of frozen remolded clay under varying freezing temperatures and sand contents were investigated through uniaxial compression tests, indicating that soil strength is significantly influenced by fractal dimensions. The findings suggest that lower freezing temperatures lead to a more dispersed soil skeleton, resulting in a higher fractal dimension for the frozen–thawed soil. Likewise, an increase in sand content enlarges the soil pores and the fractal dimension of the frozen–thawed soil. Furthermore, an increase in fractal dimension caused by freezing temperatures results in increased soil strength, while an increase in fractal dimension due to changes in sand content leads to a decrease in soil strength.

Keywords: freeze–thaw; sandy soft soil; fractal characteristics; NMR; pore distribution; uniaxial compressive strength

1. Introduction

It is essential to investigate the microstructural characteristics of sandy soft soil for several reasons [1]. The Yangtze River Delta region in China is characterized by extensive soft soil strata that have developed into sand interlayers due to repeated wave activity. The artificial ground freezing method is commonly used in subway construction through these sandy soft soil layers [2–6], as it minimizes construction disturbance, effectively isolates groundwater, and prevents sand and water ingress. However, the freeze–thaw process damages the soil's microstructure, leading to significant settlement that can affect subway operation safety. To clarify the effects of freeze–thaw action, Zhou et al. [7,8] investigated the freeze–thaw properties of soft clay under seepage conditions in composite strata. Kong et al. [9] examined the pore structures of marine soft soils under various freezing conditions. Yao et al. [10] conducted uniaxial compressive strength tests on artificially frozen soft soil, clarifying the impact of freezing temperatures on uniaxial compressive strength. The results show that freezing damage rearranges the soil's internal microstructure, resulting in diminished strength and stability, consequently leading to uneven foundation settlements [11–16].

The concept of fractals was introduced in 1975 by the American scholar Mandelbrot, who focused on the study of irregular geometric shapes characterized by their self-similarity [17,18]. Prior research has confirmed that pore distributions exhibit self-similarity, and the pore structures of porous media manifest fractal characteristics, making them amenable to analysis through fractal theory [19–24]. Currently, the characterization of pore structures and fractal dimensions is primarily conducted using techniques such as scanning electron microscopy (SEM) [25], mercury intrusion porosimetry (MIP) [26], transmission electron microscopy (TEM) [27], X-ray microtomography [28], nitrogen adsorption [29], and nuclear magnetic resonance (NMR) [30]. Zhu et al. [31] investigated the pore structure of modified sandy soil using SEM, demonstrating that multiple fractal parameters can effectively describe the variations in pore characteristics and quantify the features of the pore structure in modified sandy soils. Kong et al. [32] acquired the pore size distribution (PSD) of freeze–thaw soft soil using NMR and SEM, and subsequently employed fractal theory to clarify the fractal characteristics of the PSD. He et al. [33] performed NMR tests to examine the influence of grain size on the microstructural pore characteristics and fractal features of carbonate-based and silicate-based sands, discovering that larger grain sizes correspond to increased maximum pore sizes and fractal dimensions. Ferreiro et al. [34] used MIP and nitrogen adsorption to investigate PSD across various equivalent pore size ranges, proposing that multifractal methods are suitable for characterizing the heterogeneity of the soil pore system. The integration of multiple methods by researchers not only facilitates a comprehensive understanding of the soil's microstructure from diverse perspectives but also significantly enhances the accuracy and depth of the studies [35,36]. These studies mainly focus on the fractal characteristics of pure soft soil, providing theoretical support for subway construction using artificial ground freezing (AGF) in soft soil.

In view of the lack of research on the micro-pore structure of sandy soft soil, this study focuses on examining sandy soft soil using NMR technology to analyze the pore size distribution curves and their variations before and after freeze–thaw cycles. Fractal theory is also applied to characterize the soil's fractal properties. Additionally, unconfined compressive strength tests are conducted to determine the impact of fractal dimensions on the strength properties of frozen remolded clay under different freezing temperatures and sand contents. The research results can serve as a reliable basis for predicting the structural damage of frozen and thawed soil under freezing method construction conditions, which can help to investigate the law of the inhomogeneous settlement of metro lines in the Yangtze River Delta region.

2. Materials and Methods

2.1. Experimental Materials

Soft clay samples taken from Zijingang Campus of Zhejiang University, located in Hangzhou City, Zhejiang Province, China. The basic physical properties of the soft soil and the incorporated sand were determined through indoor geotechnical testing, as shown in Tables 1 and 2.

Table 1. Basic physical indices of soft soil.

Index	Natural Density ($g{\cdot}cm^{-3}$)	Dry Density ($g{\cdot}cm^{-3}$)	Water Content (%)	Liquid Limit (%)	Plastic Limit (%)	Cohesion (kPa)	Internal Friction Angle (°)
Mean	1.75	1.37	33.96	38.34	20.59	15.18	18.23

Table 2. Basic physical indices of sand.

Index	Particle Size Range (mm)	Particle Size Range (mm)	Maximum Dry Density ($g{\cdot}cm^{-3}$)	Minimum Void Ratio	Maximum Void Ratio	Specific Gravity
Mean	0.1~1	1.43	1.73	0.53	0.84	2.63

2.2. Experimental Design and Sample Preparation

This study examines the impacts of freezing temperature and sand content on the micro-mechanisms and mechanical properties through a single-factor experimental design, where all other parameters remain constant, while one variable is changed to assess its influence. Samples were prepared with varying sand content configurations, as detailed in Table 3. Reflecting real-world construction conditions, a freezing temperature gradient of 5 °C was established, with soil samples subjected to freezing at −5 °C, −10 °C, −15 °C, and −20 °C, to investigate the varying impacts of freezing temperatures on soil performance, as specified in Table 4. In the table, sample identifiers such as S0 and T-5 correspond to 0% sand content and a temperature of −5 °C, respectively, with other identifiers following a similar notation.

Table 3. Sand content variation experimental plan.

Sample ID	Sample Dimensions (mm × mm)	Percentage of Sand Content (%)
S0	38 × 76	0
S5	38 × 76	5
S10	38 × 76	10
S15	38 × 76	15
S20	38 × 76	20
S25	38 × 76	25
S30	38 × 76	30

Table 4. Temperature variation experimental plan.

Sample ID	Sample ID	Freezing Temperature (°C)	Freezing Duration (h)	Melting Temperature (°C)	Melting Duration (h)
WD	38 × 76	-	-	-	-
T-5	38 × 76	−5	24	20	24
T-10	38 × 76	−10	24	20	24
T-15	38 × 76	−15	24	20	24
T-20	38 × 76	−20	24	20	24

"WD" denotes samples not subjected to freezing.

Soft sandy soils with varying sand content were prepared according to the test program and sealed for one day to ensure thorough mixing and stabilization. Vaseline was uniformly applied to the inner wall of the three-valve membrane fixture, and the prepared soil samples were placed into the mold in layers to ensure consistent density and homogeneity, as shown in Figure 1. The specimen diameter and height were set to 38 mm and 76 mm, respectively. As shown in Figure 2, the three-valve membrane specimen was placed into the vacuum saturator. The cover seam was coated with petroleum jelly to prevent air leakage. The instrument was started, pumped for 2 h, then filled with water until the specimen was completely submerged, and then turned off to stand for 10 h. To ensure sample saturation, the triaxial instrument was used for re-saturation. The sample was considered saturated when the B-value exceeded 97 and was maintained for 15 min. We removed the specimen and the film apparatus to ensure the specimen's integrity. We wrapped the demolded specimen (Figure 3) with plastic film and placed it into the air freezing and thawing box to freeze for 24 h. The frozen specimen is shown in Figure 4.

As shown in Figures 5 and 6, before freezing and thawing, the SEM images of sandy soft soil demonstrate large continuous flocculent structures and typical honeycomb structures. The soil structure is dense with few pores. After freezing and thawing, the breaking of large particles and the aggregation of small particles occur simultaneously, resulting in a lamellar interlocking particle aggregate structure. The framework becomes looser, the proportion of pores increases, and the connectivity rate also rises, adding complexity to the soil structure.

Figure 1. Sample preparation.

Figure 2. Vacuum saturation.

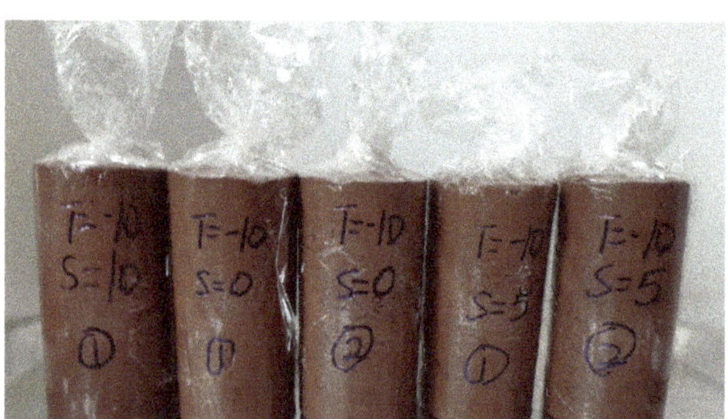

Figure 3. Demolded sample.

Figure 4. Frozen sample.

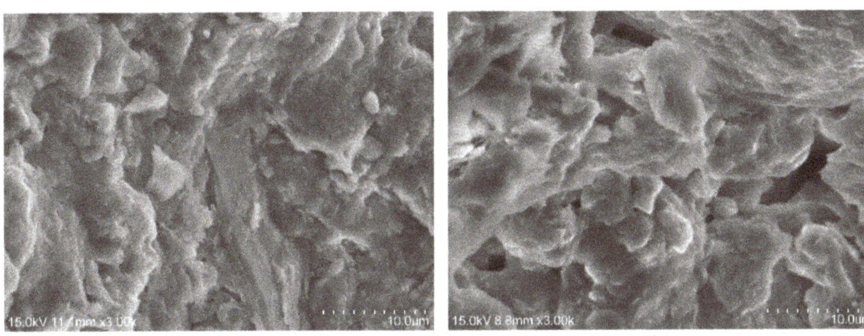

(**a**) (**b**)

Figure 5. Scanning electron microscope (SEM) images of soft soil at a magnification of 3000×: (**a**) unfrozen soil, (**b**) frozen–thawed soil.

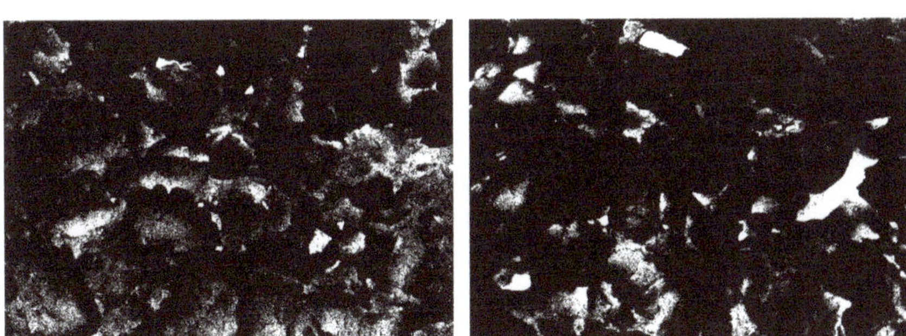

(**a**) (**b**)

Figure 6. Pore morphology of soft soil at 3000× magnification, with images processed through binarization: (**a**) unfrozen soil, (**b**) frozen–thawed soil.

2.3. Test Methods

2.3.1. Basic Principles of Nuclear Magnetic Resonance

The experimental apparatus used in this study is the MesoMR23-060H-I model low-field NMR instrument developed by Shanghai Niumag Corporation (Shanghai, China). This instrument autonomously maintains the magnetic field temperature at 35 ± 0.01 °C, with the maximum homogeneity in magnetic field dimensions being 60 × 60 mm. NMR clarifies the microstructural characteristics of substances by observing the changes in the spin states of atomic nuclei, which possess spin magnetic moments, as they absorb electromagnetic waves at specific frequencies in the presence of an external magnetic field [37].

This study primarily utilizes transverse relaxation time (T_2) values measured by nuclear magnetic resonance to estimate water content by quantifying the number of hydrogen atoms in the sample. The cutoff values in the T_2 relaxation spectra facilitate the differentiation between movable and bound fluid types in the soil [38]. For the frozen–thawed sandy soft soil discussed in this paper, the relaxation spectrum is dominated by a single peak, with the main peak occurring at less than 10 ms. Therefore, an empirical method was adopted, using the midpoint of the right half of the main peak as the T_2 cutoff value. As shown in Figure 7, the T_2 cutoff value was determined to be 2.31 ms. Fluids in pores with T_2 values greater than 2.31 ms were considered movable fluids, whereas fluids with T_2 values less than 2.31 ms were regarded as bound. The formula for T_2 relaxation is as follows:

$$\frac{1}{T_2} \approx \frac{1}{T_{2S}} = \rho_2\left(\frac{S}{V}\right)_{pore} = \rho_2\frac{\alpha}{R}, \tag{1}$$

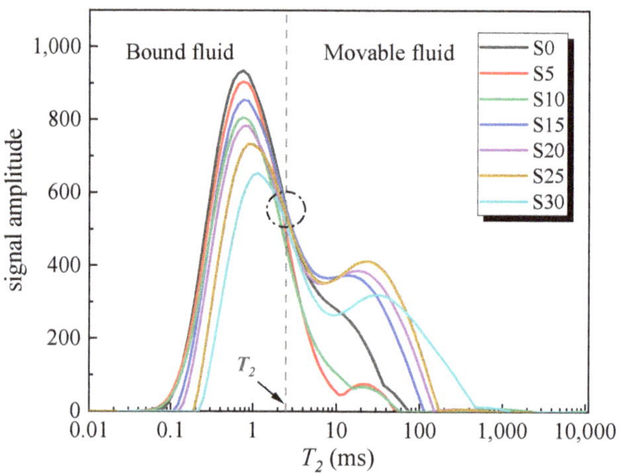

Figure 7. Determination of cutoff values for T_2 relaxation spectra (T_2 cutoff value is 2.31 ms, S0 to S30 represent sand contents of 0% to 30%).

In the equation, ρ_2 denotes the surface relaxation rate (μm/ms), which is associated with the properties of the sample's pore surfaces and binders. For the soft soil discussed here, ρ_2 is approximated at 27 μm/ms. α signifies the pore shape factor, with assigned values of 1, 2, and 3, corresponding to flat, cylindrical, and spherical pores, respectively. R denotes the equivalent pore radius, and S/V represents the pore-specific surface area.

2.3.2. Fractal Dimension Principle

Fractal dimension is a quantitative parameter used to characterize fractal features, reflecting the complexity and irregularity of fractals. Previous studies have confirmed that pore distributions exhibit self-similarity, and the pore structures of porous media demonstrate fractal features, which can be analyzed using fractal theory [39]. Currently, fractal theory is widely applied in quantifying and analyzing the geometric properties of PSD [33]. Below is the distribution of smaller pore sizes, calculated as a mass percentage:

$$M(R > R_a)/\rho_p C_m = (R/R_{max})^{3-D}, \tag{2}$$

In the equation, R represents the aperture, while R_{max} denotes the aperture that constitutes the largest percentage. M represents the mass of pores larger than the measured size R_a. C_m is a constant associated with the shape and size of pores, and ρ_p represents the density of pores under assumed conditions. Due to the fractal characteristics of soil pore distribution, the fractal dimension D is strictly confined to the range $0 < D < 3$ [40]. The fractal dimension D can be derived when the percentage of cumulative pore volume falls below a specified threshold. Given the linear relationship between the T_2 value and pore size R, and using the fractal dimension calculation formula for movable fluid space developed by Zhou et al. [39], the corresponding fractal geometric approximation for pore size R is expressed as follows:

$$S_V = (R_{max}/R)^{D-3}, \tag{3}$$

In the equation, S_V (%) represents the percentage of the total pore volume composed of pores smaller than the aperture R. D denotes the fractal dimension, and R_{max} represents the maximum aperture size. Upon a logarithmic transformation of the formula, the nuclear magnetic resonance-derived fractal dimension D model corresponding to the PSD for aperture R is as follows:

$$\lg S_V = (3 - D_b)\lg R + (D_b - 3)\lg R_{max} \text{ (Bond fluid)}, \tag{4}$$

$$\lg S_V = (3 - D_m)\lg R + (D_m - 3)\lg R_{max} \text{ (Movable fluid)}, \tag{5}$$

In the equation, D_b represents the fractal dimension of the bound fluid, while D_m denotes the fractal dimension of the movable fluid.

3. Analysis of Experimental Results

3.1. Quantitative Analysis of the Microstructure of Frozen–Thawed Sandy Soft Soil

3.1.1. Impact of Freezing Temperature on Pore Distribution

As shown in Figure 8a, as the temperature decreases, the freezing effect on pore expansion becomes more pronounced, leading to an overall increase in pore size. Concurrently, the peak value of the main peak shows a decreasing trend as the temperature lowers. Particularly in the −10 °C to −15 °C range, the peak value of the main peak after freezing drops more significantly, decreasing by approximately 20%. However, the peak value of the secondary peak exhibits an upward trend as the temperature decreases; this indicates that some small pores in the soil structure connect with nearby pores to form larger ones due to the freezing and expansion of pore water, thereby leading to a decrease in the main peak value and an increase in the secondary peak value.

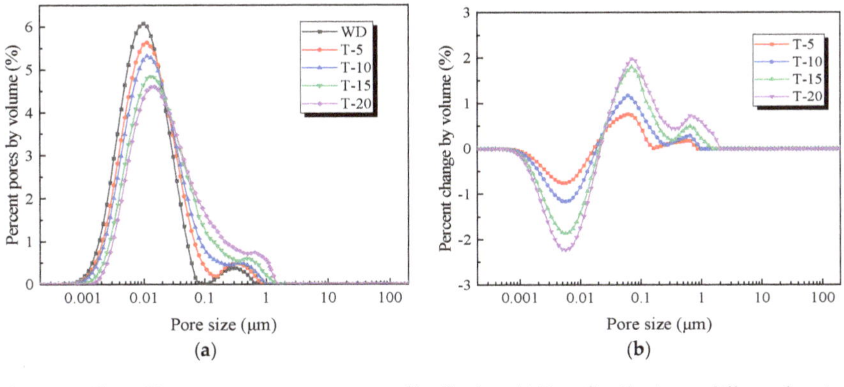

Figure 8. Effect of freezing temperature on pore distribution. (**a**) Pore distribution at different freezing temperatures. (**b**) Change in pore distribution at different freezing temperatures. ("WD" denotes samples not subjected to freezing. T-5 to T-20 represent freezing temperatures of −5 °C to −20 °C).

Considering the change in pore percentage, within the 0.001–0.025 μm range, the number of pores decreases as the temperature decreases, forming a distinct "concave area". These pores, originally in the "concave zone", shift to a larger aperture zone after freezing. Concurrently, the number of pores in the 0.025–0.4 μm range gradually increases, forming the "main convex area". According to the principle of the ice expansion of pore water, small pores expand upon freezing to form slightly larger pores. The diagram clearly shows that the "main convex area", which has increased in proportion, originates from the "concave area" which has decreased. Additionally, between 0.4 and 2 μm, a "secondary convex area" forms, potentially due to some pores from the "main convex area" further expanding during freezing or medium-sized pores merging under the effect of freezing, thereby increasing the number of pores in this larger pore size range. In conclusion, after freezing, the sandy soft soil shows a general tendency to expand. The small pore size increases by about 10 times and the percentage of larger pores grows dramatically.

3.1.2. Impact of Sand Content on Pore Distribution

According to the pore distribution curves for soft soils with varying sand contents shown in Figure 9, it is evident that the proportion of the main peak pores is negatively correlated with the sand content. At a constant temperature, the peak value of the main peak diminishes with increasing sand content, while the peak value of the secondary peak rises. This suggests that a higher sand content correlates with fewer small-sized pores and more large-sized pores, primarily due to the formation of more irregular large pore structures.

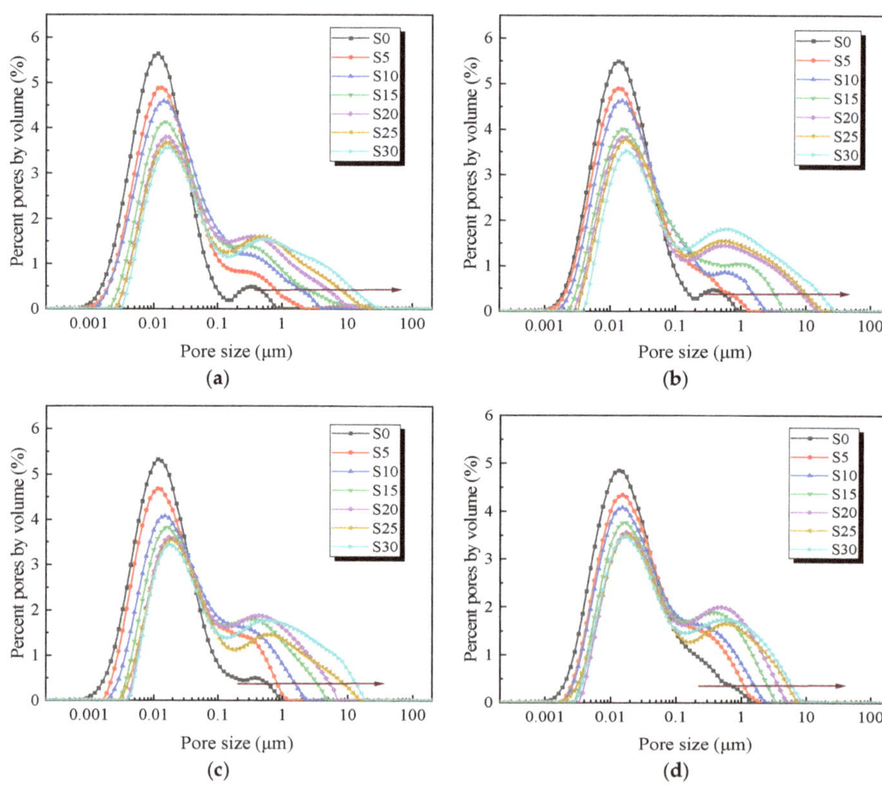

Figure 9. Changes in pore distribution of sandy soft soil at different sand contents: (**a–d**) represent temperatures of −5 °C, −10 °C, −15 °C, −20 °C, respectively. (S0 to S30 represent sand contents of 0% to 30%).

According to Figure 10, observing the changes in pore size percentage, when the temperature ranges from −5 °C to −15 °C, the curve displays a "concave area" and two "convex areas". Within the 0.001 μm to 0.05 μm "concave area", the number of pores decreases with an increase in sand content, showing a significant percentage change. Concurrently, the number of pores in the 0.05 μm to 0.4 μm range gradually increases, forming the first "convex area", and a second "convex area" emerges above 0.4 μm. At a temperature of −20 °C, the pore distribution curve exhibits distinct characteristics. Within the 0.001 μm to 0.1 μm range, a wider "concave area" forms compared to the "concave area" at −5 °C to −15 °C, suggesting that at lower temperatures, more small pores are reduced due to frost heave effects. Additionally, only one "convex area" forms above 0.1 μm, as the more intense freezing at lower temperatures results in the more pronounced expansion and merging of larger pores.

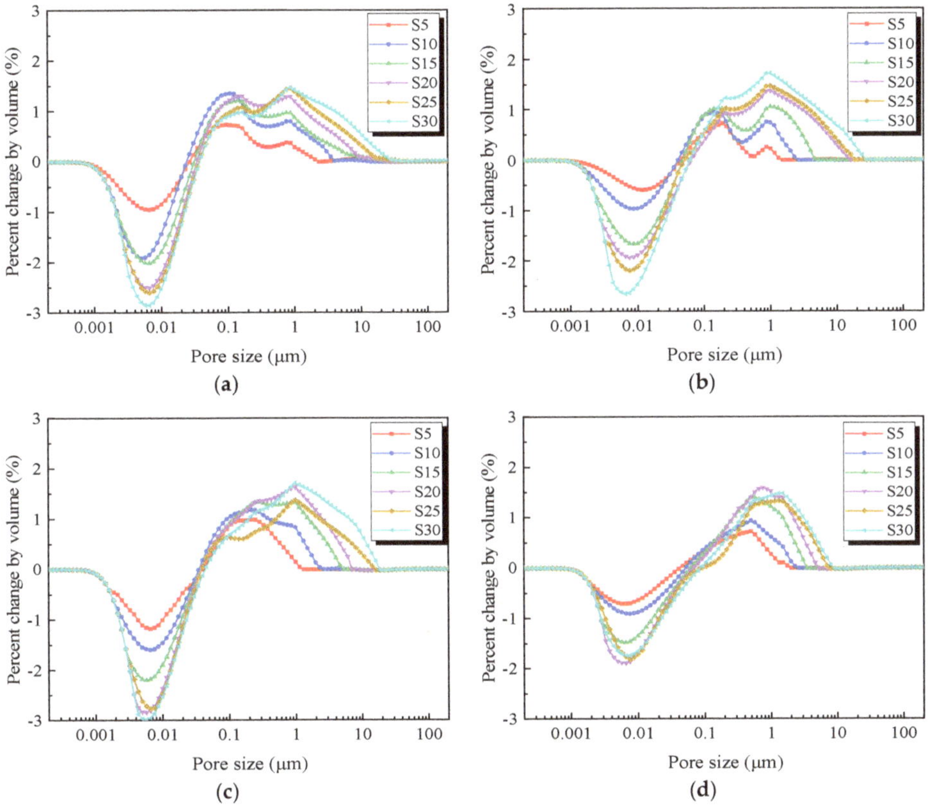

Figure 10. Changes in soft soil pore distribution relative to the initial distribution at different sand contents: (**a**–**d**) represent temperatures of −5 °C, −10 °C, −15 °C, −20 °C, respectively. (S0 to S30 represent sand contents of 0% to 30%).

3.2. Analysis of Fractal Characteristics in Frozen–Thawed Sandy Soft Soil

3.2.1. Analysis of Fractal Dimensions of Sandy Soft Soil at Various Freezing Temperatures

Freezing influences the distribution of soil pores, consequently altering the fractal dimensions of these pores. Based on Equations (5) and (6), fractal dimensions across various freezing temperatures are calculated, as summarized in Table 5. Figure 9 displays the study results on the relationship between freezing temperatures and fractal dimensions in sandy soft soil.

Table 5. Fractal dimension statistics for sandy soft soil at various freezing temperatures.

Sample ID	D_b	R^2	D_m	R^2
T-5	0.583	0.82	2.944	0.94
T-10	0.634	0.76	2.948	0.95
T-15	0.888	0.83	2.951	0.96
T-20	0.927	0.81	2.958	0.98

According to Figure 11 and Table 5, sandy soft soil displays distinct regular changes in the fractal dimensions of pore structures as revealed by nuclear magnetic resonance after undergoing various freezing temperatures. As temperatures decrease from −5 °C to −20 °C, the fractal dimensions of both bound and movable fluids gradually increase, exhibiting an upward trend. This suggests that the soil's pore structure becomes increasingly complex during the freezing process. Further comparisons reveal that the fractal dimensions of movable fluids are more significantly correlated with temperature changes, with correlation coefficients ranging from 0.94 to 0.98. Conversely, the fractal dimensions of bound fluids exhibit weaker correlations with temperature, with correlation coefficients between 0.76 and 0.83, indicating that freeze-up and thaw-sinking mainly affect the sample's larger pore structures.

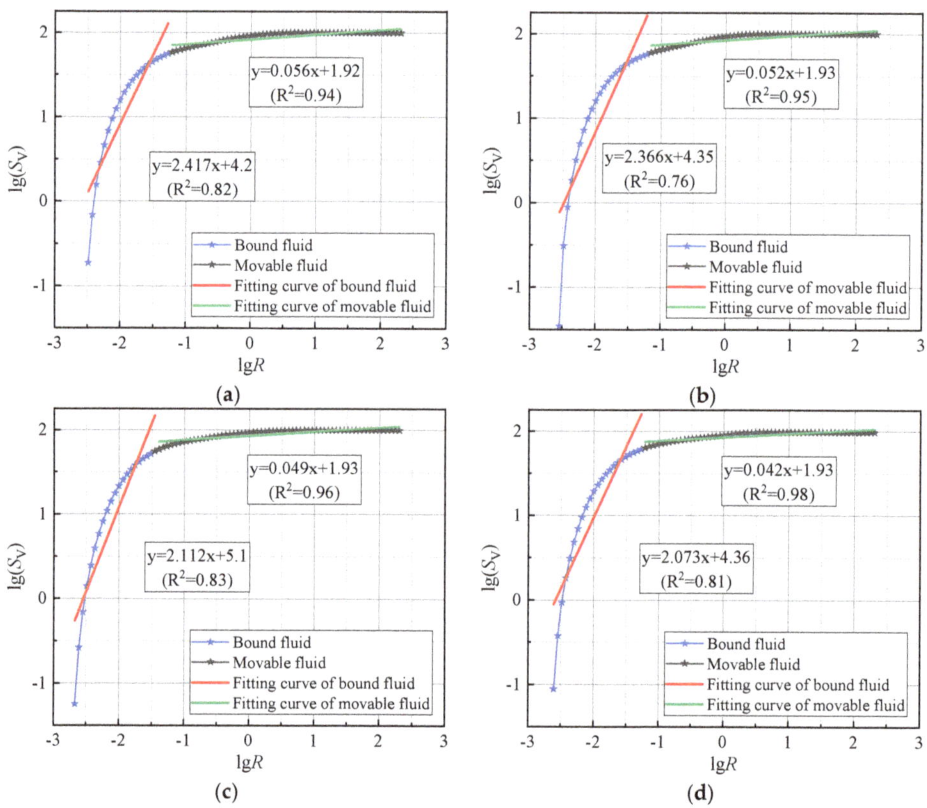

Figure 11. Pre-freezing lg(SV)-lgR curves of soft soil at different freezing temperatures: (**a**–**d**) represent temperatures of −5 °C, −10 °C, −15 °C, −20 °C, respectively.

3.2.2. Analysis of Pre-Freezing Fractal Dimensions of Soft Soil at Various Sand Contents

According to Figure 12 and Table 6, the variations in fractal dimensions of soft soils with different sand contents before freezing adhere to fractal laws. Specifically, with increasing sand content, the fractal dimensions of both bound and free fluids exhibit a consistent upward trend. This observation indicates that an increase in sand content results in the enhanced complexity of pore distribution. Comparing the fractal dimensions of bound fluid (D_b 0.82–0.89) and free fluid (D_m 0.89–0.99), it is clear that the fractal features of the free fluid are more pronounced than those of the bound fluid, primarily due to the irregular sands forming more branched and complex large void structures.

Table 6. Pre-freezing fractal dimension statistics for soft soil at various sand contents.

Sample ID	D_b	R^2	D_m	R^2
S0	0.757	0.83	2.919	0.89
S5	0.774	0.82	2.932	0.90
S10	0.848	0.84	2.944	0.93
S15	0.866	0.85	2.953	0.97
S20	0.907	0.87	2.962	0.98
S25	0.984	0.84	2.977	0.99
S30	1.013	0.89	2.985	0.99

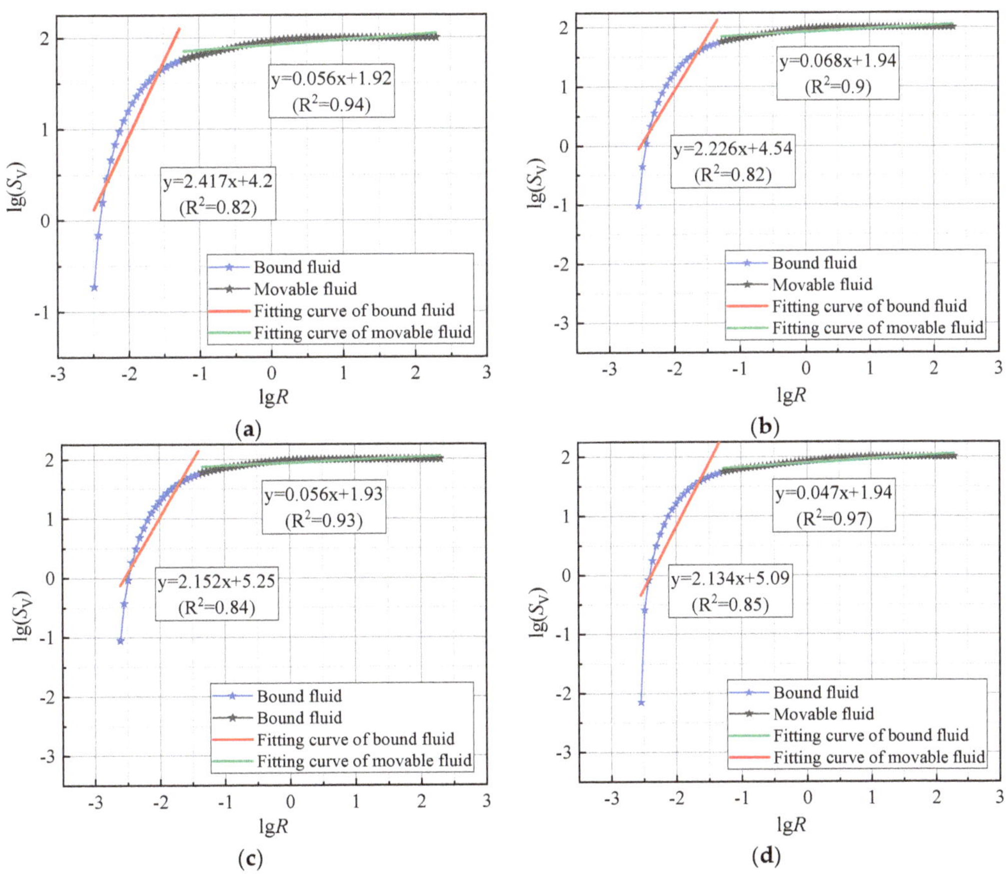

Figure 12. *Cont.*

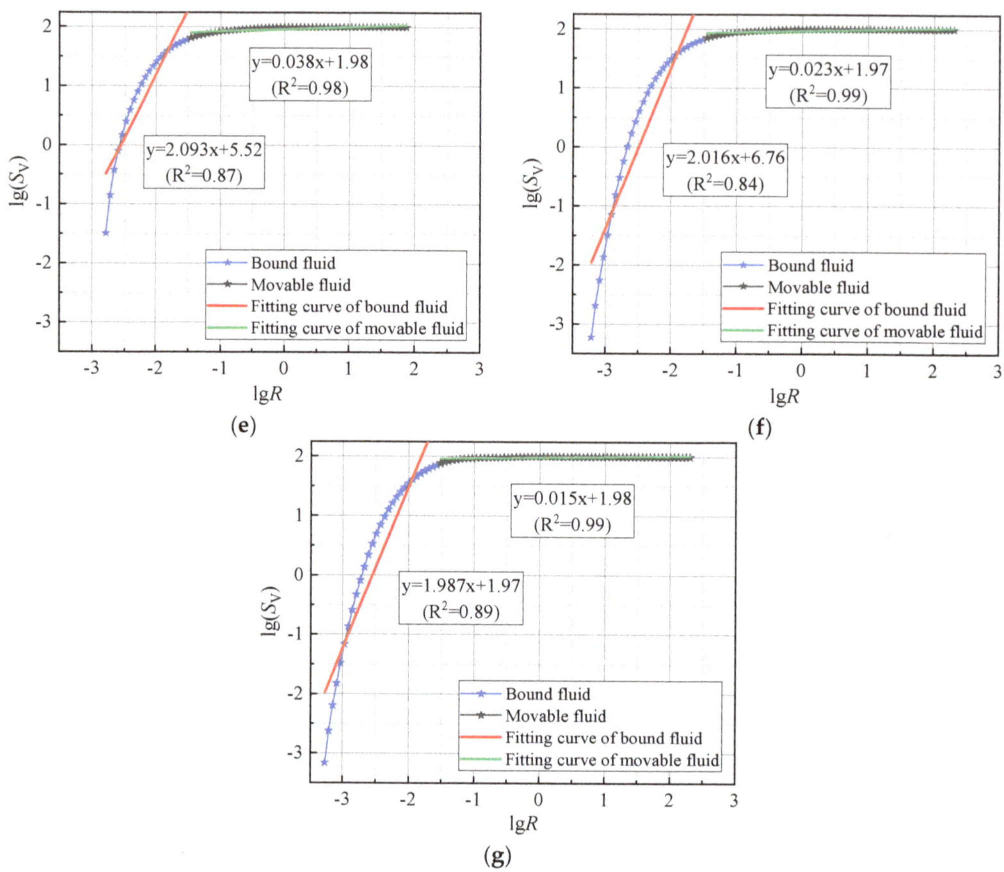

Figure 12. Pre-freezing lg(SV)-lgR curves of soft soil at different sand contents: (**a**–**g**) represent sand contents of 0%, 5%, 10%, 15%, 20%, 25%, 30%, respectively.

3.2.3. Analysis of Post-Freeze–Thaw Fractal Dimensions of Soft Soil at Various Sand Contents

According to Figure 13 and Table 7, the fractal dimensions of soft soil with varying sand contents continue to follow pre-freezing fractal laws even after undergoing freeze–thaw effects. Fractal dimensions increase gradually with rising sand content, and the fractal characteristics of movable fluid are more pronounced than those of bound fluid. Changes in fractal dimensions before and after freezing correspond to the complexity of pore distribution, illustrating the internal damage process in sandy soft soil under freeze–thaw conditions. For bound fluid, the fractal dimensions of smaller pores slightly decrease post freeze–thaw, primarily because small pore water freezes and expands into larger pores, reducing the quantity of smaller pores. Consequently, the number of larger pores increases, thereby enhancing the fractal dimensions of movable fluid post freeze–thaw.

Table 7. Post-freeze–thaw fractal dimension statistics for soft soil at various sand contents.

Sample ID	D_b	R^2	D_m	R^2
S0	0.706	0.8	2.93	0.94
S5	0.725	0.82	2.95	0.96
S10	0.775	0.8	2.96	0.97
S15	0.818	0.87	2.97	0.98
S20	0.837	0.84	2.979	0.99
S25	0.916	0.86	2.984	0.99
S30	1.035	0.87	2.994	0.99

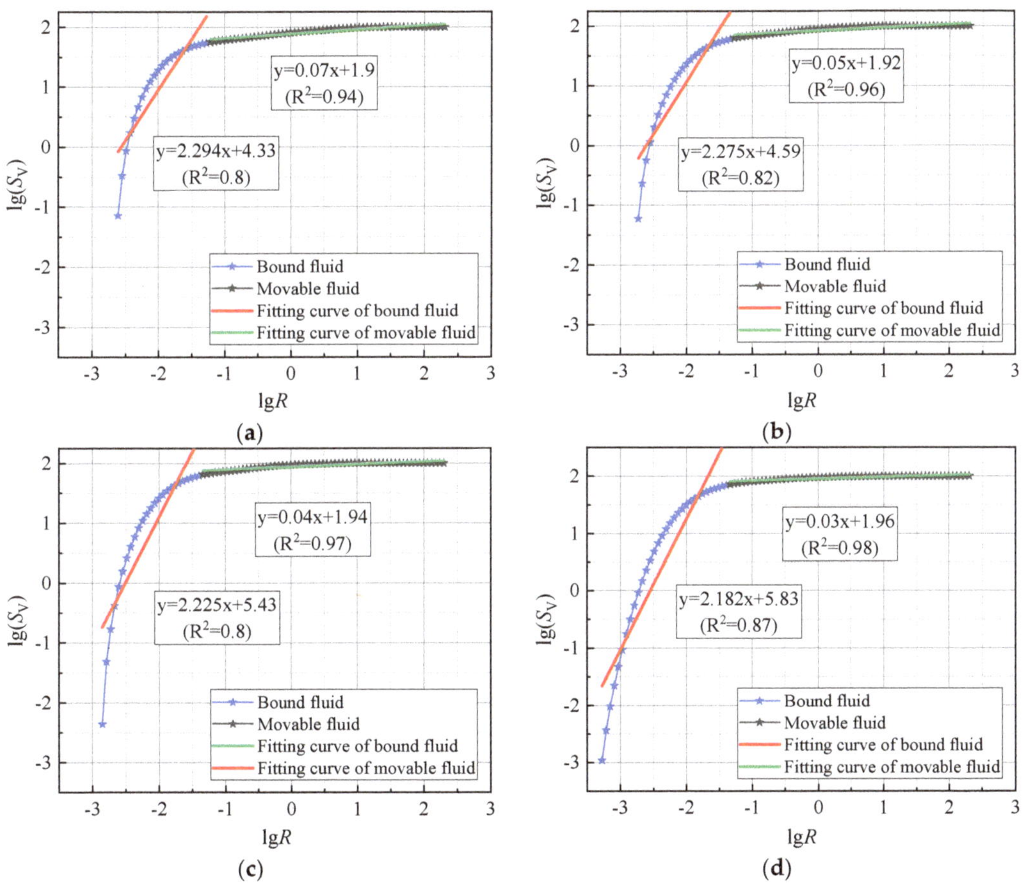

Figure 13. *Cont.*

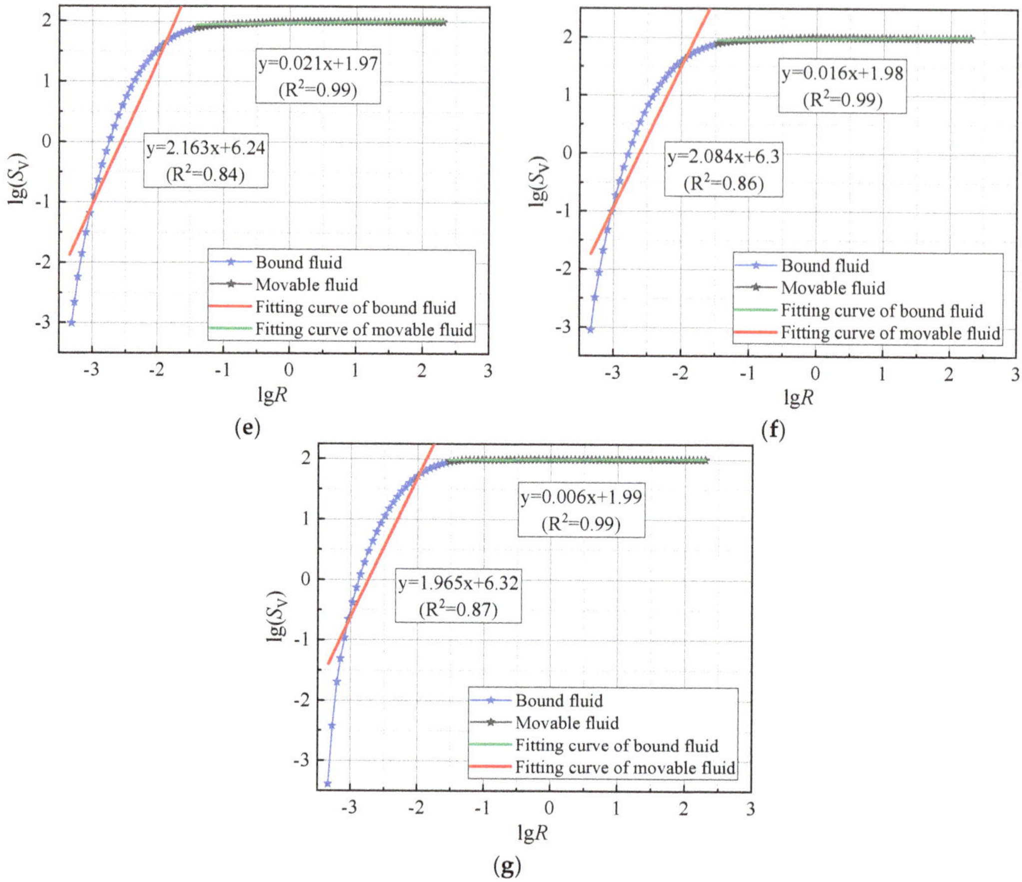

Figure 13. Post-thaw lg(SV)-lgR curves of soft soil at different sand contents: (**a**–**g**) represent sand contents of 0%, 5%, 10%, 15%, 20%, 25%, 30%, respectively.

3.3. Analysis of the Unconfined Compressive Strength of Frozen Sand–Clay Mixtures

As shown in Figure 14 and Table 8, unconfined compressive strength tests were conducted on frozen sandy soft soil with sand content rates of 0%, 5%, 15%, 20%, 25%, and 30% at freezing temperatures of −5 °C, −10 °C, −15 °C, and −20 °C. The resulting graph illustrates the relationship between unconfined compressive strength, sand content rate, and freezing temperature, described by the following equation:

$$\sigma = 1.182 - 0.116T - 0.04S,\ R^2 = 0.96 \quad (6)$$

where σ is the unconfined compressive strength, T is the freezing temperature, and S is the sand content. As the sand content in the frozen sand–clay increases, its strength exhibits a decreasing trend. This is because the increase in sand particles reduces the soil's cohesion. Furthermore, as the sand content increases, the soil's permeability improves, enhancing water mobility. This increased water mobility hinders the freezing of water, thereby reducing the soil's freezing capacity and subsequently affecting its post-freezing structural strength.

As the temperature decreases, the strength of the frozen sandy soft soil increases. This occurs because, under low-temperature conditions, the liquid phase water in the soil transitions to solid ice. The solid ice bonds the loose soil particles together, forming a much more stable frozen soil compared to non-frozen soil. As the temperature continues to decrease, the rate of phase transition of water in the soil increases, reducing its deformation capacity and making it more rigid and stable. This, in turn, leads to an increase in unconfined compressive strength.

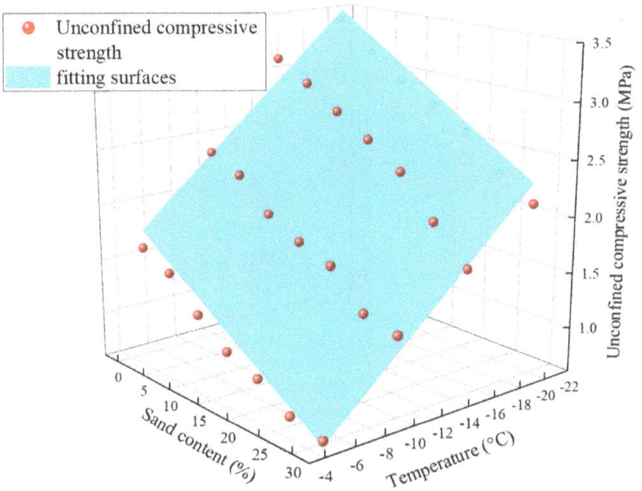

Figure 14. The relationship between compressive strength, sand content, and freezing temperature.

Table 8. Unconfined compressive strength of frozen sand–clay mixtures under various test conditions.

Sample ID	Unconfined Compressive Strength (MPa)			
	T-5	T-10	T-15	T-20
S0	1.60	2.35	3.12	3.41
S5	1.47	2.22	2.95	3.16
S10	1.20	1.95	2.76	2.95
S15	0.99	1.80	2.58	2.80
S20	0.88	1.69	2.37	2.61
S25	0.69	1.38	2.02	2.32
S30	0.62	1.31	1.71	2.13

4. Discussion

Figure 15a,b illustrate the relationship between fractal dimensions and freezing temperatures, vividly showing the extent of damage to pore structures due to freezing. Both bound and movable fluids' fractal dimensions show a notable correlation with freezing temperatures, with the correlation being more significant for movable fluid. This occurs because the freezing process entails an energy exchange and lower freezing temperatures have greater energy to break chemical and physical bonds between pore water. Consequently, lower freezing temperatures lead to a looser skeletal structure and higher fractal dimensions. Additionally, since larger pore waters have higher volumes and their water molecules exhibit greater freedom of movement, they are more susceptible to freezing than smaller pore waters, further substantiating the stronger correlation of fractal dimensions with temperature in movable fluids compared to bound fluids.

Figure 16a,b show the relationship between the fractal dimensions of sand-containing soft soils before freezing and the sand content, clearly illustrating the impact of sand content on pore structures. The fractal dimensions of both bound and free fluids exhibit a linear correlation with sand content. In soils with a higher sand content, pores are larger and distributed more irregularly and unevenly, contributing to an increase in fractal dimensions.

The fractal dimension of the free fluid has a stronger correlation with the sand content because the sand has large particles, which form more irregular and large voids.

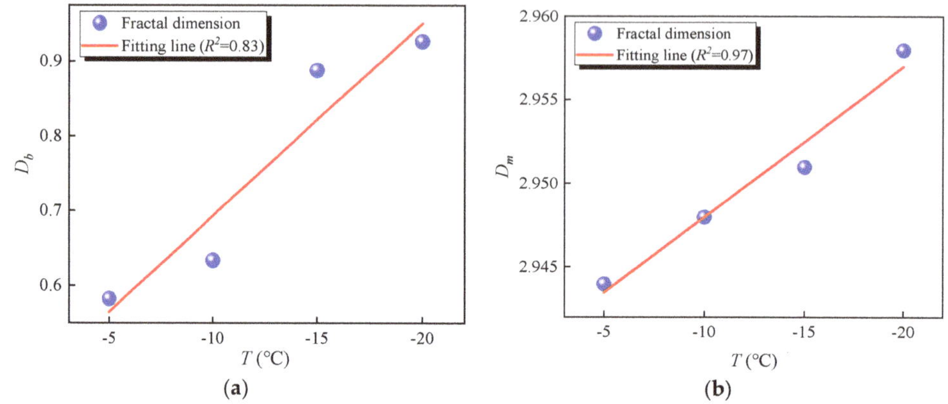

Figure 15. Fractal dimensions of sandy soft soil at different freezing temperatures: (**a**) fractal dimension for bound fluid; (**b**) fractal dimension for movable fluid.

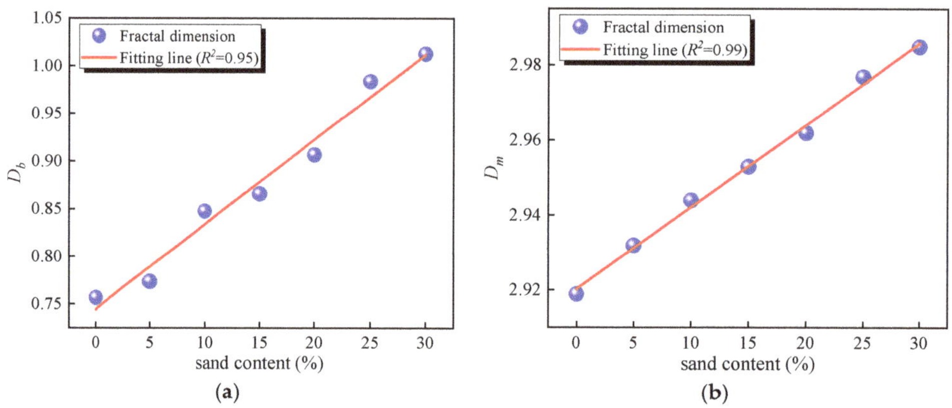

Figure 16. Fractal dimension of sandy soft soil before freezing: (**a**) fractal dimension for bound fluid; (**b**) fractal dimension for movable fluid.

Figure 17a,b show the relationship between the post-freeze–thaw fractal dimensions of sandy soft soil and sand content, showing clear correlations for both bound and movable fluids' fractal dimensions with sand content. However, these correlations are diminished compared to pre-freezing, primarily because sand resists deformation. After freezing, the specimen's freezing and thawing are "suppressed" by the sand, weakening the correlation between the fractal dimension and sand content.

Figure 18a,b depict the relationship between the fractal dimension, freezing temperature, and unconfined compressive strength of frozen–thawed sand–clay. It is evident that the fractal dimension significantly influences the unconfined compressive strength. The fractal dimensions of both bound fluid and movable fluid exhibit a linear correlation with freezing temperature. From the previously discussed conclusions, it is known that when only the freezing temperature is lowered, it leads to a larger fractal dimension and the creation of a large number of pores with an irregular distribution and a high degree of inhomogeneity, and the structurally complex pores created in this case lead to a higher degree of freezing of the soil and consequently to a greater compressive strength of the frozen soil. At this point the change in fractal dimension is positively correlated with the strength of the permafrost.

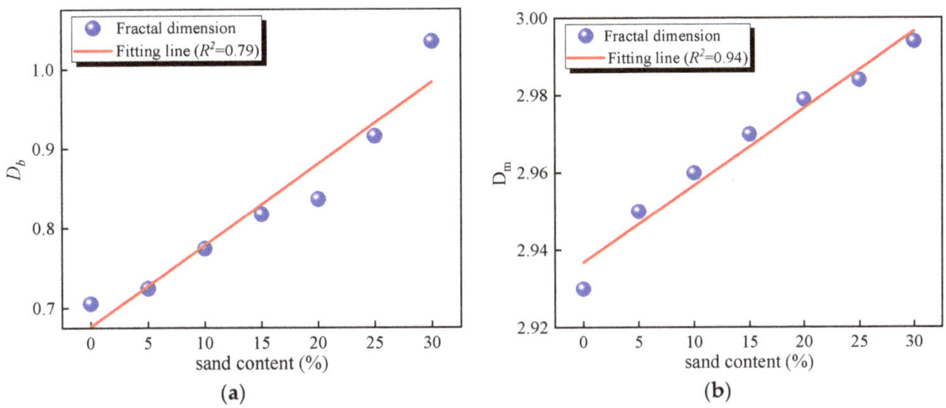

Figure 17. Fractal dimension of sandy soft soil after freeze–thaw: (**a**) fractal dimension for bound fluid; (**b**) fractal dimension for movable fluid.

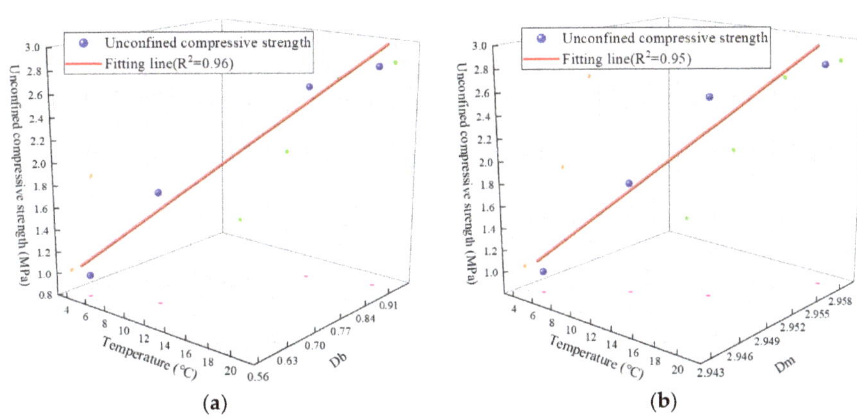

Figure 18. The relationship between compressive strength, freezing temperature, and fractal dimension. (**a**) The relationship between bound fluid fractal dimension and compressive strength. (**b**) The relationship between movable fluid fractal dimension and compressive strength.

Figure 19a,b illustrate the relationship between fractal dimension, sand content, and unconfined compressive strength of frozen–thawed sand–clay. It is clear that the fractal dimension has a significant impact on the unconfined compressive strength. The fractal dimensions of both bound fluid and movable fluid show a linear relationship with sand content. As previously discussed, it is known that when only the sand content is increased, it leads to a larger fractal dimension, and the structurally complex pores produced in this case lead to a lower degree of freezing of the soil and consequently to a lower compressive strength of the frozen soil mass. At this point, the change in the fractal dimension is negatively correlated with the strength of the frozen soil.

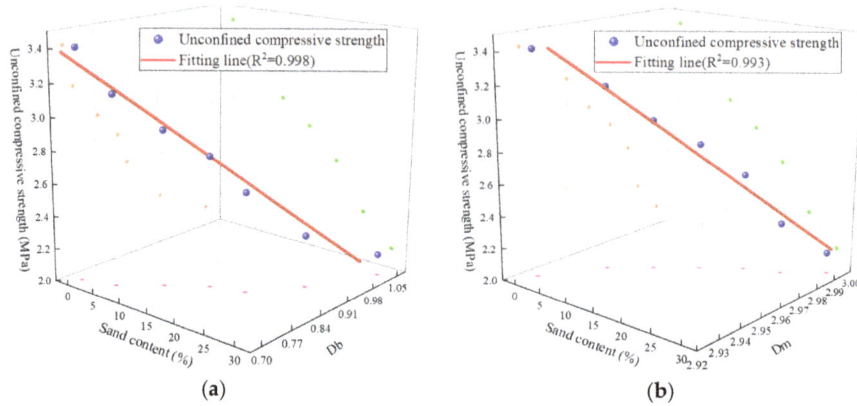

Figure 19. The relationship between compressive strength, sand content, and fractal dimension. (**a**) The relationship between bound fluid fractal dimension and compressive strength. (**b**) The relationship between movable fluid fractal dimension and compressive strength.

5. Conclusions

Artificial freezing methods are commonly used in water-rich sandy areas, but frost heave during construction can severely damage the soil's structure. Nuclear magnetic resonance technology enables the observation of microscopic pore changes and the calculation of fractal dimensions in frozen–thawed sandy soil through pore distribution curves. This reveals the complexity and regularity of pore structures, providing a new perspective on the microstructural characteristics of soft soil. Uniaxial compression tests were conducted on frozen remolded clays at various freezing temperatures and sand contents to analyze the effect of freezing on the strength of remolded clays from the perspective of fractal dimension. The following conclusions have been drawn:

(1) Pore distribution shows self-similarity, indicating fractal characteristics. Higher fractal dimensions in pore size distribution suggest more complex pore structures. Pore fluids are categorized based on the T_2 cutoff value: fluids with $T_2 < 2.31$ ms are considered bound, while those with $T_2 > 2.31$ ms are considered movable. Fractal characteristics are more pronounced in movable fluids compared to bound fluids.

(2) Both pre- and post-freeze–thaw pore distributions follow a bimodal pattern, with the main peak representing smaller pores and the secondary peak larger pores. After freezing, there is a trend towards increasing pore size, which becomes more pronounced as temperatures decrease. This indicates that as pore water expands during freezing, some smaller soil pores connect with adjacent pores to form larger pores or fissures.

(3) Variations in sand content affect pore distribution. As sand content increases, the number of smaller pores decreases, and the largest pores in the soil structure tend to expand. This trend occurs because irregular sands form more branched and complex large void structures, altering the local soil structure.

(4) Variations in sand content and freezing temperature both affect the unconfined compressive strength of frozen sand–clay. A higher sand content leads to a lower unconfined compressive strength of the frozen soil. Conversely, lower freezing temperatures result in higher unconfined compressive strength.

(5) As the freezing temperature decreases, the skeletal structure becomes more loosely arranged, leading to an increase in fractal dimension. This indicates that lower temperatures exacerbate the damaging effects of freezing on soil structure.

(6) The fractal dimensions of bound and free fluids show significant correlations with sand content both before and after freeze–thaw cycles, with these correlations being more pronounced before freezing. A higher sand content leads to larger, more irregularly and unevenly distributed pores, thereby increasing the fractal dimension.

(7) A larger fractal dimension signifies more complex pores and a looser skeletal structure when the freezing temperature changes. This indicates a higher degree of freezing and results in greater unconfined compressive strength.
(8) When sand content varies, a larger fractal dimension indicates an increase in large voids and a more complex pore structure, making it harder for water to freeze. This results in a lower degree of freezing and reduced unconfined compressive strength.

This paper investigates the strength and fractal characteristics of artificial frozen–thawed sandy soft soil, providing a reliable basis for predicting structural damage under freezing method construction conditions. It also aids in exploring the uneven settlement laws for underground lines in the Yangtze River Delta region. However, there are limitations. Firstly, this study was conducted in a closed system without external water influence. In actual engineering environments, soil typically exists in an open system, which may cause discrepancies between the experimental results and real-world conditions. Secondly, the study only conducted strength tests, making it difficult to fully understand the deformation and settlement mechanisms of artificially frozen and thawed sandy soft soil.

In future research, further studies should be conducted in an open system to better align with actual engineering conditions. Additionally, measurements of frozen–thawed soil deformation and other physical parameters should be included to better investigate the impact of freeze–thaw soil structural damage on the uneven settlement of metro lines.

Author Contributions: Conceptualization, B.K. and B.Z.; methodology, B.K. and Q.C.; validation, Y.Y. and J.Y.; formal analysis, B.K.; investigation, Y.Y. and H.H.; resources, B.K.; data curation, J.Y.; writing—original draft preparation, B.K.; writing—review and editing, B.K. and Y.Y.; visualization, Y.Y. and H.H.; supervision, Q.C.; project administration, B.Z.; funding acquisition, B.K. All authors have read and agreed to the published version of the manuscript.

Funding: This research was funded by the Zhejiang Provincial Natural Science Foundation of China, grant number LQ24A020015 and the Youth Natural Science Foundation of Zhejiang University of Science and Technology, grant number 2023QN026.

Data Availability Statement: The data are available from the corresponding author upon request.

Acknowledgments: The authors are very grateful to Jianzhong Xia from Zhejiang University of Science and Technology for his kind guidance in writing this article.

Conflicts of Interest: The authors declare no conflicts of interest.

References

1. Tomassi, A.; Milli, S.; Tentori, D. Synthetic seismic forward modeling of a high-frequency depositional sequence: The example of the Tiber depositional sequence (Central Italy). *Mar. Pet. Geol.* **2024**, *160*, 106624. [CrossRef]
2. Xue, F.C. Dynamic responses of subway tunnel in clay stratum to moving loads. *Arab. J. Sci. Eng.* **2017**, *42*, 1327–1340.
3. Ding, Z.; Zhang, X.; Yin, X. Analysis of the influence of soft soil grouting on the metro tunnel based on field measurement. *Eng. Comput.* **2019**, *36*, 1522–1541. [CrossRef]
4. Cui, Z.D.; Zhang, L.J.; Hou, C.Y. Seismic behavior of subway station in the soft clay field before and after freeze-thaw cycle. *Soil Dyn. Earthq. Eng.* **2023**, *175*, 108222. [CrossRef]
5. Liu, J.; Yang, Y.; Yang, C. Analysis and prediction of long-term settlement of metro shield tunnel in saturated sand. *Geotech. Geol. Eng.* **2021**, *39*, 5241–5252. [CrossRef]
6. Yan, Q.; Wu, W.; Zhang, C. Monitoring and evaluation of artificial ground freezing in metro tunnel construction—A case study. *KSCE J. Civ. Eng.* **2019**, *23*, 2359–2370. [CrossRef]
7. Zhou, J.; Guo, Z.; Wang, C. Analysis of Freeze–Thaw Response and Pore Characteristics of Artificially Frozen Soft Soil under Combined Formation Seepage. *Appl. Sci.* **2022**, *12*, 10687. [CrossRef]
8. Zhou, J.; Tang, Y. Practical model of deformation prediction in soft clay after artificial ground freezing under subway low-level cyclic loading. *Tunn. Undergr. Space Technol.* **2018**, *76*, 30–42. [CrossRef]
9. Kong, B.; Xia, F.; Yu, B. Pore size changes in marine soft soil under various freezing conditions. *J. Mar. Sci. Eng.* **2020**, *8*, 170. [CrossRef]
10. Yao, Y.; Zhu, Y.; Shen, D. Fuzzy random evaluation of creep model of frozen soft soil in metro tunnel construction using artificial ground freezing technique. *Sci. Rep.* **2023**, *13*, 9468. [CrossRef]
11. Fan, W.; Yang, P.; Yang, Z.J. Impact of freeze-thaw on the physical properties and compressibility of saturated clay. *Cold Reg. Sci. Technol.* **2019**, *168*, 102873. [CrossRef]

12. Fan, W.; Yang, P. Ground temperature characteristics during artificial freezing around a subway cross passage. *Transp. Geotech.* **2019**, *20*, 100250. [CrossRef]
13. Li, J.; Tang, Y.; Feng, W. Creep behavior of soft clay subjected to artificial freeze–thaw from multiple-scale perspectives. *Acta Geotech.* **2020**, *15*, 2849–2864. [CrossRef]
14. Xiao, J.; Shan, J.; Wu, N. Long-term settlement of metro viaduct piers: A case study in Shanghai soft soil. *Transp. Geotech.* **2023**, *42*, 101075. [CrossRef]
15. Zhang, R.; Cheng, X.; Li, Y. Research on the ground subsidence mechanism of cross passage caused by freezing method construction. *Adv. Civ. Eng.* **2021**, *2021*, 6622177. [CrossRef]
16. Ma, J.; Huang, K.; Zou, B. The influence of tunnel insulation measures on the temperature spatiotemporal variation of frozen soil during artificial ground freezing. *Cold Reg. Sci. Technol.* **2023**, *214*, 103942. [CrossRef]
17. Pfeifer, P.; Avnir, D. Chemistry in noninteger dimensions between two and three. I. Fractal theory of heterogeneous surfaces. *J. Chem. Phys.* **1983**, *79*, 3558–3565. [CrossRef]
18. Avnir, D.; Jaroniec, M. An isotherm equation for adsorption on fractal surfaces of heterogeneous porous materials. *Langmuir* **1989**, *5*, 1431–1433. [CrossRef]
19. Li, J.; Ostoja-Starzewski, M. Thermo-poromechanics of fractal media. *Philos. Trans. R. Soc. A* **2020**, *378*, 20190288. [CrossRef]
20. Zhao, Y.; Feng, Z.; Lv, Z. Percolation laws of a fractal fracture-pore double medium. *Fractals* **2016**, *24*, 1650053. [CrossRef]
21. Chen, X.; Yao, G.; Herrero-Bervera, E. A new model of pore structure typing based on fractal geometry. *Mar. Pet. Geol.* **2018**, *98*, 291–305. [CrossRef]
22. Zhu, J.F.; Liu, J.Z.; Yang, Y.M. Fractal characteristics of pore structures in 13 coal specimens: Relationship among fractal dimension, pore structure parameter, and slurry ability of coal. *Fuel Process. Technol.* **2016**, *149*, 256–267. [CrossRef]
23. Xu, J.; Wu, K.; Li, Z. A model for gas transport in dual-porosity shale rocks with fractal structures. *Ind. Eng. Chem. Res.* **2018**, *57*, 6530–6537. [CrossRef]
24. Wei, D.; Gao, Z.; Zhang, C. Pore characteristics of the carbonate shoal from fractal perspective. *J. Pet. Sci. Eng.* **2019**, *174*, 1249–1260. [CrossRef]
25. Yu, Q.; Dai, Z.; Zhang, Z. Estimation of sandstone permeability with SEM images based on fractal theory. *Transp. Porous Media* **2019**, *126*, 701–712. [CrossRef]
26. Sun, M.; Zou, C.; Xin, D. Pore structure evolution mechanism of cement mortar containing diatomite subjected to freeze-thaw cycles by multifractal analysis. *Cem. Concr. Compos.* **2020**, *114*, 103731. [CrossRef]
27. Wang, H.Z.; Liu, X.H.; Yang, X.J. Intergrowth of a carbon layer and fractal-like trees on 3Y-TZP in TEM observations. *Mater. Sci. Eng. A* **2001**, *311*, 180–184. [CrossRef]
28. Kulkarni, P.; Watwe, V.; Doltade, T. Fractal kinetics for sorption of methylene blue dye at the interface of Alginate Fullers earth composite beads. *J. Mol. Liq.* **2021**, *336*, 116225. [CrossRef]
29. Xu, S.; Yang, Z.; Wu, S. Fractal analysis of pore structure differences between shale and sandstone based on the nitrogen adsorption method. *Nat. Resour. Res.* **2022**, *31*, 1759–1773. [CrossRef]
30. Liu, K.; Jin, Z.; Zeng, L. Determination of clay bound water in shales from NMR signals: The fractal theory. *Energy Fuels* **2021**, *35*, 18406–18413. [CrossRef]
31. Zhu, S.; Zhen, Q.; Zhang, X. Multifractal characteristics of the pore structures of physically amended sandy soil and the relationship between soil properties and multifractal parameters. *Arch. Agron. Soil Sci.* **2020**, *66*, 1188–1202. [CrossRef]
32. Kong, B.; He, S.H.; Tao, Y. Pore structure and fractal characteristics of frozen–thawed soft soil. *Fractal Fract.* **2022**, *6*, 183. [CrossRef]
33. He, S.H.; Ding, Z.; Hu, H.B. Effect of grain size on microscopic pore structure and fractal characteristics of carbonate-based sand and silicate-based sand. *Fractal Fract.* **2021**, *5*, 152. [CrossRef]
34. Paz Ferreiro, J.; Miranda, J.; Vázquez, E.V. Multifractal analysis of soil porosity based on mercury injection and nitrogen adsorption. *Vadose Zone J.* **2010**, *9*, 325–335. [CrossRef]
35. Guo, X.; Huang, Z.; Zhao, L. Pore structure and multi-fractal analysis of tight sandstone using MIP, NMR and NMRC methods: A case study from the Kuqa depression, China. *J. Pet. Sci. Eng.* **2019**, *178*, 544–558. [CrossRef]
36. Hu, Z.; Zheng, S.; Li, Y. Synthesis of a novel porous diatomite/AlOOH composite material and its humidity control performance. *Chem. Pap.* **2022**, *76*, 3909–3918. [CrossRef]
37. Ge, X.; Fan, Y.; Liu, J. An empirical method to correct nuclear magnetic resonance porosity of tight sandstone using low-field nuclear magnetic resonance data. *AAPG Bull.* **2023**, *107*, 539–551. [CrossRef]
38. Kuptsov, K.; Griffiths, R.; Maggs, D. Technology update: Magnetic resonance-while-drilling system improves understanding of complex reservoirs. *J. Pet. Technol.* **2015**, *67*, 26–29. [CrossRef]
39. Zhou, S.D.; Liu, D.M.; Cai, Y.D. Fractal characterization of pore–fracture in low-rank coals using a low-field NMR relaxation method. *Fuel* **2016**, *181*, 218–226. [CrossRef]
40. Liang, Z.K.; Li, Z.; Jiang, Z.X. Characteristics of pore structure and fractal dimension in continental shale based on NMR experiments and SEM image analyses—A case study of Shahezi formation shale in Changling fault depression of Songliao Basin. *China J. Earth Sci. Environ.* **2020**, *42*, 313–328.

fractal and fractional

MDPI

Article

Multi-Scale Research on the Mechanisms of Soil Arching Development and Degradation in Granular Materials with Different Relative Density

Luju Liang [1,2], Yi Pik Cheng [3], Xiaozhen Fan [1,2,*], Zhi Ding [1,2] and Changjie Xu [4,5]

1 Institute of Civil Engineering, Hangzhou City University, Hangzhou 310015, China; lianglj@hzcu.edu.cn (L.L.); dingz@hzcu.edu.cn (Z.D.)
2 Key Laboratory of Safe Construction and Intelligent Maintenance for Urban Shield Tunnels of Zhejiang Province, Hangzhou City University, Hangzhou 310015, China
3 Department of Civil, Environmental and Geomatic Engineering, University College London, London WC1E 6BT, UK; yi.cheng@ucl.ac.uk
4 State Key Laboratory of Performance Monitoring Protecting of Rail Transit Infrastructure, East China Jiaotong University, Nanchang 330013, China; xucj@zju.edu.cn
5 Research Center of Coastal and Urban Geotechnical Engineering, Zhejiang University, Hangzhou 310012, China
* Correspondence: fanxz@hzcu.edu.cn

Abstract: Soil arching is significantly influenced by relative density, while its mechanisms have barely been analyzed. A series of DEM numerical simulations of the classical trapdoor test were carried out to investigate the multi-scale mechanisms of arching development and degradation in granular materials with different relative density. For analysis, the granular assembly was divided into three zones according to the particle vertical displacement normalized by the trapdoor displacement δ. The results show that before the maximum arching state (corresponding to the minimum arching ratio), contact forces between particles in a specific zone (where the vertical displacement of particles is larger than 0.1δ but less than 0.9δ) increase rapidly and robust arched force chains with large particle contact forces are generated. The variation in contact forces and force chains becomes more obvious as the sample porosity decreases. As a result, soil arching generated in a denser particle assembly is stronger, and the minimum value of the arching ratio is increased with the sample porosity. After the maximum arching state, the force chains in this zone are degenerated gradually, leading to a decrease in particle contact forces in microscale and an increase in the arching ratio in macroscale. The recovery of the arching ratio after the minimum value is also more significant in simulations with a larger relative density, as the degeneration of contact force chains is more obvious in denser samples. These results indicate the importance of contact force chain stabilities in specific zones for improving soil arching in engineering practice.

Keywords: soil arching; discrete element modeling; persistent homology; multi-scale analysis; relative density

1. Introduction

Soil arching in geotechnical engineering, investigated by Terzaghi [1], has attracted increasing attention in recent years as it is a general phenomenon in sandy soil and other granular materials. Soil arching is developed when a soil mass yields. The shear resistance developed at the interface between the yielded portion and the adjoining stationary portion will result in stress transformation and redistribution. The degree of soil arching can be assessed through the arching ratio (ρ),defined as the ratio of the stress within a certain area to the overburden stress at the same location [2]. The variation in the arching ratio with the increase in relative displacement of soil mass can be described by the ground reaction curve (GRC), as presented in Figure 1 [3]. GRC is very important in describing soil arching because it can clearly illustrate the four stages of soil arching: initial arching, maximum arching, load recovery, and the ultimate state. The applicability of the GRC to describe the load–displacement relationship in a trapdoor test has been investigated by several authors [4–6].

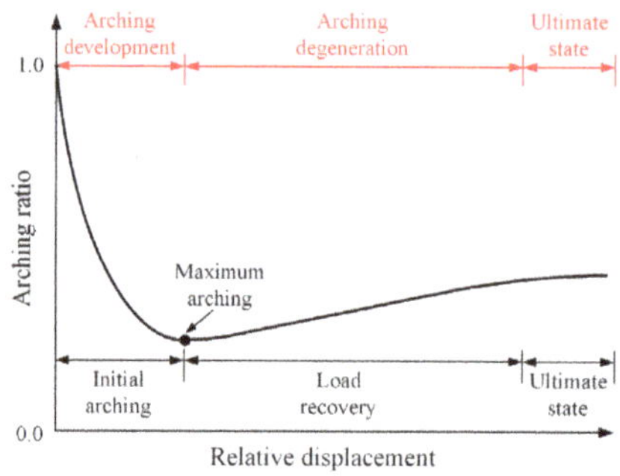

Figure 1. General ground reaction curve (GRC).

Recently, to further analyze the mechanisms of soil arching evolution as shown in GRC, many researchers have also conducted single trapdoor tests [7,8] and multiple trapdoor tests [9,10]. According to these studies, soil arching evolution is related to the progressive development of slip surfaces induced by relative displacement of soil mass. Triangular slip surfaces are observed at the maximum arching stage, while vertical slip surfaces are generally observed at the ultimate state. The test results also show that soil arching evolution is significantly influenced by the relative density (RD), which is defined as:

$$RD = (n_{max} - n)/(n_{max} - n_{min}), \quad (1)$$

where n_{max}, n_{min}, n represent the maximum porosity, the minimum porosity and the porosity of the granular materials, respectively. In the trapdoor tests with different sample densities, the arching ratio for the test at a lower relative density is higher than that for the test at higher relative density, indicating that soil arching is less mobilized in the backfill at a lower relative density [10]. PIV results of the shear strain field in trapdoor tests show that triangular slip surfaces also cannot be formed when the relative density of soil mass is low [11]. On the other hand, DEM simulation results of the trapdoor test also indicate that the stick–slip behavior of the arching ratio is more obvious for the case with a higher density at a given fill height [12,13], and the failure mechanisms of dense sand and medium dense sand in the trapdoor simulations are different [14]. However, previous studies have not explicitly explained why soil arching shows different development and degradation processes with different soil relative densities. Since soil arching is widely encountered in tunnels, pile-supported embankments, retaining structures, etc., and the relative densities of granular soil in engineering are different from each other, understanding the evolution mechanisms of soil arching in granular materials with different densities is critical for geotechnical engineering practices. Actually, the evolution of soil arching is a macroscopic reflection of the mechanical properties of soil particles in microscale. At present, the load transfer mechanisms [15] and deformation patterns [16,17] of soil arching in macroscale have been well examined. In microscale, the particle–particle interaction during the evolution of soil arching [18,19] has been considered extensively, and some effort has also been made to describe and analyze the particle behaviors in the fill with different particle sizes [20] and particle shapes [21,22], but how the particle behaviors in microscale influence the variation in the arching ratio in macroscale is still largely unknown.

In order to bridge the gap between microscale particle behaviors and macroscale arching effect, mesoscale behaviors related to the force network in soil arching should be quantitatively analyzed, because the force chains always play a key part in load transfer [23].

Recently, significant progress has been made toward developing a better understanding of force networks using a variety of tools [24–27]. In this paper, we focus on the topological analysis method—persistent homology (PH) [28,29]—for the analysis of mesoscale structures in 2D granular material. Persistent homology is a fractal method that performs multi-scale analysis on fractional complex networks and identifies their topological structures such as clusters, holes and cavities. Such an approach has been used to discuss the mesoscale structure of force networks in granular systems that were compressed [30], vibrated [31,32], sheared [33] and used for the analysis of the yielding of a granular system during pullout of a buried intruder [34]. Therefore, it allows for formulation of simple but informative measures describing the force networks, and for comparison of different networks during the evolution of soil arching.

According to the literature review presented above, a series of 2D DEM simulations of the trapdoor test were conducted to investigate the soil arching development and degradation in granular materials with different porosity. Persistent homology (PH) was adopted in this study to quantitatively analyze the features of force networks at different arching stages that related, in a broad sense, to so-called force chains. The main objects of this paper are: (i) Identifying the different development and degeneration processes of soil arching in granular materials with different relative densities; (ii) Identifying the corresponding particle behaviors in the different processes; (iii) Specifying how the particle behaviors in microscale influence the variation in arching ratio in macroscale.

2. Materials and Methods

PFC2D 5.0 is adopted in this study to simulate the trapdoor test procedure and analyze the multi-scale mechanisms of soil arching. Figure 2 presents the DEM model of the 2D trapdoor test. In this model, the trapdoor is simulated by a rigid wall at the bottom with width $B = 0.6$ m. The height of the granular sample is denoted as H. The distance between the trapdoor and the two side walls is equal to B. This boundary distance is comparable to that in the literature [12,20,21,35,36] (range from $0.5B$ to $2B$). A wider boundary distance would provide more accurate results. However, it would also increase the simulation time and have little effect on the behaviors of soil arching with different relative densities of granular material.

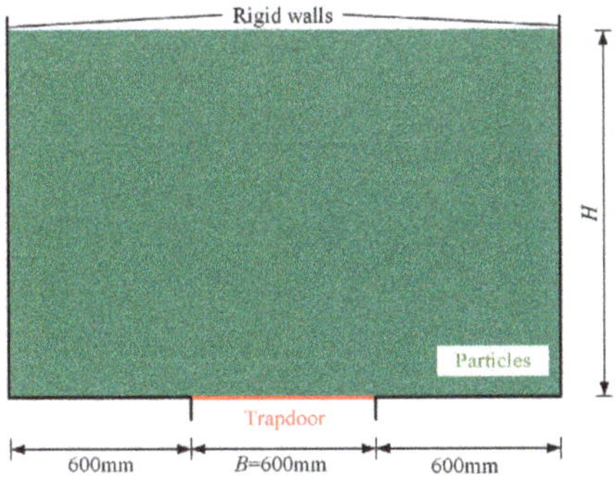

Figure 2. Schematic representation of the DEM model.

The particle size of DEM sample is based on the purely two-dimensional trapdoor test conducted by Xu et al. [37], in which Taylor–Schneebeli soil analogues with perfect disk cross-sections were used to simulate soil. Circular disks with three different diameters (3 mm, 4 mm and 5 mm) mixed at a mass ratio of 1:1:1 are randomly rained into the rectangular region (1.8 m × H) with different initial frictional coefficients ($\mu_{\text{p-p}} = 0.1, 0.3, 0.5$) layer-by-layer using the Grid-Method [38] to generate the simulation sample with different porosity ($n = 0.16, 0.18, 0.20$). After all particles have been settled down, $\mu_{\text{p-p}}$ is reset as 0.5 and the assembly is cycled to equilibrium for later simulation. During the simulation of trapdoor test, the trapdoor is moved downward and the movement speed is controlled at 1×10^{-4} m/s so that the simulation process satisfies the quasi-static condition. As listed in Table 1, six simulations with different sample porosity and buried depth ratio (H/B) are planned to be conducted in this study. For example, "H1-n0.16" represents a simulation with H/B =1 and n = 0.16.

Table 1. General configuration of DEM simulation.

Simulation Name	H1-n0.16	H1-n0.18	H1-n0.20	H2-n0.16	H2-n0.18	H2-n0.20
H/B	1	1	1	2	2	2
n	0.16	0.18	0.20	0.16	0.18	0.20
Particle number	82,000	80,000	78,000	164,000	160,000	156,000

The Hertz–Mindlin contact model between particles is adopted in this study. The micromechanical parameters of particles were determined and calibrated from numerical biaxial tests and numerical trapdoor tests. In the numerical biaxial tests, particles were randomly generated and then confined to different confining pressures (50 kPa, 100 kPa, 150 kPa) for biaxial compression test simulation. $\mu_{\text{p-p}}$ is adjusted to different values during the confining process so that the porosity of the particle assembly is controlled at 0.17 before loading. These parameters are totally the same as the model test conducted by Xu et al. [37]. The comparison between the simulation results and test results is presented in Figure 3, with the micromechanical properties listed in Table 2. In general, the numerical results are in good agreement with the experimental results. The maximum deviation between the numerical and test results of deviatoric stress and arching ratio is around 10%. A relatively larger deviation is observed in the comparison of volume strain, but these comparisons show that the micro-mechanical parameters adopted in this study can still capture the basic variation trend of the volume strain during the shear process.

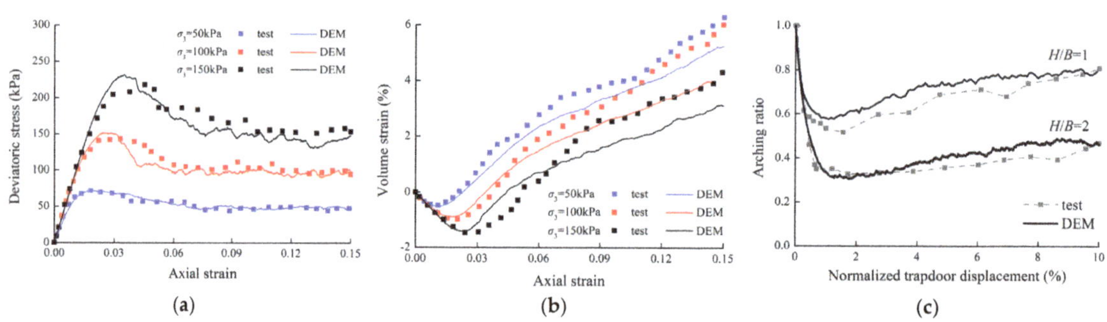

Figure 3. Comparison between the simulation results and test results [33]: (**a**) Deviatoric stress in biaxial compression test; (**b**) Volume strain in biaxial compression test; (**c**) Arching ratio in trapdoor test.

Table 2. Micro-mechanical parameters used in DEM analyses.

Parameters	Values
Particle density, ρ_p	2650 kg/m^3
Particle shear modulus, G	2.5×10^{10} N/m^2
Particle Poisson's ratio, ν	0.3
Friction coefficient between particles, $\mu_{\text{p-p}}$	0.5
Friction coefficient between walls and particles, $\mu_{\text{p-w}}$	0.0
Sample porosity, n	0.16, 0.18, 0.20

3. Results

3.1. Macroscale Results Analysis

3.1.1. GRCs with Different Sample Porosity

In order to gather insights into the multi-scale mechanisms of soil arching development and degradation, as well as the impacts of different sample porosity (n = 0.16, 0.18 and 0.20) through DEM simulations of this study, the macro phenomena of soil arching, including variation in the arching ratio and the displacement field evolution during the simulations, are firstly analyzed.

In all simulations, the trapdoor is moved downward until $\delta/B = 0.1$ (corresponding to a trapdoor displacement of 60 mm), as 0.1 is recommended as the normalized displacement (δ/B) corresponding to an ultimate arching ratio [6] according to the present trapdoor test results. Figure 4a presents the GRCs of the simulations with $H/B = 1$. The arching ratio $\rho = P/P_0$ (P is the pressure on the trapdoor and P_0 is the overburden stress on the trapdoor) of H1-n0.16 and H1-n0.18 both decrease to the minimum value at around $\delta/B = 0.02$, and then increase to a steady ultimate value with the increase in normalized trapdoor displacement. The minimum values of ρ in H1-n0.16 and H1-n0.18 are 0.55 and 0.62, respectively, while the steady ultimate values of ρ in these two simulations are almost the same (around 0.80). This indicates that the soil arching is initially gradually mobilized by the trapdoor movement, but then degenerated as the trapdoor displacement increases continuously. The minimum arching ratio increases with the increase in sample porosity, meaning that the maximum soil arching is stronger in denser samples. However, the GRC of H1-n0.20 is different. During the whole simulation, ρ continuously decreases to the steady ultimate value that is identical with the ultimate values in the other two simulations. As a result, no significant degradation process of soil arching can be observed in loose samples. This phenomenon indicates the influence of relative density on soil arching. In view of engineering, soil arching is highly correlated with the shear behaviors of soil. For loose soil, no obvious strain softening stage can be observed in the stress–strain curve, and therefore the degradation of soil arching is also insignificant in loose soil.

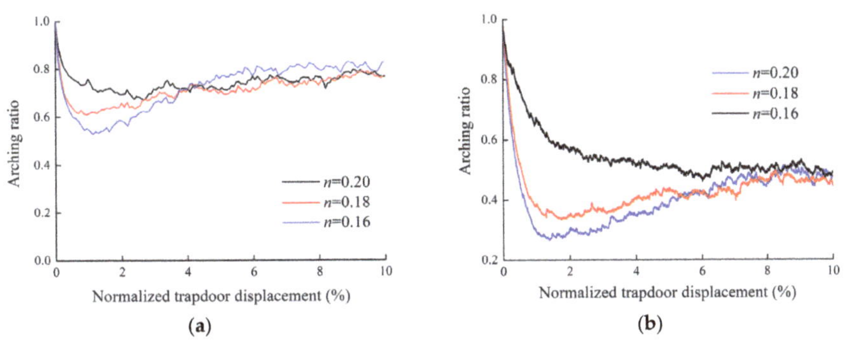

Figure 4. GRCs in different simulations: (**a**) $H/B = 1$; (**b**) $H/B = 2$.

Figure 4b presents the GRCs of the simulations with $H/B = 2$. The variation trends of the GRCs in Figure 4b are similar to that in Figure 4a. δ/B corresponding to the minimum arching ratio ρ in the scenario of $H/B = 2$ is almost identical with that when $H/B = 1$. This phenomenon means that the relative displacement needed to fully motivate the initial maximum soil arching in particle assemblies with different depths is identical. The soil arching in the simulations with $H/B = 2$ is observed to be stronger than that in the simulations with $H/B = 1$, as the minimum arching ratio on the GRCs of $H/B = 1$ is smaller than that on the GRCs of $H/B = 2$. Therefore, the relative soil movement required to fully motivate soil arching in the ultimate state when $H/B = 2$ is larger, so that the normalized trapdoor displacement corresponding to the steady ultimate value of ρ in the scenario of $H/B = 2$ is quite larger than that in the simulations with $H/B = 1$.

3.1.2. Displacement Field and the Arching Zones

The soil arching development and degradation process as reflected by GRC can be attributed to the gradual transformation of the movement patterns of soil mass within the zone of arching [14]. Although many DEM numerical simulations [21,35,36,39], PIV analyses [7,8,40] and X-ray CT scans [41] have been conducted to investigate the displacement of soil mass in a trapdoor test, the "arching zone" that dominates soil arching has not been clearly defined because the deformation patterns of soil mass at different soil arching stages are complex.

Figures 5 and 6 show the particle vertical displacement fields at different stages of the simulations with $H/B = 1$ and 2, respectively. The particle displacement is normalized by the trapdoor displacement to compare the displacement field at different stages. For analysis, the DEM sample can be artificially divided into three zones according to the vertical displacement: (a) "Zone-I" (marked as blue) with vertical displacement of particles less than 0.1δ; (b) "Zone-II" with vertical displacement of particles less than 0.9δ but larger than 0.1δ; (c) "Zone-III" (marked as red) with vertical displacement of particles larger than 0.9δ. The area ratios of the different zones denoted in these figures are calculated by dividing the particle area of a certain zone into the whole particle area of the sample. According to Figure 5, the boundaries of zone-III are similar to the triangular slip surfaces observed in the model tests [5,7,9]. The area of zone-III is influenced little by sample porosity, while zone-I and zone-II are observed to be significantly influenced by sample porosity. The area of zone-II expands with the increase in sample porosity, leading to the reduction in the area of zone-I. On the other hand, the area ratios of zone-I and zone-III are both increased with the increase in trapdoor displacement, while the area of zone-II is decreased. This phenomenon is more obvious when $H/B = 2$ (Figure 6). In the simulation results of H2-n0.16, the area of zone-II decreases by around 19% when δ increases from 10 mm to 60 mm, while these values in H2-n0.18 and H2-n0.20 are 17% and 13%, respectively. Therefore, significant shear localization is generated in zone-II as the trapdoor displacement increases. According to these results, the shear localization induced by the gradual development of particle motion in zone-II is supposed to have a significant influence on the evolution of soil arching.

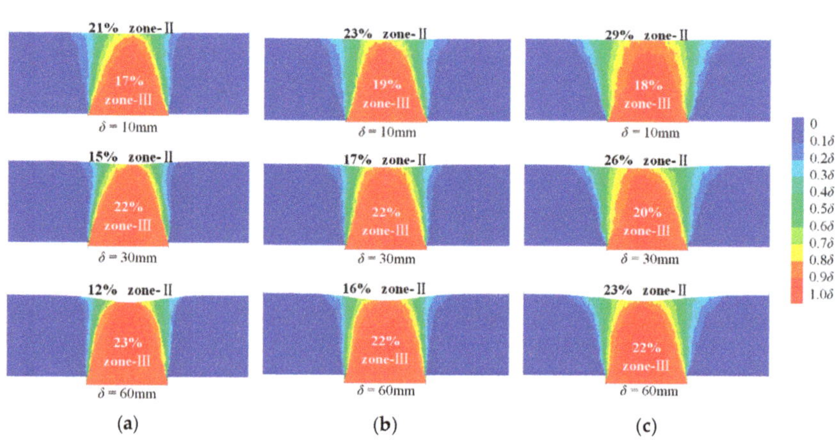

Figure 5. Normalized vertical displacement field of granular assemblies with $H/B = 1$: (**a**) $n = 0.16$; (**b**) $n = 0.18$; (**c**) $n = 0.20$.

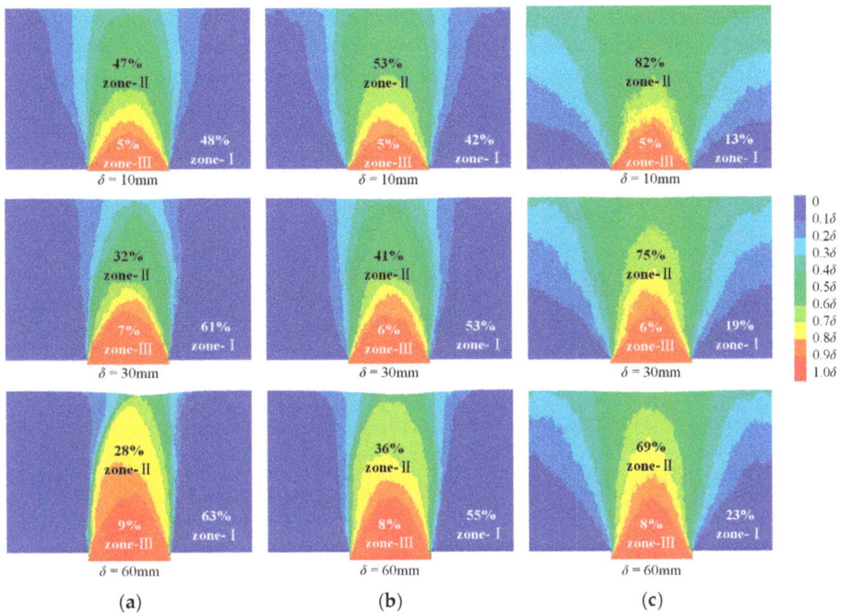

Figure 6. Normalized vertical displacement field of granular assemblies with H/B = 2: (**a**) n = 0.16; (**b**) n = 0.18; (**c**) n = 0.20.

3.2. Microscale Results Analysis

3.2.1. Coordination Number

According to the displacement field analysis, the relative displacement of particle assemblies will lead to a variation in sample porosity. However, precisely calculating the porosity in a zone of a DEM sample is quite difficult. Alternatively, the coordination number is another parameter for evaluating the variation in sample density and the evolution of microstructure in granular materials. The coordination number represents the average number of contacts per particle, and it is highly correlated to the volume strain of granular material and more sensitive than porosity [42]. The coordination number in a zone can be calculated by:

$$Z = 2C/N, \tag{2}$$

where C and N are the total contact number, and the particle number in the zone, respectively. Figure 7 shows the variation in Z in different zones during the trapdoor test simulations. According to these figures, variation in Z in zone-II is most significant among the three zones. It decreases with the trapdoor displacement during the whole process of soil arching development and degradation, and finally reaches a constant value at the ultimate state when the arching ratio is no longer varied as trapdoor displacement increases, indicating that the porosity of particle assemblies in zone-II increases continuously during the evolution of soil arching (including development and degradation). According to a previous study on the evolution of the coordination number in simulated granular materials, the reduction in the coordination number is generally induced by shear dilation [43]. Therefore, the decrease in Z in zone-II may be largely attributed to the shear dilation of particle assembly in this zone. As shear dilation is more significant in denser samples, the reduction in Z in zone-II is reduced with sample porosity. Figure 8 shows the comparison of Z in zone-II with different sample porosities. It is observed that although Z in zone-II is initially different in the simulations with different sample porosities, they all decrease to a consistent ultimate value at the end of the simulation. This ultimate constant value of Z when H/B = 2 is a little larger than that in the simulations with H/B = 1 because the stress level of the particle assembly is increased with buried depth so that a lower decrease in Z is induced by shear dilatancy when $H/B = 2$.

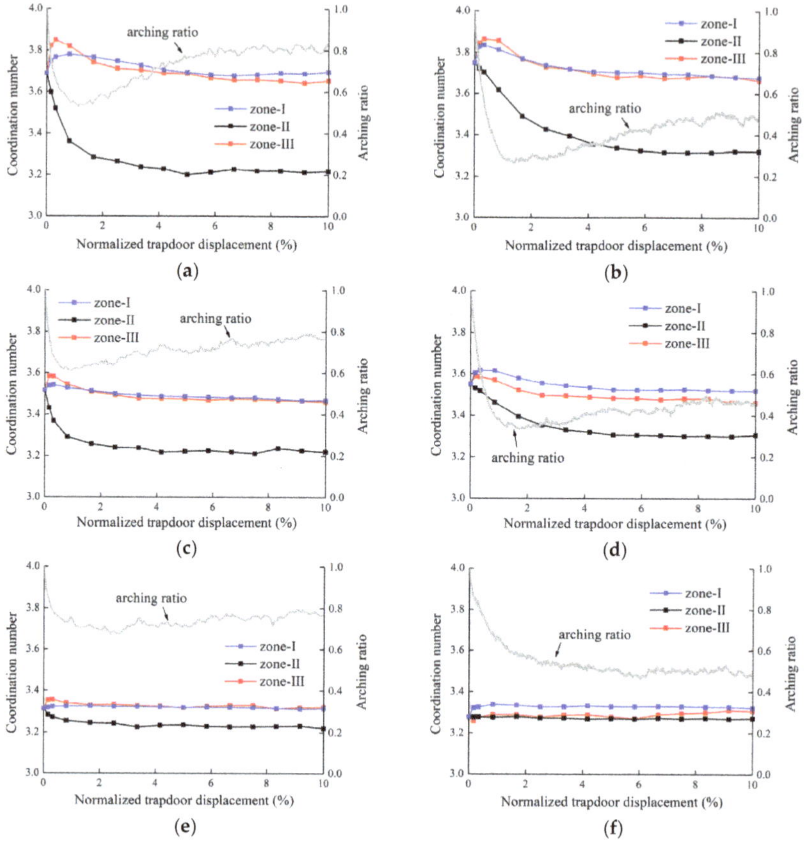

Figure 7. Evolution of average coordination number against trapdoor displacement: (**a**) H1-n0.16; (**b**) H2-n0.16; (**c**) H1-n0.18; (**d**) H2-n0.18; (**e**) H1-n0.20; (**f**) H2-n0.20.

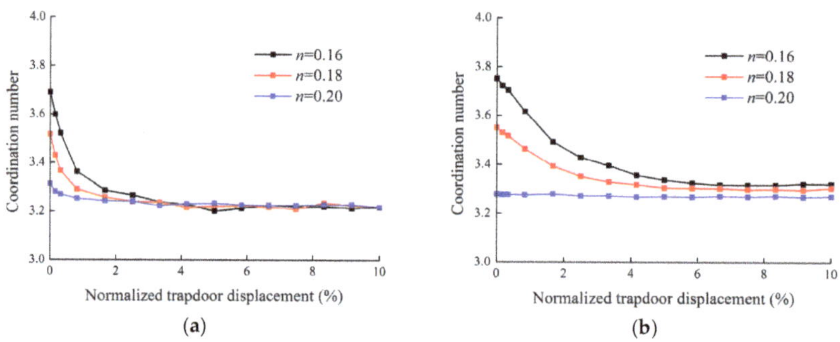

Figure 8. Comparison of average coordination number in zone-II of different simulations: (**a**) $H/B = 1$; (**b**) $H/B = 2$.

Variation in Z in both zone-I and zone-III is less significant than in zone-II. And, different to zone-II during the whole simulation process, Z values in both these two zones are decreased with the increase in sample porosity. For zone-I, the average coordination number increases during the development process of soil arching, because the upper load is transferred toward zone-I under the arching effect, leading to the compression of particle assembly in zone-I. Therefore, as soil arching is degenerated by the continuous increase in particle relative displacement, it is observed that Z decreases gradually under the lateral unloading effect and finally reaches a constant value. Especially in the simulations of H1-n0.20 and H2-n0.20, no obvious decrease in Z is observed in Figure 7e,f, because the degradation of soil arching in these two simulations is slight. The variation trend of Z in zone-III is similar to that in zone-I. The maximum value of Z in zone-III is observed to be larger than the maximum Z in zone-I because the stress level of the particle assembly in zone-III (at the bottom of the model) is much larger, compared with the stress level in zone-I at the beginning of the simulation. However, at the end of the simulation, the stress level in zone-III is decreased under the influence of soil arching so that the ultimate Z value in zone-III is less than that in zone-I.

3.2.2. Particle Contact Force

Variation in the coordination number of granular assemblies will change the contact state between particles. The particle contact force, **f**, in microscale can be resolved into two components:

$$\mathbf{f} = f_{\mathrm{n}}\mathbf{x} + f_{\mathrm{s}}\mathbf{y} \tag{3}$$

where f_{n} and f_{s} are the normal contact force and the tangential contact force between particles, respectively. The average values of f_{n} and f_{s} in a certain zone are denoted as $<f_{\mathrm{n}}>$ and $<f_{\mathrm{s}}>$, respectively.

Figure 9 shows the variation in $<f_{\mathrm{n}}>$ in different zones during the simulation. Since the analysis of tangential contact forces evolution leads to similar results with normal contact forces, it is not discussed here. As we can see in the figures, $<f_{\mathrm{n}}>$ in zone-II is supposed to be important in the development and degradation of soil arching. During the initial arching stage, $<f_{\mathrm{n}}>$ increases rapidly because of the large particle relative movement and interlocking in zone-II, leading to the sharp decrease in the $<f_{\mathrm{n}}>$ in zone-III. Because the arching ratio is correlated with the contact force acting on the trapdoor, a decrease in the arching ratio is observed in this process, indicating the gradual development of soil arching. As a result, the upper load is transferred toward two sides so that the particle contact force in zone-I increases. Then, $<f_{\mathrm{n}}>$ in zone-II decreases gradually with δ because of the continuous decrease in the coordination number as indicated in Figure 7. As a result, $<f_{\mathrm{n}}>$ in zone-I is also decreased. Meanwhile, a recovery of the arching ratio, representing soil arching degradation, is observed, according to Figure 9. The variation in $<f_{\mathrm{n}}>$ in different zones tends to be more significant as the sample porosity decreases, which is coincident with the fact that the development and degradation process of soil arching is much more obvious in granular assemblies with lower porosity. It is also observed that the decrease in $<f_{\mathrm{n}}>$ in zone-II after the maximum value when $H/B = 1$ is more significant than that in the scenario of $H/B = 2$, and therefore, degradation of soil arching in H1-n0.16 and H1-n0.18 is more obvious than that in H2-n0.16 and H2-n0.18. On the other hand, it is noted that the evolution of $<f_{\mathrm{n}}>$ in H2-n0.20 shows a different tendency comparing with other simulations. The $<f_{\mathrm{n}}>$ in the three zones all decrease gradually as δ increases. According to Figure 6c, zone-II extends to the rigid boundaries in this simulation. The different evolutionary tendency of $<f_{\mathrm{n}}>$ may be attributed to the boundary effect.

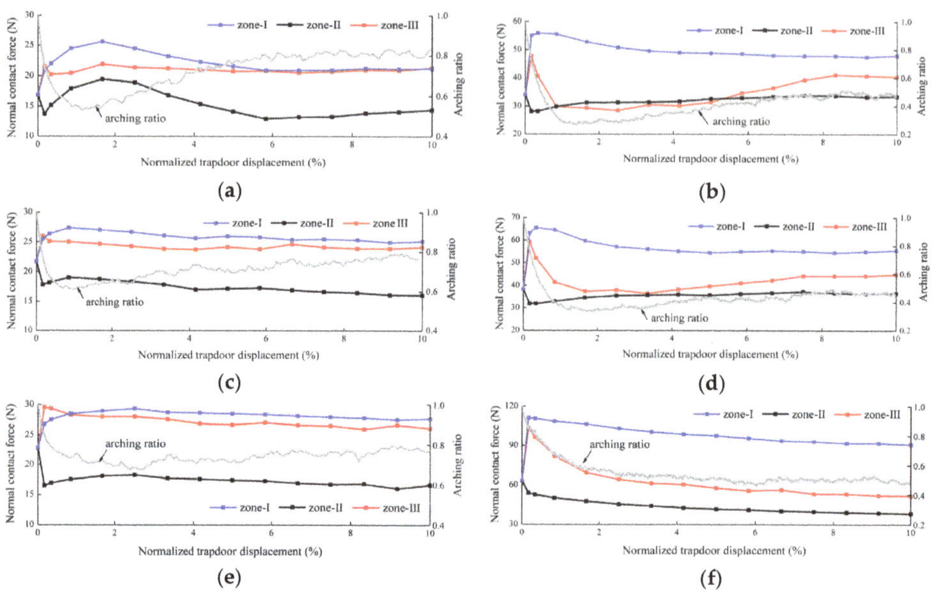

Figure 9. Evolution of the average normal contact force against the trapdoor displacement: (**a**) H1-n0.16; (**b**) H2-n0.16; (**c**) H1-n0.18; (**d**) H2-n0.18; (**e**) H1-n0.20; (**f**) H2-n0.20.

The mobilization of friction between two contact particles can be mathematically evaluated by the friction mobilization index I_m [44]:

$$I_m = |f_s|/(\mu f_n) \quad (4)$$

where μ is the friction coefficient between particles. When local failure occurred between two particles, $I_m = 1$ and particle sliding occurred at the contact. In this study, sliding is believed to occur when I_m is larger than 0.9999, and a slip ratio is defined as the number of sliding points divided by the total contact numbers. Figure 10 presents the slip ratios in different zones. According to the figures, the slip ratios of different zones in different simulations all increase sharply during the same initial stage. Then, they decrease substantially and maintain nearly constant with increases in trapdoor displacement. In all simulations, it is observed that the slip ratio in zone-II is obviously larger than that in the other two zones, and the slip ratios in zone-I and zone-III are generally identical. Therefore, the friction mobilization degree in zone-II is the highest among the three zones. The slip ratio of zone-II increases rapidly during soil arching development, indicating that the friction between particles in zone-II is fully mobilized in this process. Comparing with Figure 9, it is interesting to find that the slip ratio in zone-II reaches the maximum value at a similar trapdoor displacement, corresponding to the maximum value of $<f_n>$ and the minimum arching ratio value. Then, as soil arching is degenerated gradually, the slip ratio in zone-II also decreases and reaches a constant value at the end of the simulation. As presented in Figure 11, the maximum sliding ratios in zone-II decreased with the initial porosity. However, they tend to the same residual value at the end of the simulation. The decreases in the sliding ratio in zone-II may be attributed to the continuous decreases in the coordination number. The sliding contact fades away as the porosity increases, leading to particle rearrangement and a decrease in the sliding ratio. On the other hand, it can also be observed that the sliding ratios in zone-I and zone-III also initially increase because of the particle dislocation induced by trapdoor movement, but then they decrease rapidly to a low level (less than 30% of the sliding ratio in zone-II) with soil arching evolution and load redistribution.

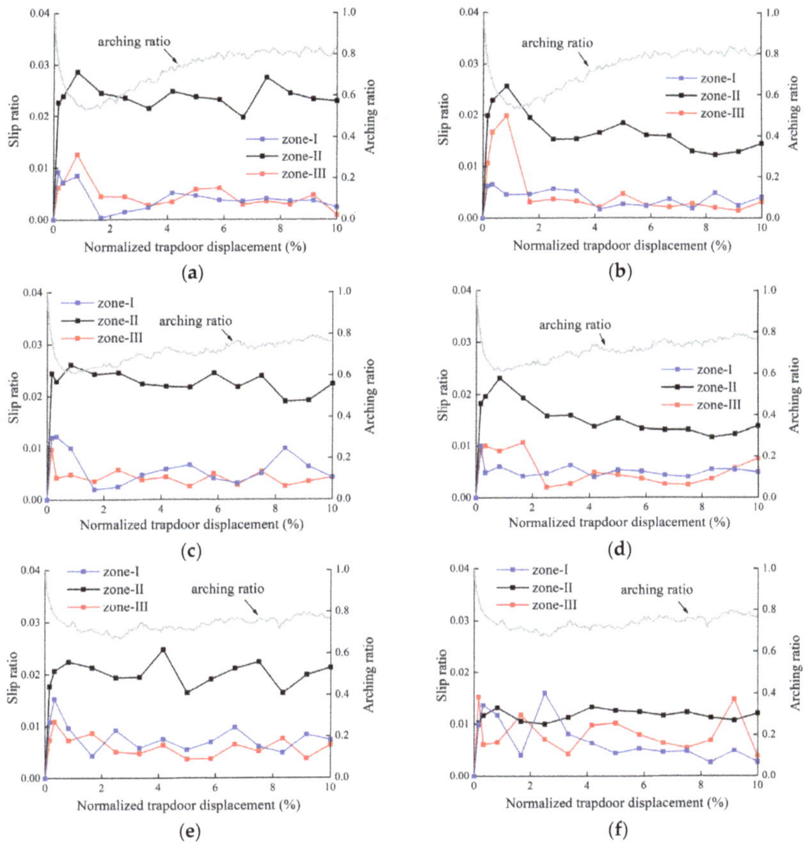

Figure 10. Evolution of slip ratio against trapdoor displacement: (**a**) H1-n0.16; (**b**) H2-n0.16; (**c**) H1-n0.18; (**d**) H2-n0.18; (**e**) H1-n0.20; (**f**) H2-n0.20.

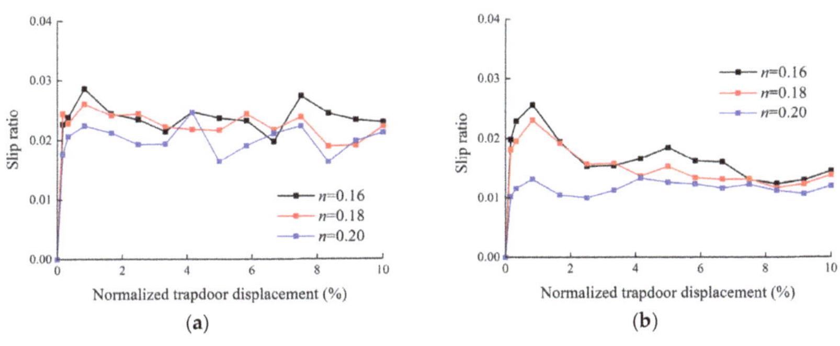

Figure 11. Comparison of sliding ratio in zone-II of different simulations: (**a**) $H/B = 1$; (**b**) $H/B = 2$.

3.3. Mesoscale Results Analysis

Variations in particle contact force and the coordination number mean that the mesoscale behaviors related to the force network, spanning roughly 10–15 particles, are also influenced by soil arching. In order to analyze the mesoscale cluster structure of force networks' evolution during soil arching development and degradation, the topological analysis method—persistent homology (PH) [23,24]—is adopted in this study. PH could be thought of as a tool for describing the weighted force networks between particles in the form of diagrams, so called persistent diagrams (PDs), which are obtained by filtering, or thresholding, the strength of the interactions between the particles. For example, considering a normal contact force network between particles in Figure 12, the numbers in Figure 12a represent the magnitude of particle contact force, and the simplest persistence diagrams in Figure 12b, called PDβ_0, encode how distinct connected components in the force network appear and then merge as the threshold θ (denoted the normal contact force value here) decreases. Each point (called a generator) in this diagram has two coordinates: 'birth' and 'death'. A birth in PDβ_0 occurs when an edge not connected to any existing edge is added to the contact force network, or a set of all contacts, and a death occurs when a newly added edge connects two existing force networks. Each point (b, d) (where b denotes the birth coordinate and d denotes the death coordinate) in PDβ_0 describes a feature of the network, and the lifespan of a point (defined as b–d) can be interpreted as the notable degree of the feature. Therefore, PDβ_0 can essentially trace how 'force chains' appear as the filtration level is decreased, or disappear as two structures merge. The reader is referred to [24] for a more in-depth presentation of this method.

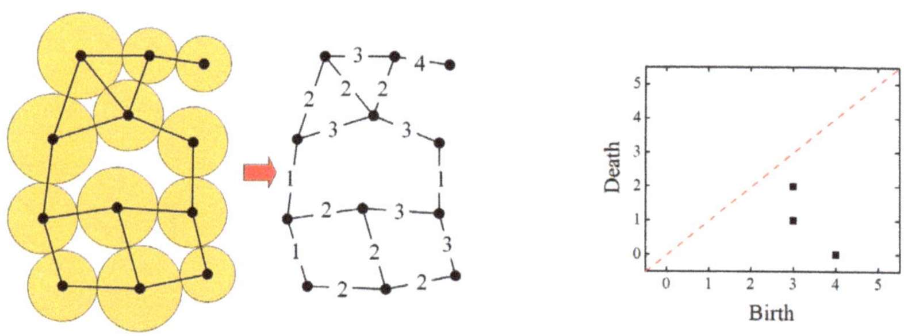

(a) **(b)**

Figure 12. Toy example illustrating contact force network and persistence diagrams: (**a**) Normal contact force network; (**b**) Persistence diagrams PDβ_0.

In order to analyze the mesoscale cluster structure of force networks' evolution during soil arching development, the persistence diagrams of the normalized force networks (all contact forces are normalized by the corresponding $<f_n>$ in different zones) are computed. according to the DEM simulation results. Figures 13 and 14 show the corresponding diagrams of the H1-n0.16 and H2-0.16 simulations, respectively. In the analysis of these results, it is important to remember that the points (generators) that are close to the diagonal represent features that persist over only a small range of thresholds and, therefore, are not significant for the purpose of identifying robust features. According to Figure 4, soil arching develops gradually when δ increases from 0 mm to around 10 mm. During this process, a lot of generators with large 'birth' coordinates can be observed away from the diagonal of PDβ_0, meaning that robust contact force chains with large normal contact force are generated. And these arched force chains are mostly within zone-II. Then, as soil arching is degenerated gradually, the robust cluster structures disappear and the generators in zone-II gradually shrink to the diagonal of PDβ_0. This phenomenon can be observed in both Figures 13 and 14.

On the other hand, information about the variations in the contact structure within zone-I and zone-III is not readily available from the point clouds in Figures 13 and 14. Hence, the average lifespans of all points within different zones are then calculated and presented in Figure 15. Generally, long lifespan represents robust force chains in the contact force network. As shown in the figures, in the simulations with n = 0.16 (H1-n0.16 and H2-n0.16), the average lifespan of generators in zone-II increases with trapdoor displacement during the initial arching stage (δ < 10 mm), meaning that robust force chains with large normal contact force are formed in this zone. Under the influence of this arched force chain, the particle contact force acting on the trapdoor is decreased, leading to the decrease in the arching ratio in this process. Then, the average lifespan of generators in zone-II decreases gradually with trapdoor displacement ($\delta \geq 10$ mm), meaning that the arched structure of the force chain is weakened and destroyed. As a result, load recovery is observed on the GRC.

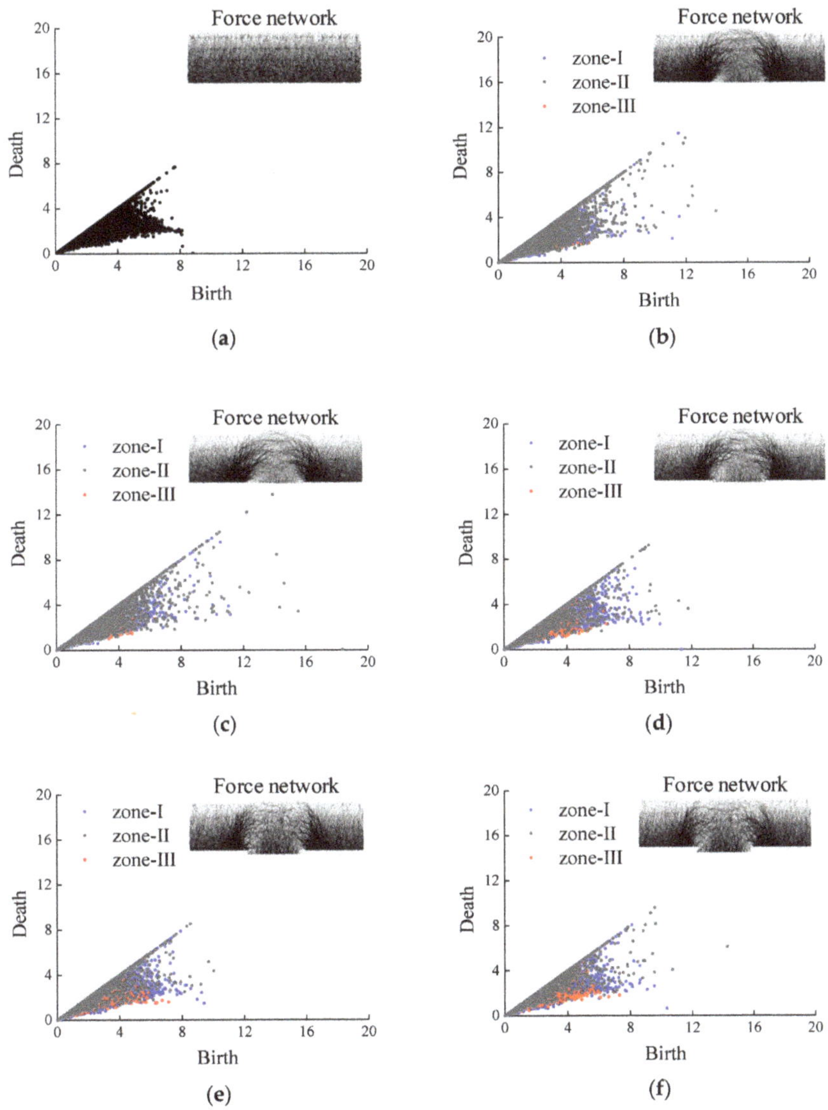

Figure 13. PDβ_0 of the contact force networks in different zones (H1-n0.16): (**a**) δ = 0 mm (K_0 condition); (**b**) δ = 5 mm; (**c**) δ = 10 mm (maximum arching); (**d**) δ = 20 mm; (**e**) δ = 40 mm; (**f**) δ = 60 mm.

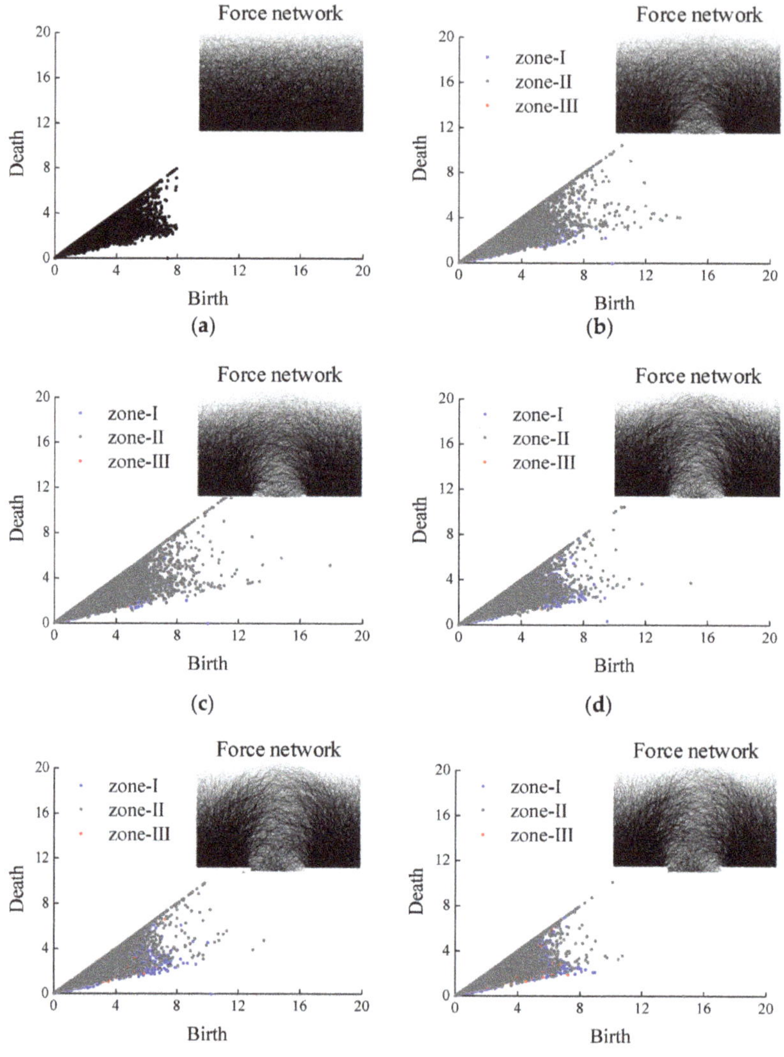

Figure 14. PDβ_0 of the contact force networks in different zones (H2-n0.16): (**a**) δ = 0 mm (K_0 condition); (**b**) δ = 5 mm; (**c**) δ = 10 mm (maximum arching); (**d**) δ = 20 mm; (**e**) δ = 40 mm; (**f**) δ = 60 mm.

Similar trends can also be observed in the simulations of n = 0.18 and n = 0.20, but the variation in the average lifespan of the generators tends to be less significant with the increase in sample porosity, which corresponds to the variation in the arching ratio and $<f_n>$ in different simulations. Figure 16 presents the comparison of the average lifespan of generators in zone-II of different simulations. The maximum average lifespan value is decreased with the initial porosity, but, similar to the average coordination number and the slip ratio, the residual values of the average lifespan at the end of the simulations with different initial porosities are identical to each other. According to these results, the variation in average lifespan may be attributed to the particle motion with increasing trapdoor displacement. In dense samples, during the initial arching stage, particle contact in zone-II decreases because of sliding and the coordination number decreases sharply, too. With this effect, particle contact force is increased, and arched force chains are formed gradually in this stage so that the average lifespan is increased. Under the shield of these arched force chains, the arching ratio is decreased. Then, as relative displacement between particles in zone-II continues to increase, particles separate with each other gradually, the coordination number continuous decreases and $<f_n>$ starts to decrease. As a result, the arched force chains are degenerated, leading to the soil arching degradation. In loose samples, a smaller decrease in the coordination number means that contact particles in zone-II would not separate further. Force chains are gradually formed in zone-II, as indicated by the average lifespan, and no obvious load recovery stage is observed in this case.

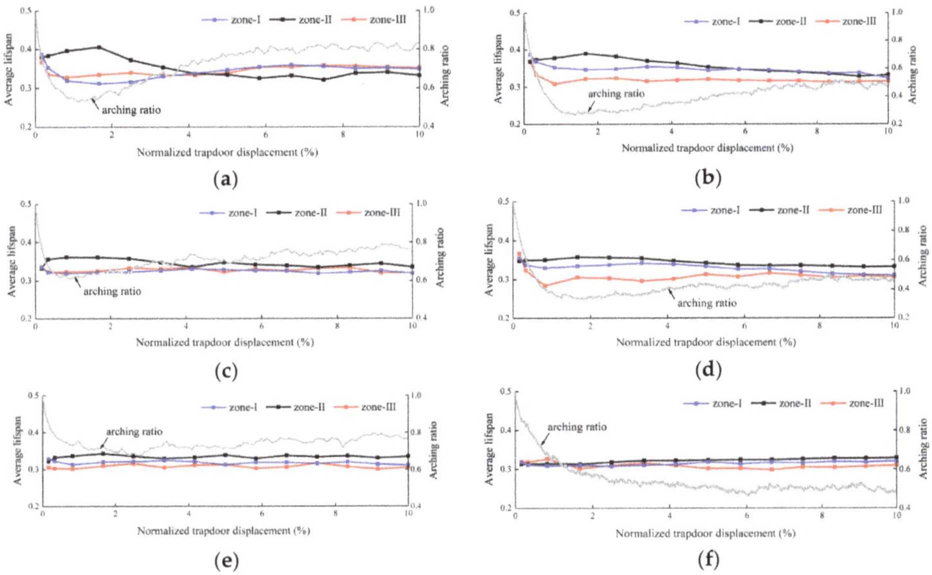

Figure 15. Variation in average lifespan of PDβ_0 in different zones during the simulations: (**a**) H1-n0.16; (**b**) H2-n0.16; (**c**) H1-n0.18; (**d**) H2-n0.18; (**e**) H1-n0.20; (**f**) H2-n0.20.

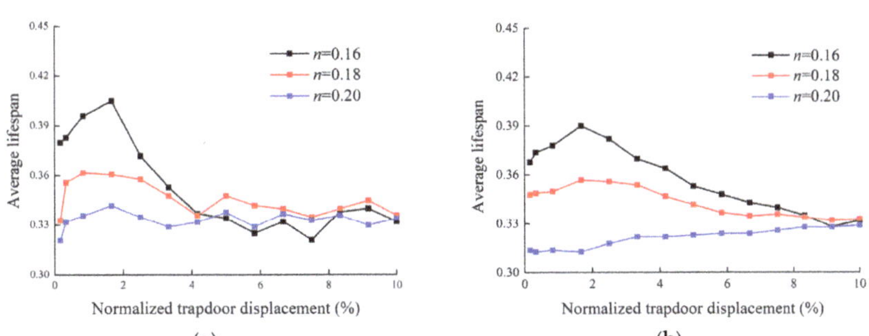

Figure 16. Comparison of average lifespan in zone-II of different simulations: (**a**) H/B = 1; (**b**) H/B = 2.

On the contrary, the average lifespan of generators in zone-I and zone-III show adverse variation trends. This phenomenon indicates that the force chains in zone-I and zone-III are degenerated when the force chains in zone-II are reinforced during the development of soil arching. Then, the force chains in zone-I and zone-III develop gradually as the force chains in zone-II are degenerated by the decreasing particle contact force when $\delta \geq 10$ mm.

4. Discussion

In this work, multi-scale analysis on the mechanisms of soil arching development and degradation in granular materials with different relative density was conducted through a series of DEM simulations of the trapdoor test. Analysis of results indicates that the evolution of force chains and particle contact forces in zone-II is important in the development and degradation of soil arching. The main results are summarized in Table 3.

Table 3. Summary of the main analysis results.

Factors	Arching State			Influence of Relative Density
	Development	**Degradation**	**Ultimate State**	
Arching ratio	Sharply decreased	Slowly increased	Constant value	The minimum value is increased with density, while the ultimate value is little influenced by density
Coordination number in zone-II	Sharply decreased	Slowly decreased	Constant value	The variation is more significant in denser samples, while the ultimate value is little influenced by density
Average particle contact force in zone-II	Sharply increased	Slowly decreased	Constant value	The variation is more significant in denser samples, while the ultimate value is little influenced by density
Slip ratio in zone-II	Sharply increased	Slowly decreased	Constant value	The maximum value is increased with density, while the ultimate value is little influenced by density
Average lifespan of force network in zone-II	Sharply increased	Slowly decreased	Constant value	The maximum value is increased with density, while the ultimate value is little influenced by density

Careful analysis of the mesoscale network structure through PH in this study provides new insight into the development and degeneration of soil arching. According to the analysis results in zone-II, before the maximum arching state (corresponding to the minimum arching ratio), robust arched force chains with large normal contact forces are generated because of the interlocking between particles. As a result, contact forces between particles in zone-II increase rapidly. After the maximum arching state, the arched force chains are degenerated gradually with the continuous increase in trapdoor displacement, leading to the continuous decrease in particle contact forces and an increase in the arching ratio. The arched force chains are more easily generated in denser samples, leading to a stronger maximum arching in the cases with higher density. However, as the slip ratio, the average particle contact force and the average lifespan of the force network all reach the same value at the ultimate state of soil arching, regardless of relative density, the ultimate arching ratio is also little influenced by porosity.

In practical engineering, such as pile embankments, soil arching will be degenerated gradually with the continuous increase in relative settlement between the pile and the sub-soil, leading to the decrease in the pile–soil stress ratio and pile efficacy, which is not favorable to the whole structure. A higher compaction degree can improve the performance of the embankment, but it should also be noticed that soil arching degeneration is more significant in denser fills. According to this study, soil arching development and degradation can be attributed to the evolution of arched force chains in specific zones. Therefore, improvement of the stability of force chains in this specific zone, such as adding geotextiles, is supposed to be considered in the measures to improve the persistence of soil arching and performance of pile-supported embankments. On the other hand, the method and results presented in this study can also provide guidance for studies of soil arching under complex conditions, such as surcharge loading, suffusion influence, etc.

However, it is noted that the DEM simulations in this paper are conducted under 2D conditions with circle particles, so it cannot capture the influence of particle shape and particle size on the multi-scale mechanisms of soil arching development and degradation in real 3D conditions, which also needs further analysis.

5. Conclusions

In this study, two-dimensional DEM numerical analysis was conducted to investigate the multi-scale mechanisms of soil arching development and degradation in granular materials with different relative density. Six DEM simulations, considering two different buried depth ratios of granular assemblies with three different porosities of particle assembly, were conducted. For analysis, the particle sample is artificially divided into three zones according to the normalized vertical displacement field: (a) "Zone-I" with vertical displacement of particles less than 0.1δ; (b) "Zone-II" with vertical displacement of particles less than 0.9δ but larger than 0.1δ; (c) "Zone-III" with vertical displacement of particles larger than 0.9δ. The evolution of the mesoscale structure of contact force networks during the evolution of soil arching is quantitatively analyzed through persistence homology. The influence of force network evolution on the macroscopic and mesoscopic phenomena of soil arching are carefully evaluated. The major conclusions are summarized according to the analysis results:

1. The porosity of particle assembly has a significant influence on the development and degradation of soil arching. According to the ground reaction curves (GRC), soil arching generated in a denser particle assembly is stronger as the minimum value of the arching ratio is increased with the initial porosity. However, the arching ratios at the end of simulations with different initial porosities are identical. The recovery of the arching ratio after the minimum value is more significant in the simulation with a denser particle assembly, which indicates a more obvious degeneration of soil arching. In macroscope, the development and degradation of soil arching can be attributed to the shear localization generated in zone-II at different arching stages according to the normalized vertical displacement fields.
2. According to the microscale analysis accounting for the coordination number and the slip ratio of contact, particle friction is mobilized and dilatancy is generated during the initial arching stage, leading to the rapid development of soil arching in a granular assembly. The granular fills in zone-II undergo the most significant shearing during the simulations. As a result, soil arching is highly correlated to the particle contact forces' evolution in zone-II. In all cases, the average normal contact forces in zone-II increase during the development of soil arching. Then, they decrease gradually after the maximum arching stage, accompanied with soil arching degradation.
3. Quantitative network analysis results indicate that the force chains show different evolution in simulations with different sample densities. The force chains in zone-II influence both the particle behaviors in microscale and the arching ratio in macroscale. Before the maximum arching state (corresponding to the minimum arching ratio), robust force chains with large normal contact forces are generated in zone-II. Stronger force chains are generated in denser samples at the maximum arching state. After the maximum arching state, the arched force chains are degenerated gradually with relative displacement between particles, leading to the decrease in normal particle contact forces in microscope and the increase in the arching ratio in macroscope. In loose samples, force chains are gradually generated with relative soil displacement until the ultimate state of soil arching, so no obvious degradation of soil arching is observed.
4. The slip ratio, the average particle contact force and the average lifespan of the force network in zone-II undergo similar evolution processes during the development and degradation of soil arching, but all reach the same value at the ultimate state of soil arching regardless of relative density. As a result, the arching ratio at the limit state of soil arching is also independent with the relative density.

Author Contributions: Conceptualization, L.L.; Project administration, X.F.; Software, L.L.; Validation, Y.P.C.; Writing—original draft, L.L.; Writing—review and editing, Z.D. and C.X. All authors have read and agreed to the published version of the manuscript.

Funding: This research was funded by National Key R&D Program of China (No. 2023YFC3009400), Natural Science Foundation of Zhejiang Province of China (No. LQ22E080009, No. LQ23E080002), National Natural Science Foundation of China (No. 52278418, No. 52308379, No. 52238009) and the open fund project of Key Laboratory of Safe Construction and Intelligent Maintenance for Urban Shield Tunnels of Zhejiang Province (No. HZCU-UST-23-07).

Data Availability Statement: The data used to support the findings of this study are available from the corresponding author upon request.

Conflicts of Interest: The authors declare no conflicts of interest.

References

1. Terzaghi, K. *Theoretical Soil Mechanics*; John Wiley and Sons: New York, NY, USA, 1943.
2. Han, J.; Gabr, M.A. Numerical analysis of geosynthetic-reinforced and pile-supported earth platforms over soft soil. *J. Geotech. Geoenviron. Eng.* **2002**, *128*, 44–53. [CrossRef]
3. Iglesia, G.R. Trapdoor Experiments on the Centrifuge: A Study of Arching in Geomaterials and Similitude in Geotechnical Models. Ph.D. Thesis, Massachusetts Institute of Technology, Cambridge, MA, USA, 1991.
4. Chevalier, B.; Combe, G.; Villard, P. Experimental and discrete element modeling studies of the trapdoor problem: Influence of the macro-mechanical frictional parameters. *Acta Geotech.* **2012**, *7*, 15–39. [CrossRef]
5. Iglesia, G.R.; Einstein, H.H.; Whitman, R.V. Investigation of soil arching with centrifuge tests. *J. Geotech. Geoenviron. Eng.* **2014**, *140*, 04013005. [CrossRef]
6. Han, J.; Wang, F.; Al-Naddaf, M.; Xu, C. Progressive development of two-dimensional soil arching with displacement. *Int. J. Geomech.* **2017**, *17*, 04017112. [CrossRef]
7. Liang, L.J.; Xu, C.J.; Chen, Q.Z.; Chen, Q.S. Experimental and theoretical investigations on evolution of soil-arching effect in 2D trapdoor problem. *Int. J. Geomech.* **2020**, *20*, 06020007. [CrossRef]
8. Burke, T.S.; Elshafie, M.Z.E.B. Arching in granular soils: Experimental observations of deformation mechanisms. *Géotechnique* **2021**, *71*, 866–878. [CrossRef]
9. Rui, R.; van Tol, A.F.; Xia, Y.Y.; van Eekelen, S.J.M.; Hu, G. Investigation of soil-arching development in dense sand by 2D model tests. *Geotech. Test. J.* **2016**, *39*, 415–430. [CrossRef]
10. Tao, F.J.; Xu, Y.; Zhang, Z.; Ye, G.B.; Han, J.; Cheng, B.N.; Liu, L.; Yang, T.L. Progressive development of soil arching based on multiple-trapdoor tests. *Acta Geotech.* **2023**, *18*, 3061–3076. [CrossRef]
11. Xu, C.J.; Liang, L.J.; Chen, Q.Z.; Luo, W.J.; Chen, Y.F. Experimental study of soil arching effect under seepage condition. *Acta Geotech.* **2019**, *14*, 2031–2044. [CrossRef]
12. Bao, N.; Wei, J.; Chen, J.F.; Wei, P. 2D and 3D discrete numerical modelling of soil arching. *J. Zhejiang Univ.-Sci. A* **2020**, *21*, 350–365. [CrossRef]
13. Liu, Q.W.; Wang, H.L.; Chen, R.P.; Yin, Z.Y.; Lin, X.T.; Wu, H.N. Effect of relative density of 2D granular materials on the arching effect through numerical trapdoor tests. *Comput. Geotech.* **2022**, *141*, 104553. [CrossRef]
14. Badakhshan, E.; Noorzad, A.; Bouazza, A.; Dafalias, Y.F.; Zameni, S.; King, L. Load recovery mechanism of arching within piled embankments using discrete element method and small scale tests. *Powder Technol.* **2020**, *359*, 59–75. [CrossRef]
15. Jenck, O.; Dias, D.; Kastner, R. Discrete element modelling of a granular platform supported by piles in soft soil—Validation on a small scale model test and comparison to a numerical analysis in continuum. *Comput. Geotech.* **2009**, *36*, 917–927. [CrossRef]
16. Rui, R.; Han, J.; van Eekelen, S.J.M.; Wan, Y. Experimental investigation of soil-arching development in unreinforced and geosynthetic-reinforced pile-supported embankments. *J. Geotech. Geoenviron. Eng.* **2019**, *145*, 04018103. [CrossRef]
17. Rui, R.; Zhai, Y.X.; Han, J.; van Eekelen, S.J.M.; Chen, C. Deformations in trapdoor tests and piled embankments. *Geosynth. Int.* **2020**, *27*, 219–235. [CrossRef]
18. Lai, H.J.; Zheng, J.J.; Zhang, J.; Zhang, R.J.; Cui, L. DEM analysis of "soil"-arching within geogrid-reinforced and unreinforced pile-supported embankments. *Comput. Geotech.* **2014**, *61*, 13–23. [CrossRef]
19. Xiong, H.; Qiu, Y.Y.; Lin, X.T.; Chen, X.S.; Huang, D.W. Multiple arching in cohesion–friction soils: Insights from deformation behavior and failure mechanisms using FEM-SPH approach. *Comput. Geotech.* **2023**, *154*, 105146. [CrossRef]
20. Liang, L.J.; Cheng, Y.P.; Xu, C.J.; Wei, G.; Ding, Z. Microscopic mechanisms of particle size effect on 2D arching effect development and degradation in granular materials. *Int. J. Multiscale Com. Eng.* **2024**, *22*, 91–108. [CrossRef]
21. Chen, R.P.; Liu, Q.W.; Wu, H.N.; Wang, H.L.; Meng, F.Y. Effect of particle shape on the development of 2D soil arching. *Comput. Geotech.* **2020**, *125*, 103662. [CrossRef]
22. Liu, Q.W.; Chen, R.P.; Wang, H.L.; Yin, Z.Y.; Wu, H.N. Effect of particle shape on soil arching in the pile-supported embankment by 3D discrete-element method simulation. *Int. J. Geomech.* **2022**, *22*, 04022027. [CrossRef]
23. Tordesillas, A.; Walker, D.M.; Liu, Q. Force cycles and force chains. *Phys. Rev. E* **2010**, *81*, 011302. [CrossRef]

24. Snoeijer, J.H.; Vlugt, T.J.H.; van Hecke, M.; van Saarloos, W. Force network ensemble: A new approach to static granular matter. *Phys. Rev. Lett.* **2004**, *92*, 054302. [CrossRef]
25. Bao, X.G.; Ma, W.Y.; Li, X. Controllability of fractional complex networks. *Fractal. Fract.* **2024**, *8*, 43. [CrossRef]
26. Babič, M.; Marinković, D.; Kovačič, M.; Šter, M.; Calì, M. A new method of quantifying the complexity of fractal networks. *Fractal. Fract.* **2022**, *6*, 282. [CrossRef]
27. Arévalo, R.; Pugnaloni, L.A.; Zuriguel, I.; Maza, D. Contact network topology in tapped granular media. *Phys. Rev. E* **2013**, *87*, 022203. [CrossRef]
28. Kramár, M.; Goullet, A.; Kondic, L.; Mischaikow, K. Evolution of force networks in dense particulate media. *Phys. Rev. E* **2014**, *90*, 052203. [CrossRef]
29. Kramár, M.; Goullet, A.; Kondic, L.; Mischaikow, K. Quantifying force networks in particulate systems. *Phys. D Nonlinear Phenom.* **2014**, *283*, 37–55. [CrossRef]
30. Kramár, M.; Goullet, A.; Kondic, L.; Mischaikow, K. Persistence of force networks in compressed granular media. *Phys. Rev. E* **2013**, *87*, 042207. [CrossRef]
31. Pugnaloni, L.; Carlevaro, C.; Kramár, M.; Mischaikow, K.; Kondic, L. Structure of force networks in tapped particulate systems of disks and pentagons. I. Clusters and loops. *Phys. Rev. E* **2016**, *93*, 062902. [CrossRef]
32. Kong, B.W.; Dai, C.X.; Hu, H.B.; Xia, J.Z.; He, S.H. The fractal characteristics of soft soil under cyclic loading based on SEM. Persistence analysis. *Fractal. Fract.* **2022**, *6*, 423. [CrossRef]
33. Gameiro, M.; Singh, A.; Kondic, L.; Mischaikow, K.; Morris, J.F. Interaction network analysis in shear thickening suspensions. *Phys. Rev. Fluids* **2020**, *5*, 034307. [CrossRef]
34. Shah, S.; Cheng, C.; Jalali, P.; Kondic, L. Failure of confined granular media due to pullout of an intruder: From force networks to a system wide response. *Soft Matter* **2020**, *16*, 7685–7695. [CrossRef]
35. Lai, H.J.; Zheng, J.J.; Cui, M.J.; Chu, J. "Soil arching" for piled embankments: Insights from stress redistribution behaviour of DEM modelling. *Acta Geotech.* **2020**, *15*, 2117–2136. [CrossRef]
36. Rui, R.; van Tol, A.F.; Xia, X.L.; van Eekelen, S.J.M.; Hu, G.; Xia, Y.Y. Evolution of soil arching; 2D DEM simulations. *Comput. Geotech.* **2016**, *73*, 199–209. [CrossRef]
37. Xu, C.; Zhang, X.Y.; Han, J.; Yang, Y. Two-dimensional soil-arching behaviour under static and cyclic loading. *Int. J. Geomech.* **2019**, *19*, 04019091. [CrossRef]
38. Duan, N.; Cheng, Y.P.; Liu, J.W. DEM Analysis of Pile Installation Effect: Comparing a Bored and a Driven Pile. *Granul. Matter* **2018**, *20*, 36. [CrossRef]
39. Liang, L.J.; Xu, C.J. Numerical and theoretical research on stress distribution in the loosening zone of the trapdoor problem. *Int. J. Numer. Anal. Met.* **2019**, *43*, 1426–1447. [CrossRef]
40. Rui, R.; van Tol, A.F.; Xia, Y.Y.; van Eekelen, S.J.M.; Hu, G. Evolution of Soil Arching: 2D Analytical Models. *Int. J. Geomech.* **2018**, *18*, 04018056. [CrossRef]
41. King, L.; Bouazza, A.; Dubsky, S.; Rowe, R.K.; Gniel, J.; Bui, H.H. Kinematics of soil arching in piled embankments. *Géotechnique* **2019**, *69*, 941–958. [CrossRef]
42. Zhao, S.; Evans, T.M.; Zhou, X.; Zhou, S. Discrete element method investigation on thermally induced shakedown of granular materials. *Granul. Matter.* **2017**, *19*, 11. [CrossRef]
43. Rothenburg, L.; Kruyt, N.P. Critical state and evolution of coordination number in simulated granular materials. *Int. J. Solids. Struct.* **2004**, *41*, 5763–5774. [CrossRef]
44. Xu, D.S.; Tang, J.Y.; Zou, Y.; Rui, R.; Liu, H.B. Macro and micro investigation of gravel content on simple shear behavior of sand-gravel mixture. *Constr. Build. Mater.* **2019**, *221*, 730–744. [CrossRef]

fractal and fractional

MDPI

Article

Fractal Operators and Convergence Analysis in Fractional Viscoelastic Theory

Xiaobin Yu and Yajun Yin *

Department of Engineering Mechanics, Tsinghua University, Beijing 100084, China; yxb18@mails.tsinghua.edu.cn
* Correspondence: yinyj@mail.tsinghua.edu.cn

Abstract: This study delves into the convergence of operators and the viscoelastic properties of fractal ladder and tree structures. It proves the convergence of fractal stiffness operators through operator algebra, revealing a fundamental connection between operator sequence limits and fractal operator algebraic equations. Our findings demonstrate that, as the hierarchical levels of these structures increase, their viscoelastic responses increasingly align with the fractional viscoelastic behavior observed in infinite-level fractal structures. We explore the similarity in creep and relaxation behaviors between fractal ladders and trees, emphasizing the emergence of ultra-long characteristic times in steady-state creep and pronounced tailing effects in relaxation curves. This research provides novel insights into the design of fractional-order viscoelastic structures, presenting significant implications for materials science and mechanical engineering.

Keywords: fractal tree; fractal ladder; fractal operator; fractional viscoelastic; relaxation; creep

1. Introduction

This paper delves into the viscoelastic mechanics of fractal structures, prompted by the growing use of polymers [1–3], rubber [4], hydrogels [5], biological tissues [6], and composite materials [7]. These advancements have spurred deeper analysis of these materials' mechanical properties and the evolution of viscoelastic mechanical theories.

Kelvin's mid-19th-century discovery of zinc's viscoelastic properties marked the beginning of viscoelastic theory [8]. Maxwell's subsequent introduction of viscosity for all bodies [9], along with the efforts of Meyer [10] and Boltzmann [11], laid the foundations of linear viscoelasticity. The early 20th century saw further advancements by Volterra [12], who developed a mathematical theory for anisotropic solids, propelling viscoelastic mechanics forward.

Viscoelastic materials are commonly divided into two broad categories: linear and nonlinear. Linear viscoelastic materials exhibit a combination of elastic and ideal viscous behaviors, acting as an intermediate state between the elastic Hookean solid and the ideal viscous Newtonian fluid [13]. In these materials, the relationship between stress and strain changes over time, yet the stress–strain relationship remains linear at any given moment. In contrast, the mechanical behavior of nonlinear viscoelastic materials is much more complex, encompassing nonlinear elasticity, non-Newtonian fluid behavior, or a combination of both. Since the mid-20th century, there has been rapid progress in the development of constitutive theories and rheology for nonlinear viscoelastic materials [14].

The complexity of nonlinear viscoelastic behavior led researchers to propose self-similar viscoelastic models, comprising infinite elements [15], such as Schiessel et al. [16,17] and Heymans et al. [18,19]. In analyzing these models, two main approaches are predominantly used: one involves studying the complex modulus of the model within the complex domain [18,19], and the other establishes stress–strain relationships at each node, solving them with the help of matrix operations [20]. Both methods, while computationally intensive, often do not directly convey structural information. Recently, scholars introduced the theories of operational calculus and operator algebra, achieving succinct and direct results [21].

Nevertheless, prior studies [13,21–26] have proceeded under the unverified assumption that the physical fractal entities under investigation exhibit self-similarity throughout their structure, thus guiding the computation of the overall response through the formulation of stiffness operator algebraic equations. Such dependence on fractal-based reasoning, however, lacks theoretical validation of its accuracy. To enhance and refine this methodology, the present paper leverages the operational calculus and operator algebra theories to rigorously demonstrate its suitability.

Moreover, the nascent stage of operator algebra theory previously left researchers unable to fully interpret operator expression outcomes, compelling them to rely on approximation techniques, such as asymptotic expansions in Ref. [23], for their research. We have advanced the theories of operational calculus and operator algebra, integrating them with integral transforms to devise the operator kernel function (OKF) method [27]. This approach has allowed us to revisit fractional viscoelastic models, develop algebraic equations for stiffness and compliance operators for self-similar fractal structures, and finally derive functional expressions.

By comparing the OKF with Boltzmann's superposition principle, we have established a link between the kernel function and quasi-static behavior, i.e., the relaxation function and the creep function, thus offering new insights into the viscoelastic behavior of fractal structures and highlighting the significant potential of fractal operators in simplifying the computational complexity associated with analyzing viscoelastic materials.

The organization of this paper is as follows: Section 2 utilizes operator algebra to prove the convergence of operators in fractal trees and ladders. Section 3 delineates the transformational relationships between fractal operators and the creep function or the relaxation function, highlighting the differences that set operator theory apart from classical viscoelasticity. Section 4 explores the discretization of continuous media, illustrating the standard approach to resolving fractal stiffness operators. Through this structured exposition, we bridge the gap between the fractal mechanics and conventional viscoelastic principles, offering a holistic framework that enriches our grasp and application of fractal operator theory within materials science.

2. Convergence Analysis of Viscoelastic Fractal Operators

2.1. Stiffness Operator Method and Compliance Operator Method

Linear viscoelastic materials, representing synergy between elasticity and perfect viscosity, serve as a transitional phase bridging the gap between the elastic behavior of Hookean solids and the ideal viscosity of Newtonian fluids, with several classic viscoelastic models illustrated in Figure 1.

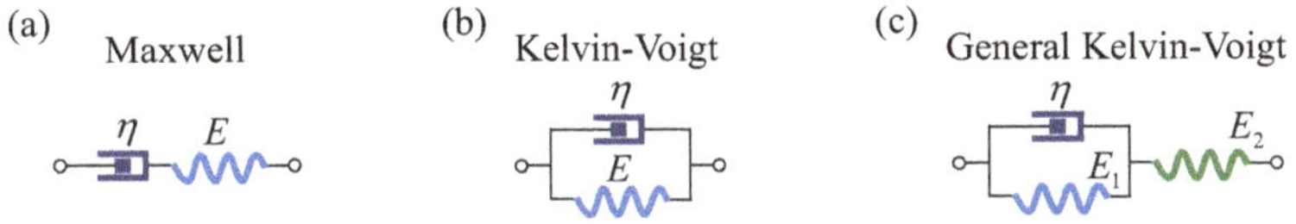

Figure 1. Some typical viscoelastic models. (**a**) The Maxwell model. (**b**) The Kelvin–Voigt model. (**c**) The General Kelvin–Voigt (GKV) model.

Establishing a structure's total constitutive relationship demands the amalgamation of stress-based equilibrium equations, strain-based compatibility equations, and the distinct constitutive relations for each element. The methodology for solving this comprehensive set of differential equations to ascertain the constitutive equation in a unidimensional framework is elaborated upon in Appendix A.

This method is viable for viscoelastic models constituted by a finite ensemble of elements. Nonetheless, with an increase in the number of elements, there is a proportional rise in the number of unknowns and equations, consequently amplifying computational complexity. In scenarios involving models with an infinite array of elements, the infinite count of unknowns renders this computational strategy untenable.

Hu et al. [13] pioneered the application of operator algebra to non-Newtonian fluid dynamics, utilizing force–electricity analogies to elucidate the stress–strain relationships inherent in viscoelastic materials. This innovative methodology has since been adopted by various researchers to explore a wide range of topics, including the viscoelastic properties of ligaments [22], the spiking and propagation of neural electrical signals [23], the complex dynamics of blood flow within the infinite elastic cavity model [25], and the intricate mechanical behaviors of bone [26]. Herein, we offer a concise review of this method.

Leveraging the concept of the force–electricity analogy, the integration of two mechanical elements in series results in a combined stiffness analogous to the parallel configuration of electrical resistors within a circuit. Similarly, their cumulative compliance corresponds to the series arrangement of resistors. Conversely, the parallel coupling of mechanical elements yields a composite stiffness akin to resistors in series, whereas their collective compliance mirrors the parallel arrangement of resistors. This correlation is visually depicted and clarified in Figure 2.

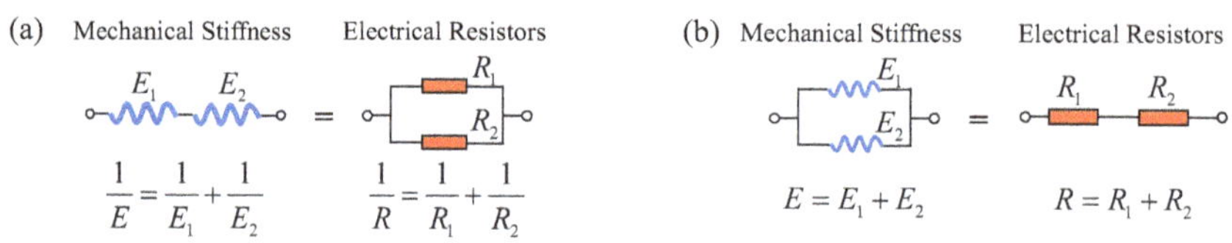

Figure 2. Schematic of the force–electricity analogy. (**a**) The stiffness interaction among mechanically series-connected elements is similar to the behavior of electrical resistors in a parallel configuration; (**b**) conversely, the stiffness of mechanically parallel-connected elements corresponds to electrical resistors arranged in series.

In the subsequent sections, we will adopt an operator-based framework where stiffness is expressed through operators, termed the stiffness operator method (SOM), and compliance is articulated in a similar manner, denoted as the compliance operator method (COM). This paper primarily focuses on thoroughly exploring and elucidating the stiffness operator methodology.

This study explores the viscoelastic behavior of models comprising energy-storing components, symbolized by springs, and energy-dissipating components illustrated by dashpots, as demonstrated in Figure 3. The stiffness operators for the springs $\boldsymbol{T}_1$ and for the dashpots $\boldsymbol{T}_2$ are represented as

$$\boldsymbol{T}_1 = E, \qquad \boldsymbol{T}_2 = \eta \boldsymbol{p}. \tag{1}$$

where $\boldsymbol{p} = \frac{\mathrm{d}}{\mathrm{d}t}$ denotes the Heaviside operator, E denotes the elastic module of the spring, and η denotes the coefficient of viscosity.

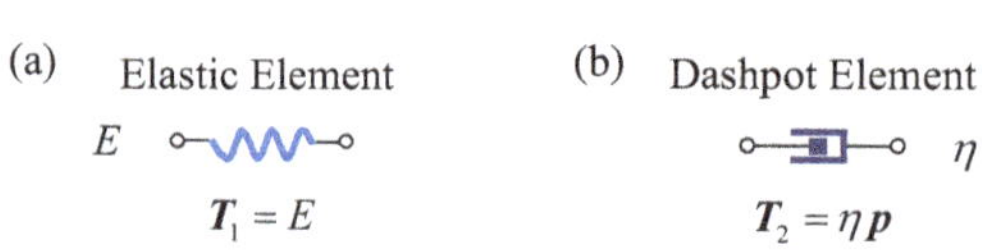

Figure 3. The stiffness operators of (**a**) elastic element and (**b**) dashpot element.

Utilizing this method enables the swift calculation of the stiffness operator for the Maxwell model as follows:

$$\boldsymbol{T}_{Maxwell} = \frac{E\eta\boldsymbol{p}}{E + \eta\boldsymbol{p}}. \tag{2}$$

The stiffness operator for the Kelvin–Voigt model is defined as

$$\boldsymbol{T}_{KV} = E + \eta\boldsymbol{p}. \tag{3}$$

Similarly, the stiffness operator for the GKV model is determined as

$$\boldsymbol{T}_{GKV} = \frac{(E_1 + \eta \boldsymbol{p})E_2}{E_1 + E_2 + \eta \boldsymbol{p}}. \quad (4)$$

Appendix A provides a guide to employing stiffness operators for determining the functional relationship between stress and strain in these models.

2.2. Algebraic Equations of Stiffness Operators for Fractal Cells

This section revisits the application of the SOM to fractal cells, as depicted in Figure 4. The figure demonstrate that a fractal tree or ladder, viewed in its entirety, matches a single fractal cell (as depicted on the left-hand side in Figure 4a,b); similarly, these structures can also be equated to a fractal element (as shown on the right-hand side in Figure 4a,b). Utilizing this equivalence, algebraic equations for stiffness operators are formulated. Solving these equations yields precise operator expressions for both fractal ladders and trees, thus providing a structured approach for analyzing their mechanical properties.

Figure 4. Schematic diagram of fractal structure cells. (**a**) Fractal ladder structure. (**b**) Fractal tree structure.

For the fractal ladder structure, see Figure 4a; based on the equivalence of compliance, we have

$$\frac{1}{\boldsymbol{T}} = \frac{1}{E} + \frac{1}{\boldsymbol{T} + \eta \boldsymbol{p}}. \quad (5)$$

In Equation (5), the left side represents the compliance operator of the fractal element, while the right side corresponds to the compliance operator of the fractal cell. Equation (5) forms a quadratic algebraic equation of operators. Solving Equation (5) and considering the condition for positive stiffness [24,28–30], we obtain the operator for the fractal ladder:

$$\boldsymbol{T} = \sqrt{\frac{(\eta \boldsymbol{p})^2}{4} + E\eta \boldsymbol{p}} - \frac{\eta \boldsymbol{p}}{2}. \quad (6)$$

For the fractal tree structure, see Figure 4b; based on the equivalence of stiffness, we have

$$\boldsymbol{T} = \frac{E\boldsymbol{T}}{\boldsymbol{T} + E} + \frac{\eta \boldsymbol{p} \boldsymbol{T}}{\boldsymbol{T} + \eta \boldsymbol{p}}, \quad (7)$$

In Equation (7), the left side represents the stiffness operator of the fractal element, while the right side corresponds to the stiffness operator of the fractal cell. Equation (7) is a quadratic algebraic equation of operators. Solving Equation (7) and considering the condition for positive stiffness [24,30] yields the operator for the fractal tree:

$$\boldsymbol{T} = \sqrt{E\eta \boldsymbol{p}}. \quad (8)$$

Remark 1. *From a mathematical perspective, both Equations (5) and (7) are expected to have two radical results [28–30]. Indeed, earlier research by Yin et al. [24] has shown that an nth-order operator algebra equation should have at least n solutions in more general circumstances. However, for the specific issue considered in this paper, the relationship between stress and strain is represented as $\sigma(t) = \boldsymbol{T}\varepsilon(t)$, indicating that an increase in positive strain applied to the structure should cause increased stress in the same direction. Given that the action of operators is realized through the* *convolution of the operator kernel function with the input, it necessitates that the structure's kernel function be positive, which also implies that the stiffness operator must be positive. Therefore, only the positive roots of Equations (6) and (8) have been retained.*

This section delves deeper into the operator expressions for fractal ladders and trees, with a focus on Equations (6) and (8). We unravel how these structures respond to a given stimulus, revealing a fundamental similarity in their behavior: both are characterized by quadratic radical operators, underscoring their non-rational nature. Yet, a striking divergence emerges in their complexity. The operator for the fractal tree is remarkably straightforward, a reflection of its symmetrical topology. In contrast, the operator for the fractal ladder exhibits greater complexity due to the disruption of this symmetry. This contrast not only highlights the distinctive architectural influences on their mechanical responses but also enriches our understanding of their intrinsic properties, offering a nuanced perspective on the dynamics of fractal-based structures.

2.3. *The Logical Foundation of the Equivalence Postulate*

In the preceding analysis, an implicit assumption of equivalence was introduced such that

$$\text{fractal elements} \approx \text{fractal trees or ladder} \approx \text{fractal cells.}$$

This perception, more intuitive than deductive, lays the groundwork for formulating algebraic equations specific to fractal operators. Closer scrutiny of these equations unveils an underlying presumption: the existence of stiffness operators for fractal ladders and trees. Our approach progresses by first positing the existence of fractal operators, then using this equivalence to derive the relevant algebraic equations, and finally solving these equations to construct the fractal operators. This methodology combines intuitive reasoning with analytical rigor to unravel the intricacies of fractal mechanics.

While this approach is standard for structures with a finite hierarchy, the premise becomes less clear when dealing with fractal structures of infinite levels. Appendix B explores the generalized Maxwell model, which is derived by arranging infinitely many Maxwell models in parallel. This approach results in paradoxes, rendering the model unsolvable. Hence, this indicates that methodologies effective for finite structures may not extend straightforwardly to infinite ones, highlighting the need for a logical and robust theoretical basis for their existence.

Our investigation primarily concentrates on two pivotal structures: fractal trees and fractal ladders. The former embody symmetric fractal topology, whereas the latter illustrate a fractal topology characterized by disrupted symmetry. This distinction not only informs our understanding of fractal mechanics but also enriches our comprehension of the diversity within fractal structures.

2.4. *Fractal Ladder*

The basic unit of the fractal ladder structure is the Maxwell element, which consists of a spring and a dashpot connected as shown in Figure 5. Due to the entire topology presenting a ladder-like configuration, when the number of structural levels approaches infinity, a fractal topology is formed; hence, the term 'fractal ladder' is used. The fractal ladder is not a traditional geometric fractal exhibiting self-similarity but rather a self-similar physical fractal composed of identical elements.

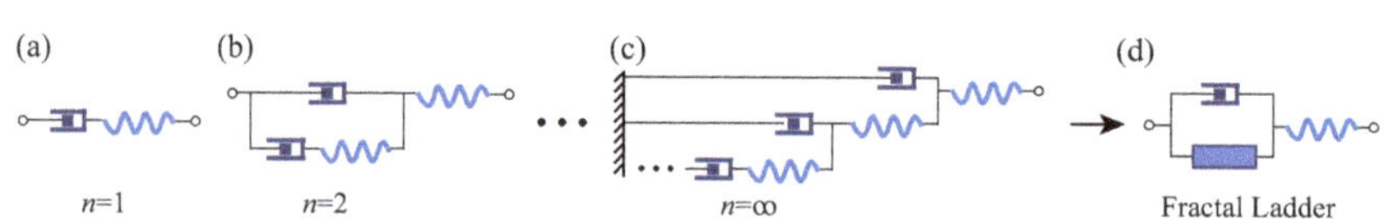

Figure 5. Schematic diagram of fractal ladder structures from level 1 and level 2 to infinite level. (**a–c**) The first, the second and the infinite level of fractal ladder. (**d**) Fractal ladder cell.

In Figure 5, according to operator algebra, the stiffness operator of adjacent nth level structures and $(n+1)$th level structures satisfies the following recursive relationship:

$$\frac{1}{\boldsymbol{T}_n + \eta \boldsymbol{p}} + \frac{1}{E} = \frac{1}{\boldsymbol{T}_{n+1}}. \tag{9}$$

Upon simplification, we can obtain

$$\boldsymbol{T}_{n+1} = \frac{(\boldsymbol{T}_n + \eta \boldsymbol{p})E}{\boldsymbol{T}_n + \eta \boldsymbol{p} + E}. \tag{10}$$

Similar to the fractal tree structure, we will now prove that the sequence of operators constituting the fractal ladder structure converges to a limit.

Proof. The stiffness operators for the first- and second-level structures are, respectively, represented as

$$\boldsymbol{T}_1 = \frac{\eta E \boldsymbol{p}}{\eta \boldsymbol{p} + E}, \qquad \boldsymbol{T}_2 = \frac{E\eta \boldsymbol{p}(2E + \boldsymbol{p}\eta)}{E^2 + 3E\eta \boldsymbol{p} + \eta^2 \boldsymbol{p}^2}. \tag{11}$$

Equation (11) can be derived to yield

$$\boldsymbol{T}_2 - \boldsymbol{T}_1 = \frac{E^3 \eta \boldsymbol{p}}{(E + \eta \boldsymbol{p})(E^2 + 3E\eta \boldsymbol{p} + \eta^2 \boldsymbol{p}^2)} > 0. \tag{12}$$

We thus proved that $\boldsymbol{T}_2 > \boldsymbol{T}_1$. Assuming $\boldsymbol{T}_n > \boldsymbol{T}_{n-1}$, we next prove that $\boldsymbol{T}_{n+1} > \boldsymbol{T}_n$. Let the function $f(x)$ be

$$f(x) = \frac{(x + \eta \boldsymbol{p})E}{x + \eta \boldsymbol{p} + E} = E - \frac{E^2}{x + \eta \boldsymbol{p} + E}. \tag{13}$$

Differentiating Equation (13) with respect to x yields

$$f'(x) = \frac{E^2}{(E + x + \eta \boldsymbol{p})^2} > 0. \tag{14}$$

Therefore, the function is monotonically increasing, implying$(x_2 - x_1)[f(x_2) - f(x_1)] > 0$. Given that $\boldsymbol{T}_n > \boldsymbol{T}_{n-1}$, it follows that

$$f(\boldsymbol{T}_n) > f(\boldsymbol{T}_{n-1}). \tag{15}$$

Combining Equations (10), (13) and (15), we arrive at

$$\boldsymbol{T}_{n+1} > \boldsymbol{T}_n. \tag{16}$$

Note that the fractal ladder, formed by connecting a spring on the outermost side in series with other structures, inherently has an equivalent stiffness lower than that of a single spring. This proposition's validity is further supported by Equation (13):

$$f(x) = E - \frac{E^2}{x + \eta \boldsymbol{p} + E} < E. \tag{17}$$

Therefore, the sequence of stiffness operators for the fractal ladder structure is a monotonically increasing and bounded sequence of operators. Consequently, this sequence of operators must have a limit, and $\lim_{n\to\infty} \boldsymbol{T}_{n+1} = \lim_{n\to\infty} \boldsymbol{T}_n = \boldsymbol{T}$. □

Remark 2. *Mikusiński rigorously defined the concept of operator sequence convergence in Ref. [30]: If there exists $\tilde{\boldsymbol{T}} \neq 0$ such that the sequence of operators $\{\boldsymbol{T}_n/\tilde{\boldsymbol{T}}\}$ has a uniformly convergent* *sequence of kernel functions, then the sequence of operators* $\{T_n\}$ *is said to converge. Furthermore, it is denoted by* $\lim_{n\to\infty} T_n = \tilde{T} \lim_{n\to\infty} T_n/\tilde{T}$. *In fact, the above proof process can be viewed as setting* $\tilde{T} = I$, *where* I *represents the identity operator.*

Taking the limit of both sides of Equation (9) yields the fractal operator algebraic Equation (5). As mentioned previously, this is a quadratic algebraic equation for the fractal operator, with a radical solution (see Equation (6)). Clearly, for the fractal ladder, the fractal operator is of a non-rational type. Its non-rationality still stems from the infinity of the fractal ladder's structural levels.

It is verifiable that a finite-level self-similar ladder structure, once the number of structural levels exceeds three, can be replaced by an infinite-level self-similar fractal ladder. Therefore, the use of fractal operators enables not only the attainment of concise and direct results but also ensures sufficient accuracy within the characteristic time.

2.5. Fractal Tree

Figure 6 illustrates the generative process of a fractal tree, depicting the evolution from a finite-level structure to an infinite-level structure. According to the operator stiffness method, the stiffness operator of the nth level structure is related to that of the $(n+1)$th level structure through the following recursive relation:

$$T_{n+1} = \frac{ET_n}{T_n + E} + \frac{\eta p T_n}{T_n + \eta p}. \tag{18}$$

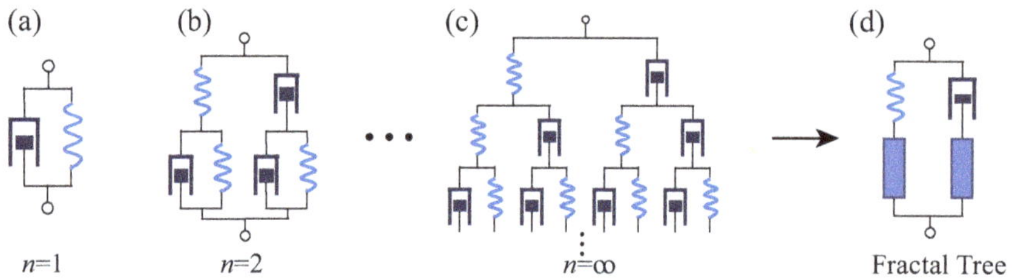

Figure 6. Schematic diagram of fractal tree structures from level 1 and level 2 to infinite level. (**a–c**) The first, the second and the infinite level of fractal tree. (**d**) Fractal tree cell.

Given that the stiffness operator and the compliance operator are inversely related, taking the reciprocal of both sides of Equation (18) yields the recursive relation for the compliance operator:

$$\frac{1}{T_{n+1}} = \frac{(T_n + E)(T_n + \eta p)}{ET_n(T_n + \eta p) + \eta p T_n(T_n + E)}. \tag{19}$$

Leveraging the principle of monotonic boundedness for operator sequences, we prove the existence of a limit for the sequence of operators, specifically $\lim_{n\to\infty} T_n = T$.

Proof. Firstly, we establish the boundedness of the operator sequence. During the loading process, the applied stress (strain) induces an increasing strain (stress), ensuring that the stiffness operator remains positive and, consequently, has a lower bound.

Utilizing mathematical induction, we will prove that the sequence of stiffness operators is decreasing. For the first-level structure, we have

$$T_1 = E + \eta p. \tag{20}$$

Substituting Equation (20) into Equation (18) yields

$$T_2 = \frac{ET_1}{T_1 + E} + \frac{\eta p T_1}{T_1 + \eta p} = (E + \eta p)\left(\frac{E}{2E + \eta p} + \frac{\eta p}{E + 2\eta p}\right). \tag{21}$$

Subtracting Equation (20) from Equation (21), we obtain

$$T_2 - T_1 = -\frac{(E + \eta p)(E^2 + E\eta p + \eta^2 p^2)}{(2E + \eta p)(E + 2\eta p)} < 0. \tag{22}$$

We thus proved that $T_2 < T_1$. Assuming $T_n < T_{n-1}$, we next prove that $T_{n+1} < T_n$. Let the function $f(x)$ be

$$f(x) = \frac{Ex}{x + E} + \frac{x\eta p}{x + \eta p}. \tag{23}$$

Differentiating Equation (23) results in

$$f'(x) = \frac{E^2}{(E + x)^2} + \frac{\eta p^2}{(\eta p + x)^2} > 0. \tag{24}$$

Therefore, the function is monotonically increasing, implying $(x_2 - x_1)[f(x_2) - f(x_1)] > 0$. Given that $T_n < T_{n-1}$, it follows that

$$f(T_n) < f(T_{n-1}). \tag{25}$$

Combining Equations (18), (23) and (25), we arrive at

$$T_{n+1} < T_n. \tag{26}$$

By immediate application of mathematical induction, it is evident that the stiffness operator decreases as the structural level n increases. Given that the stiffness operator is monotonically decreasing and bounded below, it necessarily converges to a limit as follows: $\lim\limits_{n\to\infty} T_{n+1} = \lim\limits_{n\to\infty} T_n = T$. $\square$

Since the sequence of linear operators converges and has a unique limit, taking the limit of both sides of Equation (18) simultaneously yields the fractal stiffness operator Equation (7) mentioned earlier by Guo and Yin et al. [23]. As discussed, this is a quadratic algebraic equation for the fractal operator, with a radical solution. As mentioned, the radical-type fractal operator in Equation (8) is a non-rational operator.

At this point, we can make a fundamental judgment: for finite-level self-similar structures, the operator is rational; for infinite-level self-similar fractal structures, the fractal operator is non-rational. Clearly, the non-rational nature of the fractal operator arises from the infiniteness of the structural levels. The properties of rational and non-rational operators differ significantly. As outlined in Mikusinski's work [30], kernel functions for rational operators typically manifest as elementary functions. Our earlier research confirms that kernel functions associated with non-rational operators are typically non-elementary functions.

Figure 7 illustrates the creep response of structures from levels 1 to 3 compared to an infinite-level structure. With the increase in structural levels, the multi-level structure's creep curves rapidly converge to the fractal tree within a characteristic time range, as indicated by the red dashed line. The findings suggest that, as the hierarchy of a finite-level structure extends beyond three levels, its mechanical behavior within the characteristic time can be represented by an infinite-level structure. Additionally, the analytical process for determining the behavior of an infinite-level structure proves to be simpler than for a finite-level structure, thus enhancing the model's practical applicability in viscoelastic investigations. Moreover, the level-1 Voigt model entirely lacks the long-term relaxation effect. The greater the number of structural levels, the more pronounced the long-term relaxation effect becomes. An infinite-level fractal structure exhibits the most pronounced characteristics of ultra-long relaxation time. This discovery holds significant importance for understanding the mechanical behavior of complex viscoelastic materials.

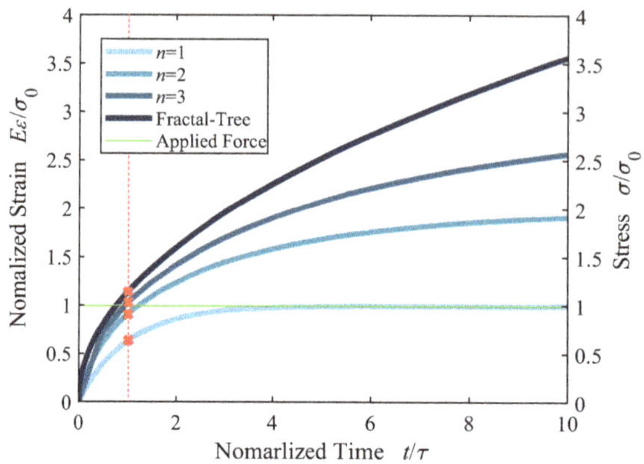

Figure 7. Creep response functions to step stress for structures at various levels. $\tau = \eta/E$ denotes the characteristic time. The red dashed line represents the response at the characteristic time $\tau = 1$.

3. Viscoelastic Response Curves of Fractal Structures

3.1. Correspondence between Operator Kernel Functions and Relaxation and Creep Functions

This section establishes a direct and intrinsic link between fractal operators and the functions governing relaxation and creep. This correlation renders fractal operators exceptionally apt for delving into the dynamics of creep and relaxation behaviors. Systems constituted by linear physical elements inherently display linearity, aligning with the Boltzmann superposition principle:

$$\sigma(t) = \varepsilon_0 Y(t) + \int_{0+}^{t} Y(t-\tilde{\tau})\frac{\mathrm{d}\varepsilon(\tilde{\tau})}{\mathrm{d}\tilde{\tau}}\mathrm{d}\tilde{\tau}. \tag{27}$$

$$\varepsilon(t) = \sigma_0 J(t) + \int_{0+}^{t} J(t-\tilde{\tau})\frac{\mathrm{d}\sigma(\tilde{\tau})}{\mathrm{d}\tilde{\tau}}\mathrm{d}\tilde{\tau}. \tag{28}$$

In the equations, $Y(t)$ represents the material's relaxation function, and $J(t)$ denotes the material's creep function. Utilizing integration by parts, Equations (27) and (28) can be rewritten in a convolution form:

$$\sigma(t) = Y(t) * \dot{\varepsilon}(t) = \varepsilon(t) * \dot{Y}(t). \tag{29}$$

$$\varepsilon(t) = J(t) * \dot{\sigma}(t) = \sigma(t) * \dot{J}(t). \tag{30}$$

Notice that both Equations (29) and (30) are convolution expressions. In fact, in Mikusinski's monograph [30], operators are defined through convolution. This implies that convolution expressions in creep and relaxation theories often correspond to a certain fractal operator. Specifically, in Equation (29), we can consider strain as the input signal and stress as the output signal, making the creep function $J(t)$ necessarily the kernel function of some fractal stiffness operator. Similarly, the relaxation function $Y(t)$ in Equation (30) is undoubtedly the kernel function of a fractal compliance operator.

Let the stiffness operator of the viscoelastic material be denoted as T and its compliance operator as $\frac{1}{T}$. Then, the viscoelastic stress–strain relationship expressed in terms of operators is

$$\sigma(t) = \boldsymbol{T}(\boldsymbol{p})\varepsilon(t), \qquad \text{or} \qquad \varepsilon(t) = \frac{1}{\boldsymbol{T}}(\boldsymbol{p})\sigma(t). \tag{31}$$

Based on the operator kernel function theory previously discussed [27], operator operations are defined through the convolution integral of their corresponding kernel functions (generalized functions), namely

$$\sigma(t) = \mathscr{L}^{-1}[\boldsymbol{T}(\boldsymbol{p})](t) * \varepsilon(t). \tag{32}$$

$$\varepsilon(t) = \mathscr{L}^{-1}[\frac{1}{\boldsymbol{T}}(\boldsymbol{p})](t) * \sigma(t). \tag{33}$$

Comparing Equations (29) and (30) with Equations (32) and (33) yields

$$\mathscr{L}^{-1}[\boldsymbol{T}(\boldsymbol{p})](t) = \dot{Y}(t), \quad \text{or} \quad Y(t) = \int_{0+}^{t} \mathscr{L}^{-1}[\boldsymbol{T}(\boldsymbol{p})](\tilde{\tau})\mathrm{d}\tilde{\tau}. \tag{34}$$

$$\mathscr{L}^{-1}[\frac{1}{\boldsymbol{T}}(\boldsymbol{p})](t) = \dot{J}(t), \quad \text{or} \quad J(t) = \int_{0+}^{t} \mathscr{L}^{-1}[\frac{1}{\boldsymbol{T}}(\boldsymbol{p})](\tilde{\tau})\mathrm{d}\tilde{\tau}. \tag{35}$$

Equations (34) and (35) indicate that, once the stiffness or compliance operator of a structure is known, the OKF can be obtained through the inverse Laplace transform, thereby deriving the relaxation or creep function. It is important to note that the above operations only require manipulation of the stiffness or compliance operator, without involving the input and output signals.

As mentioned, the stiffness and compliance operators characterize the intrinsic properties of self-similar fractal structures themselves. By understanding the intrinsic properties within the physical fractal space, one can determine the creep and relaxation functions. This facilitates research into creep and relaxation phenomena.

3.2. Comparison of Mechanical Behavior between Several Classical Viscoelastic Models and Fractal Tree, Fractal Ladder Structures

As discussed earlier, we now summarize the relaxation and creep functions, along with the response curves under cyclic loading, for classical viscoelastic models and fractal tree and ladder structures in Figure 8.

Figure 8a–e display the schematic diagrams and corresponding stiffness/compliance operator expressions for the classical Maxwell model, Voigt model, generalized Kelvin–Voigt model, fractal tree model, and fractal ladder model. Comparing the creep response to step stress in Figure 8f–j, it is evident that fractal models significantly differ from classical models. Upon instantaneous loading, the fractal ladder structure's end has a series spring element that produces instantaneous elastic deformation, similar to the Maxwell and three-parameter GKV models. In contrast, the fractal tree model and Voigt model, lacking an independent load-bearing structure, do not exhibit instantaneous strain. During the steady creep stage, models with a finite number of elements rapidly stabilize in deformation rate, indicating transfer of all viscous forces to the elastic el'ements. In comparison, the elements of the infinite-level fractal model continuously transmit stress, leading to ongoing creep with a characteristic time far exceeding that of finite-level structures.

Figure 8k–o show the stress relaxation behavior to step strain. These curves reveal that the behavior of fractal structures is significantly different from finite-level models. In terms of instantaneous behavior, the fractal tree model and Voigt model exhibit a sharp increase in stress, whereas the Maxwell, GKV, and fractal ladder models, which contain independent load-bearing elastic elements, predominantly absorb instantaneous strain through these elements. For long-term relaxation, fractal structures exhibit characteristics of decaying over time and eventually tending to zero, with a gradually decreasing rate of decay, similar to the Maxwell model. Notably, fractal structures more distinctly exhibit the tailing or dragging phenomenon, prevalent in rock creep tests, indicating that fractional viscoelastic models are more effective in describing such materials.

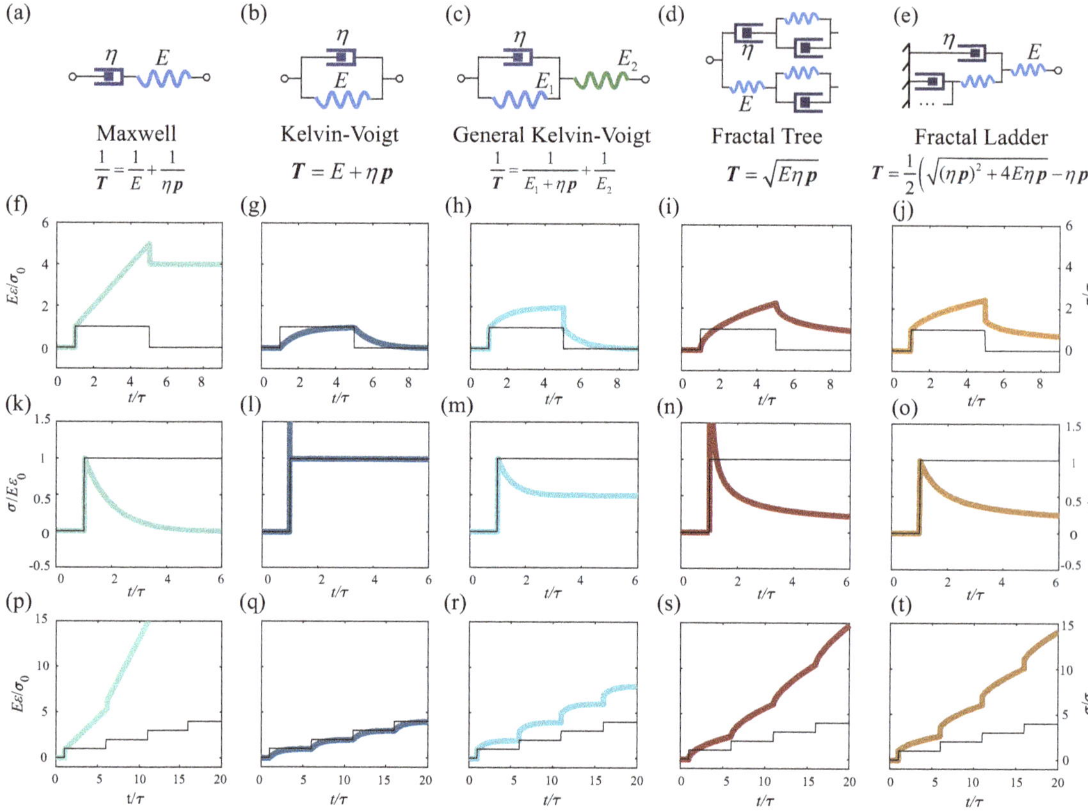

Figure 8. Quasi-static responses of classical viscoelastic models and fractional viscoelastic models. (**a–e**) Schematic diagrams and operator expressions of classical linear viscoelastic models and fractal viscoelastic models. (**f–j**) Creep function curves of the structures, with black line segments indicating applied step stress, loaded at time $t/\tau = 1$ and unloaded at time $t/\tau = 5$. The horizontal axis represents dimensionless time, while the vertical axis represents the dimensionless creep function. (**k–o**) Relaxation function curves, with black line segments indicating applied step strain, loaded at time $t/\tau = 1$. The vertical axis is dimensionless stress. (**p–t**) Creep curves under multiple loading cycles, with black line segments indicating applied step stress, loaded at time $t/\tau = 1$.

Figure 8p–t present the response curves of the models under multiple loading conditions. From these curves, it can be observed that, due to their prolonged characteristic times, fractal models cannot disperse external forces promptly within one loading cycle and thus enter the next cycle. This characteristic indicates that the mechanical behavior exhibited by fractal structures under periodic loading more closely resembles rheological properties. This behavior reveals the uniqueness of fractal structures in handling dynamic loads, which is significant for understanding the response of such structures in practical applications.

4. Discretization of Continuous Viscoelastic Bodies into Fractal Topological Structures

Discrete fractal models, built from physical elements, showcase the intriguing ability to transition between discretization and continuity. Peng et al. [25] investigated the infinite elastic cavity model of arterial blood flow, transforming the continuous vascular wall into a myriad of microelastic cavities, crafting a physical fractal model that mirrors blood flow dynamics. Similarly, Mario Di Paola et al. [31] explored fractional viscoelastic theory, adopting the strategy of breaking continua into discrete segments. This section, as illustrated by Figure 9, elucidates the process of discretizing an elastic body with distributed viscous constraints and illustrates how this approach leads to the fractalization of force transmission pathways, thereby offering a novel perspective on modeling complex physiological and material behaviors.

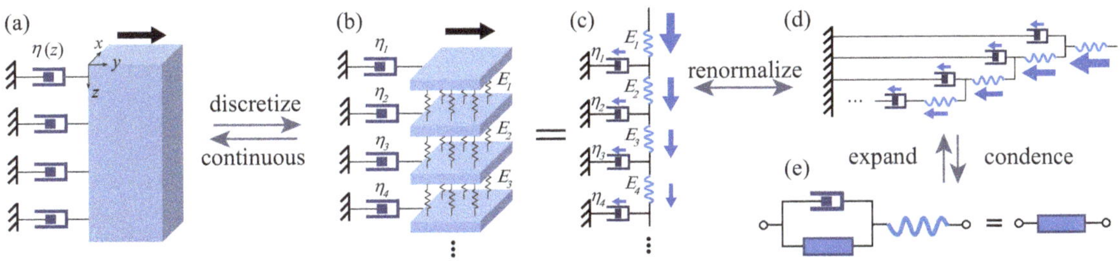

Figure 9. Schematic diagram of the discretization process from a continuous model to a physical fractal model. (**a**) Elastic body subjected to distributed viscous constraints. (**b**) Discretization of continuous structure. (**c**) Schematic diagram of force transmission paths. (**d**) Renormalization to fractal ladder structure. (**e**) Equivalent relationship between fractal ladder cells and elements.

Figure 9a,b depict an elastic body under lateral forces at its upper boundary, experiencing shear deformation that initiates force transmission from top to bottom. By dividing the body vertically (z-direction) into thin slices (each with thickness Δz), it transitions from a continuous entity (Figure 9a) to a discrete system (Figure 9b), where slices are linked by elastic elements and to boundaries by viscous elements, allowing force to propagate through springs and dashpots. This setup leads to the multi-level self-similar model in Figure 9c: applied forces traverse internal elastic components, bifurcating at each level—one path through dashpots to boundaries and the other to subsequent elastic layers. This repetitive sequence crafts a layered mesh of elastic and viscous connections. Drawing on this model, Schiessel and Blumen [16,17], Heymans et al. [18,19], and Mario Di Paola et al. [31] introduced a pivotal transformation relationship, correlating the body's internal elastic constants with the external viscous constraints' viscosity coefficients by

$$E_k = \frac{1}{2k-1}\frac{\Gamma(\beta)}{\Gamma(1-\beta)}\frac{\Gamma(k+1-\beta)}{\Gamma(k-1+\beta)}E_0. \tag{36}$$

$$\eta_k = 2\frac{\Gamma(\beta)}{\Gamma(1-\beta)}\frac{\Gamma(k+1-\beta)}{\Gamma(k+\beta)}\eta_0. \tag{37}$$

This leads to the derivation of a fractional relationship between stress and strain:

$$\sigma(t) = \frac{\mathrm{d}^\beta}{\mathrm{d}t^\beta}\varepsilon(t). \tag{38}$$

It is important to note that the constraint relations between physical constants (Equations (36) and (37)) are artificially designed with the purpose of deriving the fractional viscoelastic constitutive as in Equation (38). Such an operation is not natural. Below, we provide a more intuitive explanation.

In this scenario, we imagine a structure where each level is composed of identical elements, ensuring uniform elastic constants and equal viscosity coefficients for all viscous constraints. By applying the principle of self-similarity, we can transform the model in Figure 9c into a fractal ladder configuration, as shown in Figure 9d. Remarkably, this fractal ladder mirrors the topology of the ladder in Figure 5 exactly. Moving forward, we abstract the fractal cell from the ladder, and, by developing and resolving the algebraic equation for the fractal stiffness operator (mirroring Equation (5)), we arrive at the fractal operator expression (aligned with Equation (6)). This illustrates that the fractal ladder, while an abstract and idealized structure, accurately represents the physical reality of viscoelastic bodies. Thus, the fractal operator, a conceptual entity derived from logical reasoning, ubiquitously characterizes the viscoelastic domain.

Although the fractal operators on the fractal ladder and the fractional derivatives in the viscoelastic constitutive equations (Equation (38)) appear different, this discrepancy is only superficial. At their essence, they are fundamentally identical; that is, the fractal operator (Equation (6)) inherently operates as a fractional operator. This unity is further evidenced when analyzing Figure 8i,j, where the elimination of the outermost independent spring element from the fractal ladder aligns the creep and relaxation behaviors of both models closely. Such parallels emphasize the inherent similarity across these mathematical formulations.

In Section 2, we detailed the convergence attributes of the fractal ladder structure. Focusing on the imagery in Figure 9, as $\Delta z \to 0$, the level of the structure moves towards infinity, i.e., $n \to \infty$. At this juncture, the discrete model transitions from discrete to a coherent fractal ladder form. On the other hand, as $\Delta z \to 0$, the discrete model again converges to the continuous form depicted in Figure 9a. This transformative process, described as continuum $\to$ discrete $\to$ fractalize, fascinatingly facilitates the fractalization of the force transmission imagery and unveils the source of fractional-order effects in viscoelastic bodies.

The derivation of fractional-order effects in viscoelastic substances is traced back to fractal operators. The intrinsic non-rationality of radical-type fractal operators signifies a root cause of fractional-order dynamics. Appendix C shows that the kernel functions for radical-type fractal operators are typically non-elementary functions. This suggests that accurately depicting fractional-order effects and the characteristic tail effect in viscoelastic bodies requires the use of non-elementary functions. This approach not only encapsulates the sophisticated dynamics of these systems but also offers a fresh perspective on understanding and modeling viscoelastic behavior.

5. Conclusions

This paper delves into the convergence of operators in the viscoelastic theories pertaining to fractal ladder and tree structures, highlighting the robustness and predictability of these models in complex mechanical settings. We have rigorously demonstrated that the sequences of stiffness operators for both structures are monotonically bounded, thereby establishing definitive limits. This groundwork allows us to assert that, beyond a third-level hierarchy, the mechanical behavior of finite-level structures can effectively be represented by an infinite-level fractal framework. This insight offers a significant reduction in computational complexity, streamlining the analyses that involve structural stiffness operators.

During steady-state creep and relaxation, forces within fractal structures are transmitted in a tiered manner, resulting in characteristically prolonged durations. This behavior, prevalent in rock rheology, underscores the superiority of fractal models in capturing the nuanced rheological properties of materials.

Furthermore, we explore the transformative linkage between continuous medium models and their discrete fractal counterparts. Increased granularity in structural subdivision and level augmentation leads to fractalization of internal force pathways, culminating in a seamless transition to a fractal model. This process not only bridges the gap between continuous and fractal representations but also reinforces fractal operator theory as a potent tool for addressing intricate mechanical challenges, promising wide-ranging applications in the field.

This paper investigates the utilization of SOM and COM to characterize the quasi-static mechanical properties of materials, such as relaxation and creep. In real-world engineering applications, materials frequently undergo cyclic alternating loads, highlighting the importance of understanding the dynamic mechanical responses of viscoelastic materials. In our future work, we aim to further develop SOM and COM, broadening their application to the dynamic mechanical characterization of materials. This initiative is directed towards providing new insights into the dynamic mechanical behaviors of viscoelastic materials.

Author Contributions: Conceptualization, X.Y. and Y.Y.; methodology, X.Y.; validation, X.Y.; investigation, X.Y.; writing—original draft preparation, X.Y.; writing—review and editing, Y.Y.; supervision, Y.Y.; funding acquisition, Y.Y. All authors have read and agreed to the published version of the manuscript.

Funding: This research was partially supported by the National Natural Science Foundation of China (grant No. 12050001).

Data Availability Statement: No new data were created or analyzed in this study. Data sharing is not applicable to this article.

Acknowledgments: We thank Zelin Liu for the helpful discussion on this topic.

Conflicts of Interest: The authors declare no conflicts of interest.

Appendix A

Focusing on the three-parameter General Kelvin–Voigt (GKV) model as a case study, the structure is formed by paralleling a viscous element with an elastic element, followed by their serial connection to another elastic element. To ascertain the structure's comprehensive constitutive relationship, it involves the simultaneous solution of two stress-based equilibrium equations:

$$\begin{cases} \sigma = \sigma_2 \\ \sigma = \sigma_\eta + \sigma_1 \end{cases}, \tag{A1}$$

two strain-based compatibility equations:

$$\begin{cases} \varepsilon_\eta = \varepsilon_1 \\ \varepsilon = \varepsilon_1 + \varepsilon_2 \end{cases}, \tag{A2}$$

and the constitutive relationships of each element:

$$\begin{cases} \sigma_1 = E_1\varepsilon_1 \\ \sigma_2 = E_2\varepsilon_2 \\ \sigma_\eta = \eta\frac{\mathrm{d}\varepsilon_\eta}{t} \end{cases}. \tag{A3}$$

Addressing this set of seven differential equations with eight unknowns unravels the constitutive equation for a one-dimensional system:

$$\sigma(t) = E_2\left[\delta(t) - \frac{E_2}{\eta}\mathrm{e}^{-\frac{t(E_1+E_2)}{\eta}}\right] * \varepsilon(t). \tag{A4}$$

While this approach is viable for models with a finite number of elements, the rapid increase in the number of elements escalates both the number of unknowns and the computational complexity. For models composed of an infinite number of elements, leading to an endless count of unknowns, this computational strategy becomes unfeasible.

Based on the stiffness operator from Equation (4) and the operator kernel function method proposed by Yu et al. [27], the formula for converting the stress–strain relationship of the GKV model, as represented by operators, into a functional relationship is the following:

$$\sigma(t) = \mathscr{L}^{-1}[\boldsymbol{T}_{GKV}(\boldsymbol{p})] * \varepsilon(t) = \mathscr{L}^{-1}\left[\frac{(E_1 + \eta \boldsymbol{p})E_2}{E_1 + E_2 + \eta \boldsymbol{p}}\right] * \varepsilon(t). \tag{A5}$$

Simplifying Equation (A5) yields the same result as Equation (A4). This process does not require solving any differential equations, and all steps are algebraic manipulations.

Based on the outcomes of Equations (34) and (35), this approach does not require assuming any form of input to directly determine the model's relaxation function:

$$Y(t) = \int_{0+}^{t} \mathscr{L}^{-1}\left[\frac{(E_1 + \eta \boldsymbol{p})E_2}{E_1 + E_2 + \eta \boldsymbol{p}}\right](\tilde{\tau})\mathrm{d}\tilde{\tau}, \tag{A6}$$

and the creep function:

$$J(t) = \int_{0+}^{t} \mathscr{L}^{-1}\left[\frac{E_1 + E_2 + \eta \boldsymbol{p}}{(E_1 + \eta \boldsymbol{p})E_2}\right](\tilde{\tau})\mathrm{d}\tilde{\tau}. \tag{A7}$$

Throughout this process, we consistently engage in various algebraic operations rather than solving systems of linear differential equations.

Appendix B

Consider the generalized Maxwell model derived from an infinite parallel connection of Maxwell elements, as illustrated in Figure A1.

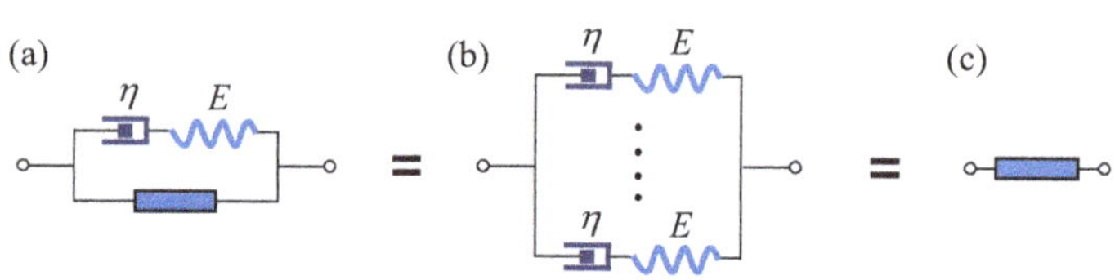

Figure A1. Schematic diagram of general Maxwell model. (**a**) Fractal cell. (**b**) General Maxwell model. (**c**) Fractal element.

Utilizing a calculation method analogous to that for fractal trees and fractal ladders, the model must fulfill the following stiffness operator algebraic equation:

$$\frac{1}{\frac{1}{E} + \frac{1}{\eta p}} + \boldsymbol{T} = \boldsymbol{T}. \tag{A8}$$

Thus, we have

$$\frac{E\eta p}{E + \eta p} = 0. \tag{A9}$$

The paradox presented by Equation (A9) arises because, as the number of levels in the structure increases, the stiffness of the generalized Maxwell model also increases. When the structure approaches an infinite number of levels, the stiffness becomes infinitely large, failing to converge to a specific stiffness operator $\boldsymbol{T}$. Therefore, the aforementioned method is no longer applicable.

Appendix C

The expression for the stiffness operator of the fractal ladder structure is the following:

$$T=\sqrt{\frac{{T_2}^2}{4}+T_1T_2}-\frac{T_2}{2}=\sqrt{\frac{(\eta p)^2}{4}+E\eta p}-\frac{\eta p}{2}=\frac{\eta}{2}\left(\sqrt{p^2+\frac{4p}{\tau}}-p\right), \tag{A10}$$

where $\tau=\frac{\eta}{E}$ represents the relaxation characteristic time. The kernel function corresponding to the fractal operator is

$$T=\mathscr{L}^{-1}[T(p)]=-\frac{Ee^{-\frac{2t}{\tau}}}{t}I_1(\frac{2t}{\tau})+E\delta(t), \tag{A11}$$

wherein $I_n(\frac{2t}{\tau})$ denotes the nth-order modified Bessel function and $\delta(t)$ represents the Dirac function. Similar kernel functions also appear in the study of stress relaxation effects in the hemodynamics of small arteries by Peng et al. [25]. Applying a step strain $\varepsilon(t)=\varepsilon_0H(t)$ to the fractal structure yields the stress expression as follows:

$$\begin{aligned}\sigma(t)&=T\varepsilon(t)\\&=-\left[\frac{Ee^{-\frac{2t}{\tau}}}{t}I_1(\frac{2t}{\tau})+E\delta(t)\right]*[\varepsilon_0H(t)]\\&=E\varepsilon_0\mathrm{e}^{-\frac{2t}{\tau}}\left[I_0(\frac{2t}{\tau})+I_1(\frac{2t}{\tau})\right],\end{aligned} \tag{A12}$$

where $H(t)$ denotes the Heaviside step function.

The expression for the compliance operator of the fractal ladder structure is

$$\frac{1}{T}=\frac{2}{\eta}\frac{1}{\sqrt{p^2+\frac{4p}{\tau}}-p}=\frac{\tau}{2\eta p}\left(\sqrt{p^2+\frac{4p}{\tau}}+p\right)=\frac{1}{2Ep}\left(\sqrt{p^2+\frac{4p}{\tau}}+p\right). \tag{A13}$$

Equation (A13) can be further simplified such that

$$\frac{1}{T}=\frac{1}{\eta E}\frac{1}{p}\frac{\eta}{2}\left(\sqrt{p^2+\frac{4p}{\tau}}-p+2p\right)=\frac{1}{\eta E}\frac{1}{p}(T+\eta p)=\frac{1}{\eta E}\frac{1}{p}T+\frac{1}{E}. \tag{A14}$$

The kernel function corresponding to the compliance operator is

$$\frac{1}{T}=\frac{1}{E}\delta(t)+\frac{1}{\eta}\mathrm{e}^{-\frac{2t}{\tau}}\left[I_0(\frac{2t}{\tau})+I_1(\frac{2t}{\tau})\right]. \tag{A15}$$

Applying step stress $\sigma(t)=\sigma_0H(t)$ to the fractal ladder results in the creep response as follows:

$$\begin{aligned}\varepsilon(t)&=\frac{1}{T}\sigma(t)\\&=\left\{\frac{1}{E}\delta(t)+\frac{1}{\eta}\mathrm{e}^{-\frac{2t}{\tau}}\left[I_0(\frac{2t}{\tau})+I_1(\frac{2t}{\tau})\right]\right\}*[\sigma_0H(t)]\\&=\frac{\sigma_0}{4\eta}\mathrm{e}^{-\frac{2t}{\tau}}\left[-\mathrm{e}^{\frac{2t}{\tau}}\frac{\eta}{E}+\left(6t+\frac{\eta}{E}\right)I_0(\frac{2t}{\tau})+6tI_1(\frac{2t}{\tau})\right]+\frac{1}{E}\sigma_0\\&=\frac{\sigma_0}{4\eta}\mathrm{e}^{-\frac{2t}{\tau}}\left[\left(6t+\frac{\eta}{E}\right)I_0(\frac{2t}{\tau})+6tI_1(\frac{2t}{\tau})\right]+\frac{3\sigma_0}{4E}.\end{aligned} \tag{A16}$$

References

1. Knauss, W.G.; Igor, E.; Lu, H. *Mechanics of Polymers: Viscoelasticity;* Springer: Boston, MA, USA, 2008. [CrossRef]
2. Gargallo Ligia, R.D. *Physicochemical Behavior and Supramolecular Organization of Polymers;* Springer: Dordrecht , The Netherlands, 2009; pp. 43–162. [CrossRef]
3. Yang, L.; Yang, L.; Lowe, R.L. A viscoelasticity model for polymers: Time, temperature, and hydrostatic pressure dependent Young's modulus and Poisson's ratio across transition temperatures and pressures. *Mech. Mater.* **2021**, *157*, 103839. [CrossRef]
4. Leblanc, J.L. A Multiparametric Approach of the Nonlinear Viscoelasticity of Rubber Materials. In *Non-Linear Viscoelasticity of Rubber Composites and Nanocomposites: Influence of Filler Geometry and Size in Different Length Scales;* Ponnamma, D., Thomas, S., Eds.; Springer International Publishing: Cham, Switzerland, 2014; pp. 273–300. [CrossRef]
5. Oyen, M.L. Mechanical characterisation of hydrogel materials. *Int. Mater. Rev.* **2014**, *59*, 44–59. [CrossRef]
6. Xu, Q.; Engquist, B.; Solaimanian, M.; Yan, K. A new nonlinear viscoelastic model and mathematical solution of solids for improving prediction accuracy. *Sci. Rep* **2022**, *10*, 2202. [CrossRef]
7. Zhao, W.; Li, N.; Liu, L.; Leng, J.; Liu, Y. Mechanical behaviors and applications of shape memory polymer and its composites. *Appl. Phys. Rev.* **2023**, *10*, 011306. [CrossRef]
8. Christensen, R. *Theory of Viscoelasticity*, 2nd ed.; Academic Press: Cambridge, MA, USA, 1982; pp. 1–364. [CrossRef]
9. Maxwell, J.C. On the dynamical theory of gases. *Phil. Trans. R. Soc.* **1867**, *157*, 49–88. [CrossRef]
10. Meyer, O.E. Theorie der elastischen Nachwirkung. *Ann Phys.* **1874**, *227*, 108–119. [CrossRef]
11. Boltzmann, L. Zur Theorie der elastischen Nachwirkung. *Ann Phys.* **1878**, *241*, 430–432. [CrossRef]
12. Volterra, V. *Theory of Functionals and of Integral and Integro-Differential Equations;* Dover: New York, NY, USA, 1959.
13. Hu, K.X.; Zhu, K.Q. Mechanical analogies of fractional elements. *Chin. Phys. Lett.* **2009**, *26*, 108301. [CrossRef]
14. Schapery, R.A. Nonlinear viscoelastic solids. *Int. J. Solids Struct.* **2000**, *37*, 359–366. [CrossRef]
15. Bauwens, J.C. Two nearly equivalent aproaches for describing the non-linear creep behavior of glassy polymers. *Colloid Polym. Sci.* **1992**, *270*, 537–542. [CrossRef]
16. Schiessel, H.; Blumen, A. Mesoscopic Pictures of the Sol-Gel Transition Ladder Models and Fractal Networks. *Macromolecules* **1995**, *28*, 4013–4019. [CrossRef]
17. Schiessel, H.; Metzler, R.; Blumen, A.; Nonnenmacher, T.F. Generalized viscoelastic models Their fractional equations with solutions. *J. Phys. A Math. Gen* **1995**, *28*, 6567–6584. [CrossRef]
18. Heymans, N.; Bauwens, J.C. Fractal rheological models and fractional differential equations for viscoelastic behavior. *Rheol. Acta* **1994**, *33*, 210–219. [CrossRef]
19. Heymans, N. Hierarchical models for viscoelasticity: Dynamic behaviour in the linear range. *Rheol. Acta* **1996**, *35*, 508–519. [CrossRef]
20. Deseri, L.; Paola, M.D.; Zingales, M.; Pollaci, P. Power-law hereditariness of hierarchical fractal bones. *Int. J. Numer. Meth. Biomed. Eng.* **2013**, *29*, 1338–1360. [CrossRef]
21. Yin, Y.J.; Peng, G.; Yu, X.B. Algebraic equations and non-integer orders of fractal operators abstracted from biomechanics. *Acta Mech. Sin.* **2022**, *38*, 521488. [CrossRef]
22. Guo, J.Q.; Yin, Y.J.; Ren, G.X. Abstraction and operator characterization of fractal ladder viscoelastic hyper-cell for ligaments and tendons. *Appl. Math. Mech.-Engl. Ed.* **2019**, *40*, 1429–1448. [CrossRef]
23. Guo, J.Q.; Yin, Y.J.; Hu, X.L.; Ren, G.X. Self-similar network model for fractional-order neuronal spiking: Implications of dendritic spine functions. *Nonlinear Dyn.* **2020**, *100*, 921–935. [CrossRef]
24. Yin, Y.J.; Guo, J.Q.; Peng, G.; Yu, X.B.; Kong, Y.Y. Fractal Operators and Fractional Dynamics with 1/2 Order in Biological Systems. *Fractal Fract.* **2022**, *6*, 378. [CrossRef]
25. Peng, G.; Guo, J.Q.; Yin, Y.J. Self-similar functional circuit models of arteries and deterministic fractal operators: Theoretical revelation for biomimetic materials. *Int. J. Mol. Sci.* **2021**, *22*, 12897. [CrossRef]
26. Jian, Z.M.; Peng, G.; Li, D.A.; Yu, X.; Yin, Y. Correlation between Convolution Kernel Function and Error Function of Bone Fractal Operators. *Fractal Fract.* **2023**, *7*, 707. [CrossRef]
27. Yu, X.B.; Yin, Y.J. Operator Kernel Functions in Operational Calculus and Applications in Fractals with Fractional Operators. *Fractal Fract.* **2023**, *7*, 755. [CrossRef]
28. Anderson, W.; Duffin, R. Series and parallel addition of matrices. *J. Math. Anal. Appl.* **1969**, *26*, 576–594. [CrossRef]
29. Uchiyama, M. Operator means and matrix quadratic equations. *Linear Algebra Appl.* **2021**, *609*, 163–175. [CrossRef]
30. Mikusiński, J. *Operational Calculus;* Pergamon Press: Oxford, UK, 1959.
31. Paola, M.D.; Zingales, M. Exact mechanical models of fractional hereditary materials. *J. Rheol.* **2012**, *56*, 983–1004. [CrossRef]

fractal and fractional

Article

A Fractal-Based Quantitative Method for the Study of Fracture Evolution of Coal under Different Confining Pressures

Ancheng Wang and Lei Wang *

State Key Laboratory of Mining Response and Disaster Prevention and Control in Deep Coal Mines, Anhui University of Science and Technology, Huainan 232001, China; 2021200245@aust.edu.cn
* Correspondence: 2006020@aust.edu.cn; Tel.: +86-17718267822

Abstract: To study the dynamic crack evolution process of loaded coal from the perspective of fractals, we carried out in situ industrial CT scanning tests of loaded coal under different confining pressures, visualizing loaded coal fracturing. Combined with fractal theory, the temporal and spatial evolution law of coal cracks is described quantitatively. The results provide two findings: (1) from the perspective of two-dimensional images and three-dimensional space, the evolution characteristics of cracks in coal under different confining pressures were basically the same in each loading stage. During the loading stages, the cracks exhibited a change rule of a slow reduction, initiation/development, rapid increase, expansion, and penetration. (2) The fractal dimension of coal was calculated by introducing fractal theory, and its change law was in good agreement with the dynamic changes of the cracks, which can explain the influence of the confining pressure on the loaded coal. The fractal dimension showed three stages: a slight decrease, a stable increase, and then a significant increase. The larger the confining pressure, the more obvious the limiting effect. Thus, our approach provides a more accurate method for evaluating the spatial and temporal evolution of cracks in loaded coal. This study can be used to predict the instability failure of loaded coal samples.

Keywords: fractal theory; fractal dimension; CT scanning; image analysis; fissure evolution; confining pressure

1. Introduction

Coal is a heterogeneous natural porous medium and geological material, which is mainly composed of a coal matrix, a large number of randomly distributed natural fissures and defects, and other minerals. The random distribution of these components determines the structural characteristics of the coal. The analysis of the dynamic evolution process of internal crack initiation, development, and expansion to the penetration of coal under external triaxial loads can help us better understand the failure process of coal fracturing [1–3], which is of great practical significance for the safe production of coal mines.

In the actual excavation process of projects, the confining pressure is constantly changing [4]. By setting different confining pressure levels in the analysis, rock stress conditions at different occurrence depths can be simulated. In recent years, scholars have conducted significant research on the mechanical properties of coal rock mass under different confining pressure conditions. For example, Jiang has shown, through triaxial compression tests, that the confining pressure and buried depth have a significant effect on the physical and mechanical properties and failure modes of shale [5]. Alejano conducted a triaxial compression test on granite and studied the influence of confining pressure on the peak strength and deformation modulus of massive granite [6]. The above research focused on the macroscopic mechanical properties of coal rock mass. Furthermore, the macroscopic fracturing of rock is closely related to the spatial and temporal evolution of its microstructure.

Rock fracture is a process of crack initiation, propagation, and evolution until penetration. Studying the crack propagation process and instability failure mechanism of cracked coal rock mass is helpful for understanding the stress variation law of the surrounding rock in coal roadways and working faces. Scanning electron microscopy, nuclear magnetic resonance, and computed tomography (CT) scanning are the primary methods used to describe the microscopic pore structure of coal rock mass [7–12]. CT scanning technology (also known as computer tomography recognition technology) can enable the non-destructive scanning of a test sample, and allow for the obtaining of a CT image sequence of the sample for display in the form of high-resolution digital three-dimensional images after three-dimensional reconstruction. With the steady advancement of industrial CT, its application in coal rock mass is becoming more extensive. Using CT technology to study the fracture structure of coal rock mass failure has unique advantages [13–15] and can facilitate the visualization of coal rock mass during loading. For example, Wang used CT data combined with three-dimensional visualization software for the quantitative characterization of the pore and fracture structure of coal and established a three-dimensional model as well as a simplified model of the pore network and its topological structure, effectively describing the pore size, pore volume, porosity, and other properties of coal [16]. Kumari carried out CT scanning tests on granite during loading and studied the propagation path of granite cracks; their results show that the crack propagation path is mainly controlled by the stress state and the heterogeneity of the rock matrix [17].

The evolution process of cracks in rock masses under loading is dynamic and involves the initiation, propagation, evolution, and penetration of cracks. Using CT scanning technology, we can observe the spatial distribution of cracks in rock masses during loading but cannot quantitatively study the rock mass fracture structure at different stages of loading. There is a lack of digital indicators reflecting the differences in the rock mass structure. The emergence of fractal theory helps solve this problem.

Fractal theory is often used to study the laws behind various complex, disorderly, and chaotic phenomena in nature, and fractal dimensions are the most important concept within fractal theory. Because the fracture distribution in coal rock mass shows good self-similar fractal characteristics, fractal dimensions can be used to quantitatively describe the degree of fracture evolution, which could facilitate the application of fractal theory in the field of rock mechanics. By extracting the fractal information from the scanning results, we can describe the development and distribution of fractures in the loaded rock mass effectively, evaluate the complexity of the overall fracture structure inside the rock mass accurately, strengthen our understanding of the evolution of rock mass pores and fracture structures, and visualize and finely and quantitatively characterize the loaded coal rock mass.

Xie [18], Zhao [19], and others have studied the fractal characteristics of cracks in coal rock mass. Zhang used industrial CT to observe and scan the rock fracture process in different stages and constructed a three-dimensional fracture model of rock with which to study the dynamic propagation and evolution process of internal cracks when the loaded rock is deformed and destroyed. The author found that the fracture propagation process can be effectively quantified using characteristic parameters, such as fracture volume, surface area, and fractal dimension [20]. To further verify the reliability of the fractal dimension in three dimensions, Wang used CT to scan six coal samples with different pore structures and reconstruct them three-dimensionally. The fractal dimension D_f of the total pore structure was calculated using the three-dimensional box-counting method. The results show that the three-dimensional fractal dimension can accurately describe the fractal characteristics of coal. The larger the D_f, the greater the porosity of the coal samples [21]. However, previous CT scanning experiments have not used real-time scanning during the loading process. Therefore, to study the temporal and spatial evolution laws of internal cracks in coal, one must capture the dynamic damage and failure process of loaded coal samples in real time. More importantly, a method for accurately describing the evolution of internal cracks in coal is required.

This study assesses the whole process of the in situ failure of coal samples under four confining pressure states using an industrial CT scanning system. We analyze the evolution law of internal cracks within coal samples from loading to failure and propose a new method with which to describe the quantitative crack evolution of coal samples during the loading process based on fractal theory. As a result, the fractal characteristics of cracks in loaded coal samples are detailed.

2. Test Scheme and System

2.1. Sample Introduction

The coal samples used in the test were core drilling samples taken from the same area (Huainan mining area), having the same internal structure, a uniform texture, no obvious joints, cracks, etc., and very similar mechanical properties. After cutting and coring, the coal samples were processed into a standard cylinder of Ø25 × 50 mm. To achieve evenly stressed coal samples during the test and high accuracy of the test results, the upper and lower surfaces of each sample were polished to obtain flatness within 0.05 mm. The non-parallelism of the two ends was less than 0.02 mm, thus meeting the requirements of the International Society of Rock Mechanics test [22]. In addition, basic physical indicators, such as quality, were measured, and the samples with basic physical indicators exceeding the average value of 10% were removed. The processed coal samples are shown in Figure 1.

Figure 1. Coal samples.

2.2. Testing Equipment

This research used the V | tome | x L300 high-precision CT triaxial loading test system to carry out triaxial and scanning testing, as shown in Figure 2a. This comprised an industrial CT scanning system and a triaxial loading system that could facilitate the uniaxial and triaxial loading of the sample, with a maximum axial pressure of 100 kN and a maximum confining pressure of 30 MPa.

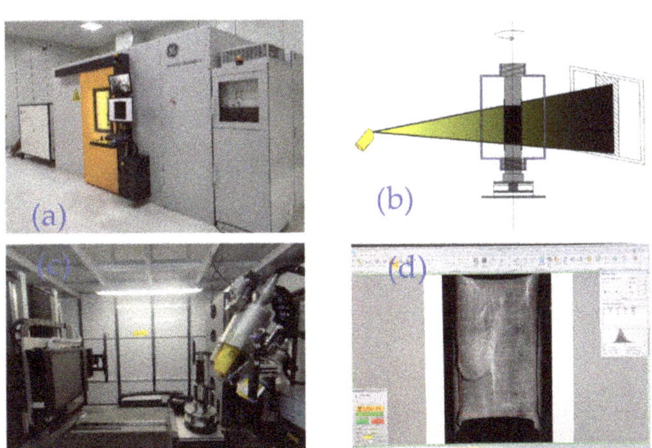

Figure 2. High-precision CT three-axis loading experimental system. (**a**) Industrial CT system, (**b**) CT scanning principle, (**c**) internal structure of the CT system, and (**d**) Phoenix Datos | x2 interface.

The industrial CT scanning system can display the internal structure and defect status of the measured object clearly, accurately, and intuitively in the form of two- and three-dimensional images without damage to the object. The scanning principle involves use of the high penetration capacity of X-rays to irradiate the different layers of the sample. Because some photons are absorbed by the sample, the light intensity is attenuated, and this information is received by the detector. After the data are converted by the computer, a digital image of the coal sample can be reconstructed. The micro-focus CT scanning system is equipped with two X-ray tubes—a high-power micron-scale ray tube and a high-resolution nano-scale ray tube. Its unique micro-focus ray source scanning system can effectively reduce ghosting and achieve a good display. The specific parameters are listed in Table 1.

Table 1. CT system parameters.

Ray Tube Model	System Parameter	Maximum Power	Maximum Tube Voltage	Detail Resolution
Micron-sized ray tube	Value	500 W	300 kV	≤2 μm
Nano-scale ray tube		15 W	180 kV	≤0.5 μm

The principle of industrial CT scanning is shown in Figure 2b. Figure 2c shows the internal structure of the industrial CT scanning system. Here, the sample was installed and fixed on the base holder for X-ray detection and scanning. The CT scan data to be reconstructed were opened using the CT data setting software Phoenix Datos | x2 (version 2.6.1-RTM) (Figure 2d). Firstly, we determined the region of interest for digital reconstruction. Next, to achieve the best compensation effect and derive the clearest three-dimensional digital image of the coal core, we performed the following series of processes on the scanned image: geometric correction, beam hardening correction, and reverse color processing.

2.3. Test Scheme

The triaxial compression testing of coal samples under different confining pressures was carried out. According to the natural state of coal as found at the considered depths in the mining area, as well as data from previous research [23,24], the confining pressures used in the triaxial compression test were 5.0, 10.0, 15.0, and 20.0 MPa. The loaded coal sample was wrapped with a special leather sleeve and placed in the loading device. An initial axial load of 2500 N was applied to fix the coal sample. Then, different confining pressures were applied (5, 10, 15, 20 MPa). After the confining pressure had stabilized, the axial pressure was controlled by displacement, and the loading rate was 0.3 mm/min. During the test, a constant confining pressure was maintained, and the axial pressure was gradually increased until the coal sample was destroyed.

We determined the scanning point in the CT scanning test stage based on the full stress–strain curve of the whole process of deformation of the raw coal under compressive loading. The deformation process is divided into the in situ stage I, initial pore compaction stage II, stable development of elastic deformation fracture III, unstable fracture development stage IV, post-fracture stage V, and residual stage VI. Internal failure of the coal sample was assessed via scanning at six points. The experimental conditions were as follows: a voltage of 170 kV, current of 180 μA, resolution of 25 μm, exposure time of 1000 ms, projection number of 1500, and scanning time of 94 min. In addition, the same scanning conditions were maintained for six repetitions (the first scan to the sixth scan corresponded to A–F). The specific test steps are as follows:

(1) CT system initialization. The system is heated, and the ray tube is vacuumized. Then, the mechanical shaft is reset. The specimen is fixed to the specimen holder with an

electronic jack, and then the holder is placed in the axial pressure loading cylinder. The assembly is installed and fixed to the CT mechanical turntable, and the pipeline is connected for exhaust extraction.

(2) The CT scanning and triaxial loading parameters are set, with those such as voltage, current, exposure time, and the number of pictures set according to the imaging effect, resolution, and gray value. The triaxial loading equipment is set up with the loading parameters required for this experiment. Multiple scans can be performed before the coal sample breaks. During the scanning process, the pressure remains unchanged, and the number of scans is associated with the stress–strain curve for adjustment.

(3) At the end of the experiment, the data are processed. Firstly, the mechanical parameters are exported from the three-axis software database for analysis. Secondly, the original image is corrected using the CT data reconstruction software Phoenix Datos | x2 (version 2.6.1-RTM). Finally, the image is reconstructed in three dimensions using the image processing software.

(4) The above steps are repeated, carrying out triaxial tests with confining pressures of 10, 15, and 20 MPa.

3. Analysis of Fracture Evolution Law

3.1. Qualitative Analysis of Two-Dimensional Fracture Evolution Characteristics

The CT scanning system can collect real-time scanning information of each section layer of the coal sample, from top to bottom. In determining the fracture propagation characteristics of coal under different confining pressures, we applied CT scanning technology to reveal the spatial fracture morphological characteristics of the coal samples under different loading conditions and the influence of intermediate principal stress on their failure. Figure 3 shows the stress–strain curve of the raw coal under different confining pressures and the two-dimensional CT images of different deformation stages, in which the gray represents the coal matrix and white and black indicate the presence of minerals and pores/fissures, respectively.

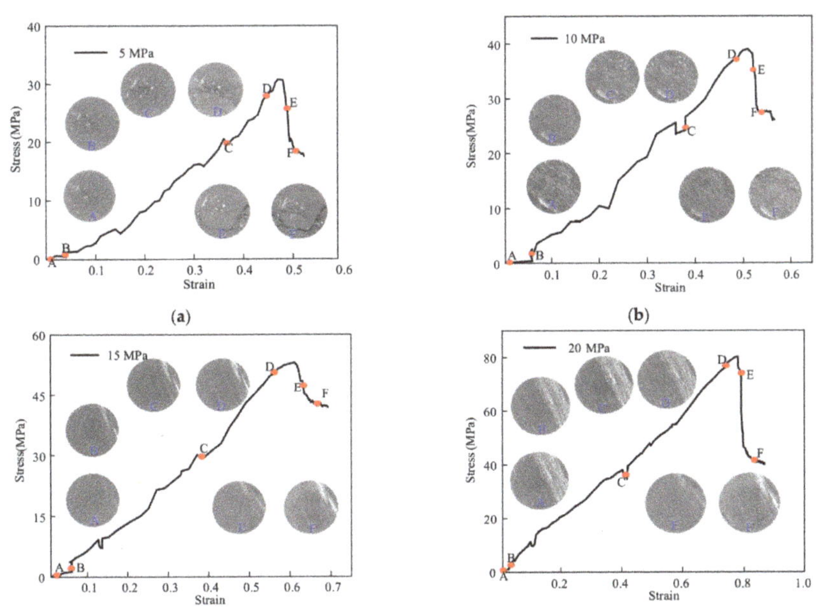

Figure 3. Two-dimensional CT images of stress–strain curves at different deformation stages under different confining pressures. (**a**) 5 MPa; (**b**) 10 MPa; (**c**) 15 MPa; (**d**) 20 MPa.

In the triaxial compression experiment, under different confining pressures, the internal crack trends in the coal sample are roughly the same—slowly decreasing at first and then gradually increasing. In the initial state, the fracture distribution of the sample under different confining pressures is anisotropic. As the confining pressure is applied, some of the original fractures are closed; furthermore, when a greater confining pressure is set, the closure effect is more obvious. When the axial stress begins to increase, internal cracks develop in the coal sample. As the axial stress continues to increase, the main crack propagation process is accompanied by the initiation of new cracks. The cracks are gradually connected with those surrounding them, and coal ruptures gradually develop from a state of diffuse and disorderly damage to local and orderly damage. Under the confining pressure of 5 MPa, the final damage degree of the sample is the greatest. The main crack is an oblique bending crack that runs through the sample. This is located in the lower part of the slice section, oriented from left to right. In summary, the greater the confining pressure, the greater the limitation of fracture development.

3.2. *Quantitative Analysis of Two-Dimensional Fracture Evolution Characteristics*

The term "fractal" was first proposed and used by Mandelbrot in the 1970s. The fractal dimension is the most important concept in fractal geometry. "Fractal" is a mathematical term, and its theory has fractal characteristics as its main research focus. It uses the fractal dimension to describe irregular and regular patterns in nature and reveals the laws of self-similarity, which nature follows at deep levels. Many scholars have developed different formulae for calculating the fractal dimension according to the basic theory [25,26]. Among these, the most widely used is the box-counting dimension, also known as the box dimension, which is a calculation method used for measuring the fractal dimension over distance. The principle of the box-counting dimension is to imagine the fractal S on a uniformly divided grid and calculate the minimum number of squares required to cover the fractal S completely. By continuously refining the grid size, the number of squares required at different scales can be reviewed. The fractal dimension of coal samples is assessed via the box-counting dimension. The approximate calculation of the box-counting dimension D_b is

$$D_b = -\lim_{s \to 0} \frac{\lg N(s)}{\lg(s)} = \frac{\lg N(s)}{\lg(1/s)},$$

where, in R^2 space, $N(s)$ is the minimum number of squares of length s required to cover this non-empty subset. The slope of the linear regression equation formed by plotting $\lg N(s)$ with $\lg(1/s)$ is the fractal dimension, which obeys the following linear equation:

$$\lg N(s) = \lg(c_1) + D_b \lg(1/s)$$

where, c_1 is a constant.

Figure 4 shows a two-dimensional fracture network diagram, which can be used in a box-counting method for determining the fractal dimension. To obtain the fractal dimension of the fracture distribution inside the coal sample, we binarized the CT scan image. After binarizing using the ImageJ (version 1.53c) software program—because of the instability of the CT detector when collecting photons, the negative working conditions of the transmission cable, or the extreme aging of the bulb tube—concentric ring artifacts that do not belong to the coal itself might arise in some scan images. Replacing the adjacent gray values and removing the ring artifacts from the image better visualizes the fracture structure that developed during the loading process of the selected coal sample section. The processed CT image strongly retains the fracture distribution characteristics inside the coal sample. After binarization and artifact removal, only black and white remain on the CT image, where black represents the fracture and all other features are white.

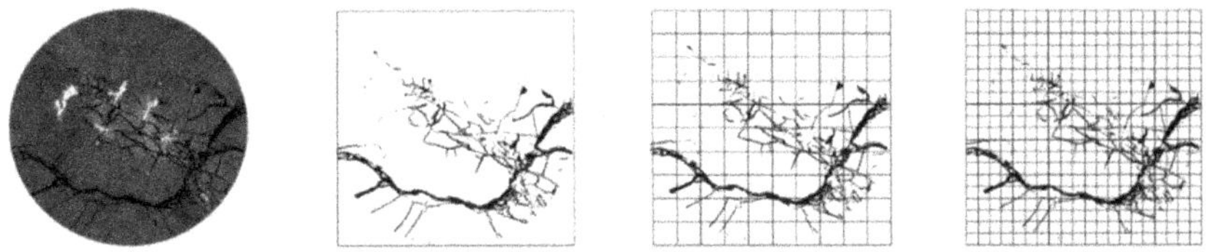

a b c d

Figure 4. Schematic diagram of the two-dimensional fracture network and the box-counting method used for assessing the fractal dimension. (**a**) Coal sample original slice, (**b**–**d**) Refinement process of grids.

ImageJ (version 1.53c) software was used to calculate the data of two-dimensional CT cross-section images taken for different stages of the coal sample. According to their specific sizes (2 × 2, 4 × 4, 8 × 8, 16 × 16, 32 × 32, and 64 × 64 mm), the processed slice images were divided into several small grids, that is, the grid was semi-divided and refined into a denser square grid. The number of grids that the cracks pass through at different scales was counted, and the box-counting dimension D_b of the internal cracks of the coal sample were obtained by fitting the linear relationship between the double logarithmic coordinates.

In the coal, the fractal dimension of the two-dimensional section is between 1 and 2. The closer the fractal dimension is to 1, the fewer pores and fissures on the coal slice surface, the smaller the area, and the simpler the distribution of pores and fissures in the section. The closer the fractal dimension is to 2, the greater the number of pores and fissures on the coal slice surface, the larger the area, and the more complex the distribution of pores and fissures on the section. The calculation results regarding the fractal dimension of the coal sample section cracks that develop during the loading process are shown in Table 2. With the increase in fracture development, the fractal dimension becomes larger. During this time, in the case of the same fracture area, the fractal dimension of the fracture structure differs. The more complex the fracture structure, the larger the fractal dimension. Therefore, the fractal dimension can better describe the characteristics of the fracture structures of coal samples [27]. Moreover, the fractal dimension can be used to effectively characterize the complexity of different fracture structures in surrounding fracture areas. In two dimensions, the larger the fractal dimension, the more sufficient the fracture development and the more complex the distribution, and vice versa.

Table 2. Calculation results of two-dimensional fractal dimension of coal sample.

Confining Pressure (MPa)		**5**	**10**	**15**	**20**
Fractal dimension	First scan	1.3631	1.2817	1.1865	1.1148
	Second scan	1.3188	1.2113	1.0475	1.0217
	Third scan	1.3531	1.2217	1.0868	1.0275
	Fourth scan	1.4104	1.2709	1.1391	1.0807
	Fifth scan	1.5237	1.3728	1.2468	1.1541
	Sixth scan	1.5869	1.4842	1.3491	1.3258

The results show that the overall development processes of cracks in the coal samples under different confining pressures were roughly the same, with obvious stage characteristics, as shown in Figure 5. Taking the analysis of all layers subjected to 5 MPa confining pressure as an example, this process can be divided into four stages:

Stage I—Initially, several primary cracks of different sizes and lengths are present in the coal samples (as shown in Figure 5a). At this time, the two-dimensional fractal dimension is 1.3631. With the increase in confining pressure, some primary cracks gradually close (as shown in Figure 5b), and the cracks decrease. At this time, the fractal dimension decreases to 1.3188.

Stage II—The primary cracks gradually close until the confining pressure reaches its specified value, and then, with the increase in axial pressure, the coal sample begins to fail. The primary crack shows a trend of expanding with the increase in axial pressure (as shown in Figure 5c,d). The fractal dimension increases to 1.3531, as shown in Figure 5c, and the crack develops slightly further. With the increase in axial pressure, obvious penetrating cracks develop, as shown in Figure 5d. At this time, the fractal dimension begins to increase from 1.4104 because of the gradual expansion of cracks.

Stage III—Under the external force, the cracks continue to evolve and extend and intersect with each other to form larger penetrating cracks. New cracks appear at the edge of the coal sample. At the same time, because of the distribution of pores and microcracks, the direction of crack propagation deviates temporarily. The main crack gradually expands along the edge and at the center of the coal sample. The coal sample deforms macroscopically, and its internal cracks begin to constitute a complex network. As the axial pressure increases, the crack expands and evolves (as shown in Figure 5e). At this time, the fractal dimension becomes larger because of the crack penetration, reaching 1.5237.

Stage IV—At this time, the post-peak residual stress stage commences, and the loaded coal sample undergoes complete deformation and failure. The internal macro-cracks converge and eventually form a complex crack network (Figure 5f). At this time, the fractal dimension maximizes at 1.5869.

Throughout the fracture process, the two-dimensional fractal dimension of the coal sample shows an overall trend of a small decrease to a large surge, which effectively reflects the overall fracture expansion process of the coal sample under confining pressure.

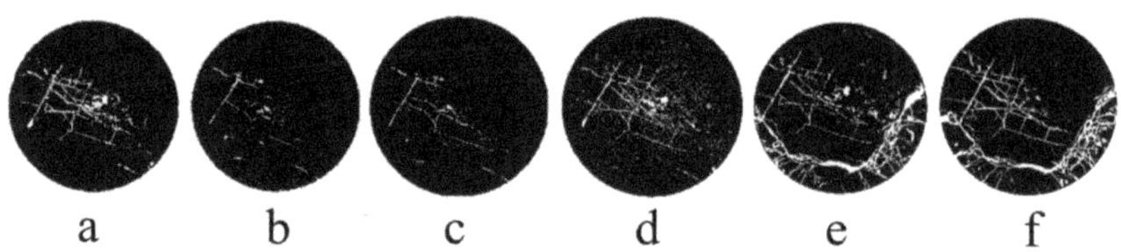

Figure 5. Coal sample slice binary graphics at 5 MPa from (**a**,**b**) Stage I, (**c**,**d**) Stage II, (**e**) Stage III, and (**f**) IV.

The internal fracture structures of each coal sample in the in situ state are different. In the initial compaction stage, cracks are almost invisible in the binary images under higher confining pressures, and the cracks become almost completely closed under high confining pressures. With the increase in axial pressure, isolated pores and micro-cracks are gradually generated inside the coal, and the spatial distribution of cracks and pores intensifies the inhomogeneity of stress distribution inside the coal rock mass, which, in turn, significantly influences the crack development in the coal rock until penetration and complete deformation and failure. The fractal dimension is calculated from the compaction stage to the final failure. Under the confining pressure of 5 MPa (Figure 5), the sixth scan's fractal dimension is 1.20 times that of the compaction stage, and the sixth scanning fractal dimension under the confining pressure of 10 MPa (Figure 6a) is 1.22 times that of the second scan's fractal dimension. The sixth scan's fractal dimension under a confining pressure of 15 MPa (Figure 6b) is 1.28 times that measured in the second scan, and the sixth scan's fractal dimension under a confining pressure of 20 MPa (Figure 6c) is 1.29 times that derived in the second scan. The increase in confining pressure is beneficial to the improvement in the ability of the sample to resist external loads. The internal crack structure network in coal under the application of low confining pressure is more complex; that is, the damage is more serious.

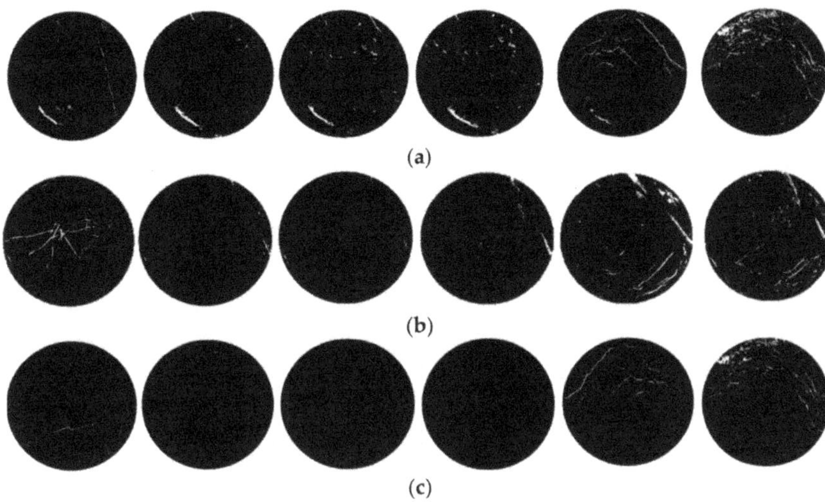

Figure 6. Binarized graph of coal sample slices at confining pressures of (**a**) 10, (**b**) 15, and (**c**) 20 MPa.

3.3. Three-Dimensional Fracture Evolution Characteristics

The CT image of a coal sample reflects the two-dimensional development of cracks to a certain extent, but only local information regarding the cracks inside the coal. To identify the complete distribution of cracks in space more clearly, study the crack evolution process in the coal sample, and characterize the three-dimensional morphology of cracks, the original two-dimensional slice must be reconstructed in three dimensions, and the cracks need to be extracted and analyzed using image analysis and processing software.

Figure 7 shows the results regarding the evolution and distribution of three-dimensional cracks in the coal samples from the first to the sixth scan under different confining pressures, where the blue area represents the cracks. The image obtained via scanning was processed, including selecting the region of interest (ROI) and filtering out the noise [28]. Using the professional CT image software VG Studio Max (version 3.3.0.165821), the noise level in each original CT image could be reduced, along with its adverse effects on image quality and the subsequent quantitative analysis. In addition, the original CT image was imported into the image processing software for rapid data acquisition, three-dimensional model reconstruction, and data processing analysis.

As shown in Figure 7, the structural diagram of coal samples obtained via three-dimensional reconstruction reflects the development of the fracture network clearly and effectively. Under triaxial compression, the coal shows obvious characteristics in five stages—fracture closure, the initial formation of fracture, the stable growth of fracture and local penetration, the accelerated growth of fracture, and comprehensive penetration and failure—which are consistent with the findings of the two-dimensional analysis of the fracture.

Before reaching peak stress, energy gradually accumulates inside the coal. When the stress concentration reaches the critical value for crack initiation, the crack expands rapidly, the fracture scale and the energy released by the fracture increase, and the fracture volume increases sharply. After the peak stress is reached, brittle failure quickly occurs as a result of the overall instability of the coal, and the volume of the stress fracture shows no obvious change. The stage characteristics of each coal sample under different confining pressures are basically the same. From the in situ state to the compaction stage, the fracture volume of the coal sample is obviously reduced. From the initial formation of the fracture to the complete destruction of the sample, the fracture volume of the coal sample increases gradually but significantly, and the growth trend changes from slow to rapid. A three-dimensional assessment thus more effectively characterizes the evolution of the overall fracture network in coal samples under different confining pressures.

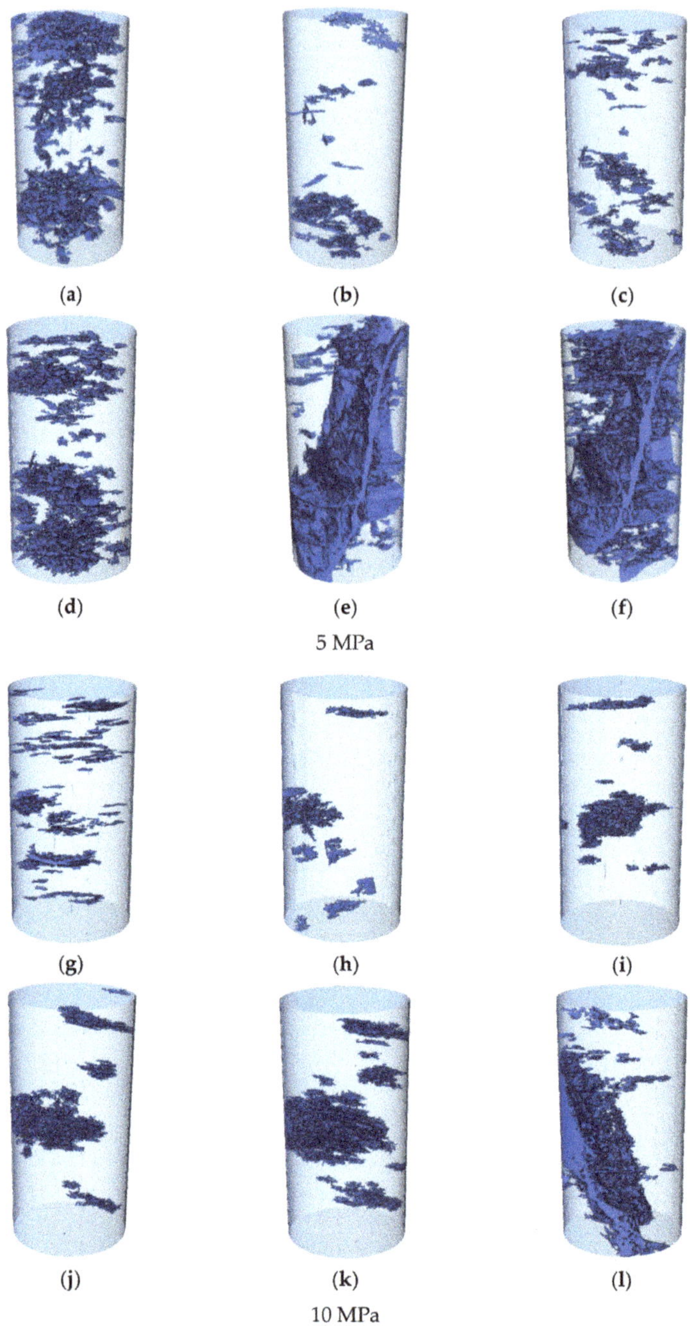

Figure 7. *Cont.*

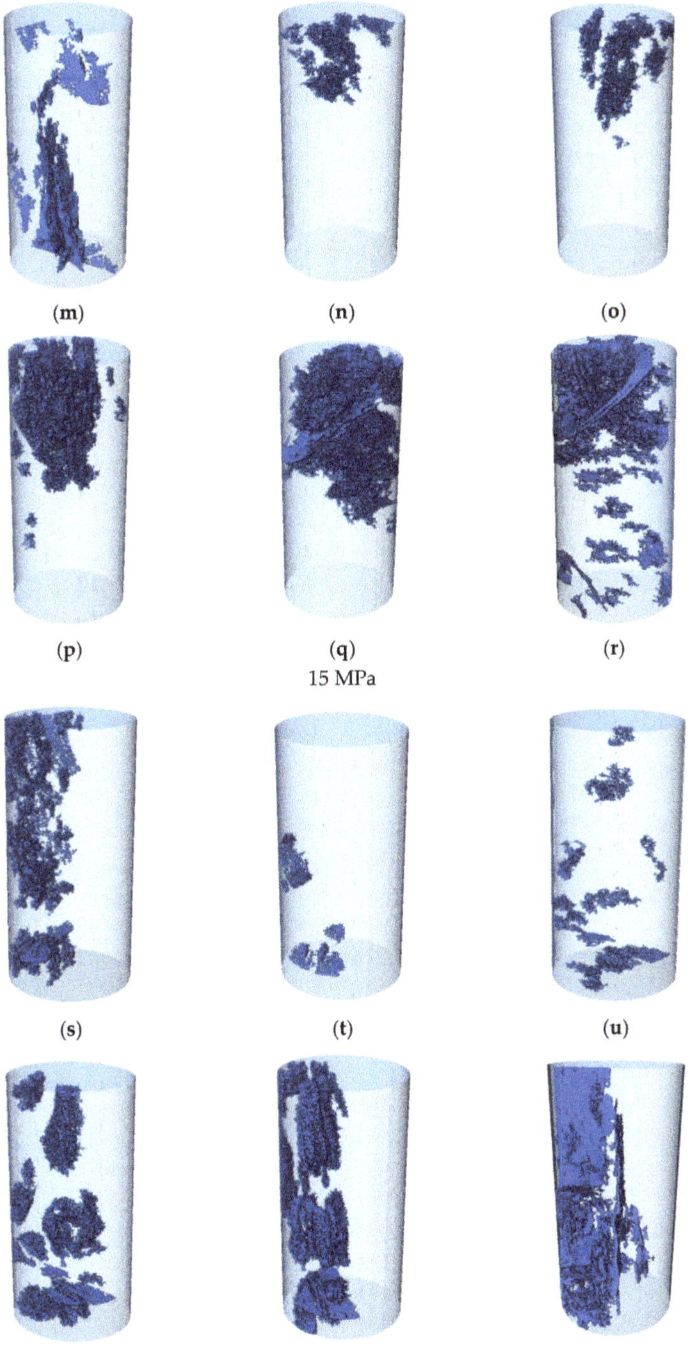

(**m**) (**n**) (**o**)

(**p**) (**q**) (**r**)

15 MPa

(**s**) (**t**) (**u**)

(**v**) (**w**) (**x**)

20 MPa

Figure 7. Fracture morphology and distribution during coal sample loading for confining pressures of (**a–f**) 5 MPa, (**g–l**) 10 MPa, (**m–r**) 15 MPa, and (**s–x**) 20 MPa.

3.4. Quantitative Analysis of Three-Dimensional Fracture Evolution Characteristics

Under loading, with differences in stress, the internal pore cracks within coal samples undergo a process of continuous change. The relationship between the internal crack volume in coal and the stress offers the most intuitive reflection of the dynamic evolution process of the cracks in loaded coal. The in-depth quantitative study of the general digital indicators extracted from the three-dimensional crack images can effectively strengthen our intuitive understanding of the evolution of pore crack structures in coal and rock. To further characterize the law of crack propagation in coal, the parameters of crack volume were calculated and extracted, as shown in Figure 8.

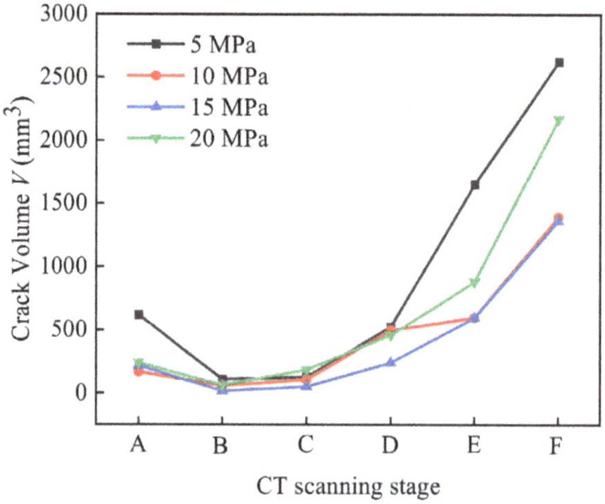

Figure 8. Relationship between fracture volume and confining pressure.

The research shows that the fractal dimension method of calculating the three-dimensional fracture structure in a material is similar to that of the two-dimensional fracture structure. Each three-dimensional image is covered with a series of grids of decreasing size. The three-dimensional fracture network is covered with a cube, which is constructed to cover the point set. By changing the side length ε of the small cube box, several small cube boxes can be formed, and the corresponding box number $N(\varepsilon)$ containing the point set can be calculated [29]. After multiple scale changes, a series of data for ε and $N(\varepsilon)$ can be obtained. The least-squares method is used to fit the scatter plot of the relationship between $\lg(1/\varepsilon)$ and $\lg N(\varepsilon)$, and the slope of the fitted line gives the three-dimensional fractal dimension. The box-counting dimension D is defined as

$$D = -\lim_{\varepsilon \to 0} \frac{\lg N(\varepsilon)}{\lg(\varepsilon)} = \lim_{\varepsilon \to 0} \frac{\lg N(\varepsilon)}{\lg(1/\varepsilon)}$$

The three-dimensional fractal dimensions of the fracture, when the specimens were destroyed under four confining pressures, were calculated. Table 3 shows that the compaction effect in the initial stage becomes more significant with the increase in confining pressure, and that the fractal dimension of the fracture structure in the coal specimen decreases gradually with the closure of the fracture. The reason for this is that the confining pressure reduces the internal fracture surface of coal–rock specimens; that is, the degree of damage is weakened and the corresponding fractal dimension D_t is reduced, indicating that an increase in confining pressure additionally and effectively inhibits the development of internal cracks in coal–rock.

Table 3. Calculation results of the three-dimensional fractal dimension of coal samples.

Confining Pressure (MPa)		5	10	15	20
Three-dimensional fractal dimension	First scan	2.2046	2. 1770	2.1312	2.1058
	Second scan	2.1056	2.0714	2.0221	2.0016
	Third scan	2.1429	2.1252	2.1093	2.0672
	Fourth scan	2.2046	2.1785	2.1633	2.1162
	Fifth scan	2.3277	2.3004	2.2578	2.2041
	Sixth scan	2.4141	2.3982	2.3327	2.2984

Coal is a natural, heterogeneous, porous material containing a large number of naturally occurring micropores. Particularly when the coal is forced into a stress state, its pores and fissures expand and extend with the external force and the accumulation of energy, a process that is complex and random. The three-dimensional fractal dimensions of the samples at each scanning stage under different confining pressures are shown in Figure 9.

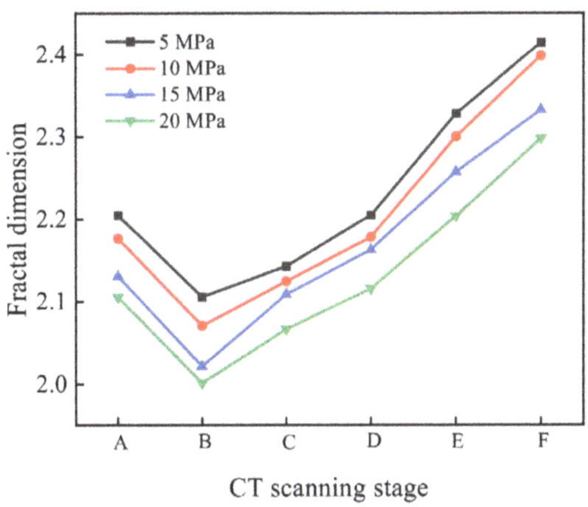

Figure 9. Relationship between three-dimensional fractal dimension and confining pressure.

In the above study, triaxial compression experiments of coal under different confining pressures were performed, with six scans taken at different stages, and the results are shown in Figure 9. Under the same confining pressure, the three-dimensional fractal dimension shows a trend of decreasing first and then increasing; however, with the increase in confining pressure, the fractal dimension values at the same stage are different, though they are all between 2 and 3. The closer the fractal dimension is to 2, the fewer the pores and fissures inside the coal, the smaller their volume, and the simpler their distribution. The closer the fractal dimension is to 3, the greater the number of pores and fissures inside the coal, the larger their volume, and the more complex their distribution.

The damage deformation processes of loaded coal under different confining pressures are similar in each stage. The damage process can be roughly divided into the following stages: compaction, elastic deformation, plastic yield, post-peak failure, and residual. In this study, the range of fracture surfaces inside the coal samples in their initial states is not large. Different numbers and scales of micro-cracks and pores inside the coal exhibited a point-like discrete distribution, and the degree of pore fracture development was low. The total crack volumes in the samples before being subjected to 5, 10, 15, and 20 MPa were 618.83 (Figure 7a), 169.87 (Figure 7g), 220.75 (Figure 7m), and 244.19 (Figure 7s) mm^3, respectively.

During the compaction stage, under the continuous application of confining and axial pressure, the initial cracks in the coal gradually closed, and the volume of the coal decreased continuously. At this stage, the fracture volume and fractal dimension of the coal took their lowest values in the whole compression process. The fracture volumes were

108.30 (Figure 7b), 61.40 (Figure 7h), 20.42 (Figure 7n), and 65.51 (Figure 7t) mm^3, and the fractal dimensions were 2.1056, 2.0714, 2.0221, and 2.0016 for the samples before being subjected to 5, 10, 15, and 20 MPa, respectively. The deformation generated at this stage is plastic and cannot be restored after axial compression unloading. After the second scan, the compaction stage ended, the elastic deformation stage began, and the fracture volume and fractal dimension of the coal increased.

In the third scan, the change in the three-dimensional fractal dimension was smaller than that in the compaction stage, and the degree of damage was lower. The main reason for this was that the third scan focused on the elastic deformation stage of coal. In this stage, the stress–strain curve presents a mostly linear relationship under the action of axial compression, and the mechanical properties are relatively stable. We see a minor initiation of internal pores and cracks, with almost no damage. In the fourth scan, the elastic deformation stage basically ended, and the plastic yield stage was about to begin. Compared with the third scan, the increase in fractal dimension was still not large. The damage in this stage is still elastic. A small number of pores and cracks were generated, and the coal was less damaged.

The fifth scanning point was located after the stress peak point, which entailed the reaching of the compressive strength of the coal. A large number of the micropores in the coal began to expand and interconnect, and macroscopic cracks appeared. The crack volumes in the coal samples increased rapidly, reaching 1651.89 (Figure 7e), 601.37 (Figure 7k), 601.69 (Figure 7q), and 884.88 (Figure 7w) mm^3 for the samples before being subjected to 5, 10, 15, and 20 MPa, respectively. The growth rates of the three-dimensional fractal dimension were also large, rising to 2.3277, 2.3004, 2.2578, and 2.2041, respectively. At this time, the coal gradually lost its bearing capacity and became damaged; however, because of the interaction between internal friction within the coal and confining pressure, the coal retained some strength after reaching the peak of the stress–strain curve.

With the further development of deformation, in the sixth scan, the fracture volume and fractal dimension of the coal under different confining pressures were maximized. At this time, the coal samples were almost completely destroyed, and their bearing capacities were basically lost. The macroscopic fractures expanded across a large range, and some fractures began to penetrate the fracture surface. Soon after this, the fracture surfaces began to slip. With the accumulation of energy inside the coal, this slippage of the fracture surface was accompanied by a large release of energy, resulting in a small-scale rock burst phenomenon, scattering broken coal blocks around.

In summary, the evolution characteristics of loaded coal under different confining pressures are consistent. They can be separated into three stages of development—slight decrease, stable increase, and significant increase—determined by the confining pressure. The ranges of variation in each are different. The fractal dimension can be used to elucidate the differences in fracture structure in the loaded coal under different confining pressures, i.e., changing the confining pressure can reduce the damage suffered by the coal.

4. Discussion

In this paper, to explore the damage- and failure-related mechanical properties of coal under different confining pressures, triaxial compression tests under confining pressures of 5, 10, 15, and 20 MPa were performed. The stress–strain curve is shown in Figure 10. The mechanical parameters were also extracted, as shown in Table 4. The peak strain, peak stress, and elastic modulus of the coal are proportional to the confining pressure produced under loading. The increase in confining pressure improves the compressive strength of coal.

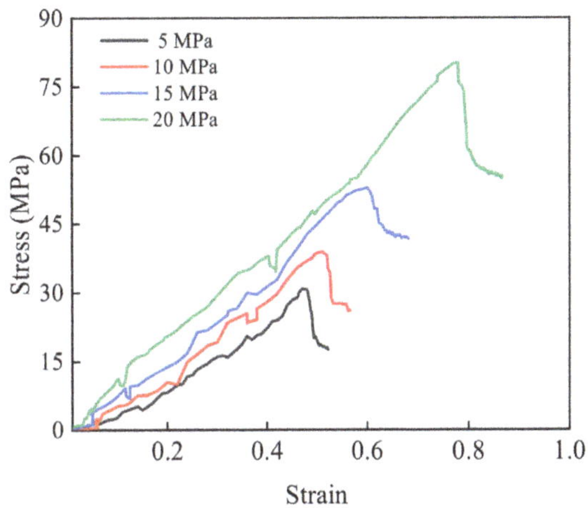

Figure 10. Stress–strain curves under different confining pressures.

Table 4. Triaxial compression test results.

Confining Pressure (MPa)	Peak Strength (MPa)	Peak Strain	Elastic Modulus (MPa)
5	30.79	0.47	64.13
10	38.99	0.51	76.59
15	52.93	0.60	82.08
20	80.24	0.78	92.06

The conventional triaxial compression test is an essential experimental method used to verify the mechanical parameters of coal and rock. Using different strength theories, we can obtain different strength parameters of coal and rock failure. At present, four strength theories are widely used in underground engineering. In this paper, the main failure mode of coal studied under different confining pressures is shear. The Mohr–Coulomb yield criterion is valuable in the maximum shear stress theory.

Figure 11 shows a Mohr stress circle diagram. According to the Mohr–Coulomb strength criterion [30], in this study, setting different initial confining pressures of $\sigma 2$ (the intermediate principal stress) = $\sigma 3$ (the minor principal stress) during the loading process of coal destroys coal rock mass when the shear stress in a certain section reaches the failure value. At this time, if the size and direction of the minimum principal stress are known, then the size and direction of the maximum principal stress in the coal rock mass can be inferred accordingly. The functional relationship of the Coulomb strength curve is

$$\tau = c + \sigma \cdot \tan \alpha$$

where c is the cohesion between the internal materials of the coal, α is the internal friction angle of the coal, the radius of the Mohr circle is $(\sigma 1 - \sigma 3)/2$, and the coordinate of the center of the Mohr circle is $[(\sigma 1 + \sigma 3)/2, 0]$. When a certain section of the coal sample reaches a critical value, the Mohr circle is tangential to the Coulomb strength curve; that is, the distance from the center of the Mohr circle to the Coulomb strength curve is equal to the radius of the Mohr circle. At this time, according to the formula for the distance d from the point to the straight line

$$d = \left| \frac{kx_0 - y_0 + c}{\sqrt{k^2 + 1}} \right|$$

where, k is the slope of the slash, and the circle center coordinates are (x_0, y_0). Additionally, we can infer that $\sigma 1$ and $\sigma 3$ have the following relationship:

$$\sigma_1 = \frac{1+\sin\alpha}{1-\sin\alpha}\sigma_3 + \frac{2c\cos\alpha}{1-\sin\alpha}$$

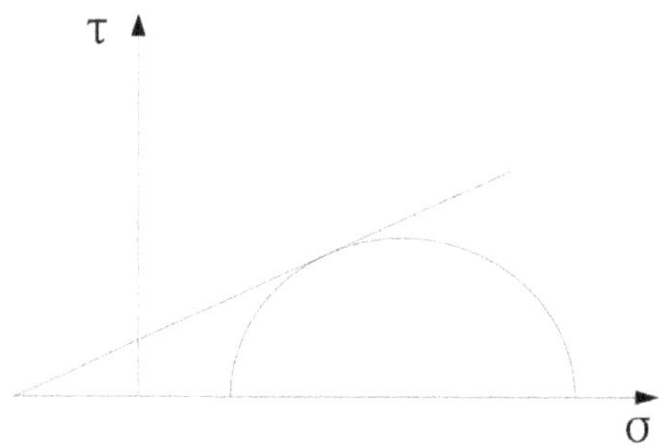

Figure 11. Mohr's stress circle.

Thus, with the increase in confining pressure, $\sigma 1$ and $\sigma 3$ become linear, and the slope of the curve increases. Additionally, a certain proportional relationship exists between the elastic modulus of coal and the confining pressure. The development of confining pressure also helps to improve coal's mechanical parameters and ability to resist damage [31]. The lateral application of confining pressure closes the primary cracks in the coal sample to a certain extent and strengthens the structure within the coal.

5. Conclusions

(1) The coal samples tested under different confining pressures all showed stage characteristics. The five stages are as follows: compaction, elastic, plastic, post-peak failure, and residual. The internal pores and fissures of coal underwent closure in the initial compaction stage under loading, followed by a gradual increase in the elastic stage and a sudden increase in the failure stage. Under the confining pressure, the coal's ability to resist failure was enhanced, and the development of fractures in coal samples was limited. The greater the confining pressure, the greater the limitation to fracture development.

(2) Via two-dimensional images and three-dimensional analyses, the dynamic changes in cracks emerging in loaded coal under different confining pressures can be quantitatively analyzed. The method for accurately quantifying the evolution of internal cracks within coal samples via fractal theory has good reliability. Based on the changes in the fractal dimension and fracture volume, the two- and three-dimensional fracture evolution characteristics shown within the scanning results are consistent. For coal under different confining pressures, the fracture evolution law is highly consistent, showing a small decrease first and then a large increase. The application of confining pressure can reduce damage in coal.

(3) In the triaxial compression tests, the main failure mode of the coal samples was shear failure. Using the Mohr–Coulomb yield criterion, the relationship between mechanical parameters and confining pressure can be obtained. The results of the experiment verify the theory, confirming that the confining pressure improves the mechanical properties of the coal to a large extent. In addition, greater confining pressures result in an increased compressive strength and elastic modulus of the coal.

In this paper, the whole process of coal mass failure is elucidated through laboratory tests. Based on fractal theory, we propose a method for quantifying the fracture structure of coal. Furthermore, when this method is extended to catastrophic circumstances at an engineering site, it can be effectively used to evaluate the failure state of the coal mass and facilitate the prediction of the possible extent of the disaster. In addition, confining pressure refers to the pressure exerted by the rock mass surrounding deep coal; it is intrinsic to the stress environment of coal rock mass. At different depths of coal, the confining pressure differs, increasing with depth. This paper aims to provide a theoretical basis for projecting instability disasters in coal rock mass at different depths or in different environments; this substantiates the engineering application value of this method.

Author Contributions: Conceptualization, A.W.; writing—original draft preparation, A.W.; funding acquisition, L.W.; supervision, L.W.; writing—review and editing, L.W. All authors have read and agreed to the published version of the manuscript.

Funding: This study is supported by the Anhui Province Science and Technology Major special projects (No. 202203a07020010).

Data Availability Statement: The data presented in this study are available upon request from the corresponding author.

Conflicts of Interest: The authors declare that they have no known competing financial interests or personal relationships that could have appeared to influence the work reported in this paper.

References

1. Ghamgosar, M.; Erarslan, N.; Williams, D.J. Experimental Investigation of Fracture Process Zone in Rocks Damaged Under Cyclic Loadings. *Exp. Mech.* **2017**, *57*, 97–113. [CrossRef]
2. Jing, Y.; Armstrong, R.T.; Ramandi, H.L.; Mostaghimi, P. Coal cleat reconstruction using micro-computed tomography imaging. *Fuel* **2016**, *181*, 286–299. [CrossRef]
3. Zhao, Y.; Sun, Y.; Liu, S.; Chen, Z.; Yuan, L. Pore structure characterization of coal by synchrotron radiation nano-CT. *Fuel* **2018**, *215*, 102–110. [CrossRef]
4. Brzovic, A.; Villaescusa, E. Rock mass characterization and assessment of block-forming geological discontinuities during caving of primary copper ore at the El Teniente mine, Chile. *Int. J. Rock Mech. Min.* **2007**, *44*, 565–583. [CrossRef]
5. Jiang, G.; Zuo, J.; Li, Y.; Wei, X. Experimental investigation on mechanical and acoustic parameters of different depth shale under the effect of confining pressure. *Rock Mech. Rock Eng.* **2019**, *52*, 4273–4286. [CrossRef]
6. Alejano, L.R.; Arzúa, J.; Bozorgzadeh, N.; Harrison, J.P. Triaxial strength and deformability of intact and increasingly jointed granite samples. *Int. J. Rock Mech. Min.* **2017**, *95*, 87–103. [CrossRef]
7. Van Stappen, J.F.; McBeck, J.A.; Cordonnier, B.; Pijnenburg, R.P.; Renard, F.; Spiers, C.J.; Hangx, S.J. 4D Synchrotron X-ray Imaging of Grain Scale Deformation Mechanisms in a Seismogenic Gas Reservoir Sandstone During Axial Compaction. *Rock Mech. Rock Eng.* **2022**, *55*, 4697–4715. [CrossRef]
8. Liu, H.; Wang, L.; Zhao, H.; Li, S. Dynamic characteristics and deterioration mechanism of coal under distinct initial gas pressure. *Geomech. Geophys. Geo-Energy Geo-Resour.* **2023**, *9*, 111. [CrossRef]
9. Roslin, A.; Pokrajac, D.; Zhou, Y. Cleat structure analysis and permeability simulation of coal samples based on micro-computed tomography (micro-CT) and scan electron microscopy (SEM) technology. *Fuel* **2019**, *254*, 115579. [CrossRef]
10. Yu, X.; Kemeny, J.; Li, J.; Song, W.; Tan, Y. 3D Observations of Fracturing in Rock-Backfill Composite Specimens Under Triaxial Loading. *Rock Mech. Rock Eng.* **2021**, *54*, 6009–6022. [CrossRef]
11. Zaidi, M.; Ahfir, N.; Alem, A.; Taibi, S.; El Mansouri, B.; Zhang, Y.; Wang, H. Use of X-ray computed tomography for studying the desiccation cracking and self-healing of fine soil during drying–wetting paths. *Eng. Geol.* **2021**, *292*, 106255. [CrossRef]
12. Huang, J.G.; Xu, K.M.; Guo, S.B.; Guo, H.-W. Comprehensive Study on Pore Structures of Shale Reservoirs Based on SEM, NMR and X-CT. *Geoscience* **2015**, *29*, 198–205.
13. Ju, Y.; Zhang, Q.; Zheng, J.; Wang, J.; Chang, C.; Gao, F. Experimental study on CH_4 permeability and its dependence on interior fracture networks of fractured coal under different excavation stress paths. *Fuel* **2017**, *202*, 483–493. [CrossRef]
14. Wildenschild, D.; Sheppard, A.P. X-ray imaging and analysis techniques for quantifying pore-scale structure and processes in subsurface porous medium systems. *Adv. Water Resour.* **2013**, *51*, 217–246. [CrossRef]
15. Shang, R.; Wang, L.; Liu, H.; Zhu, C.; Li, S.; Chen, L. The Influence of Dip Angle of Rock Bridge on Mechanical Properties and Fracture Characteristics of Fractured Coal Body at Three-Dimensional Scale. *Rock Mech. Rock Eng.* **2023**, *56*, 8927–8946. [CrossRef]
16. Wang, G.; Shen, J.; Liu, S.; Jiang, C.; Qin, X. Three-dimensional modeling and analysis of macro-pore structure of coal using combined X-ray CT imaging and fractal theory. *Int. J. Rock Mech. Min.* **2019**, *123*, 104082. [CrossRef]
17. Kumari, W.G.P.; Ranjith, P.G.; Perera, M.S.A.; Li, X.; Li, L.H.; Chen, B.K.; Isaka, B.A.; De Silva, V.R. Hydraulic fracturing under high temperature and pressure conditions with micro CT applications: Geothermal energy from hot dry rocks. *Fuel* **2018**, *230*, 138–154. [CrossRef]
18. Xie, H.P. *Theory to Fractal and Rock Mechanics*; Science Press: Beijing, China, 2005.
19. Zhao, Y.S. *Mine Rock Fluid Mechanics*; China Coal Industry Publishing House: Beijing, China, 1994.

20. Zhang, Y.B.; Xu, Y.D.; Liu, X.X.; Yao, X.L.; Wang, S.; Liang, P.; Sun, L.; Tian, B.Z. Quantitative characterization and mesoscopic study of propagation and evolution of three-dimensional rock fractures based on CT. *Rock Soil. Mech.* **2021**, *42*, 2659–2671.
21. Wang, G.; Qin, X.; Shen, J.; Zhang, Z.; Han, D.; Jiang, C. Quantitative analysis of microscopic structure and gas seepage characteristics of low-rank coal based on CT three-dimensional reconstruction of CT images and fractal theory. *Fuel* **2019**, *256*, 115900. [CrossRef]
22. Zhou, Y.X.; Xia, K.; Li, X.B.; Li, H.B.; Ma, G.W.; Zhao, J.; Zhou, Z.L.; Dai, F. Suggested methods for determining the dynamic strength parameters and mode-I fracture toughness of rock materials. *Int. J. Rock Mech. Min.* **2012**, *49*, 105–112. [CrossRef]
23. Li, Y.; Yang, S.; Liu, Z.; Sun, B.W.; Yang, J.; Xu, J. Study on mechanical properties and deformation of coal specimens under different confining pressure and strain rate. *Theor. Appl. Fract. Mech.* **2022**, *118*, 103287. [CrossRef]
24. Sun, Y.H. *Modeling and Numerical Simulation of Fractured Low Permeability Sandstone Reservoir*; Tianjin Science and Technology Press: Tianjin, China, 2010.
25. Zhang, R.; Ai, T.; Li, H.; Zhang, Z.; Liu, J. 3D reconstruction method and connectivity rules of fracture networks generated under different mining layouts. *Int. J. Min. Sci. Technol.* **2013**, *23*, 863–871. [CrossRef]
26. Peng, R.D.; Xie, H.P.; Ju, Y. Computation method of fractal dimension for 2-D digital image. *J. China Univ. Min. Technol.* **2004**, *33*, 19–24.
27. Fu, X.; Qin, Y.; Xue, X.; Li, G.; Wang, W. Research on Fractals of Pore and Fracture-Structure of Coal Reservoirs. *J. China Univ. Min. Technol.* **2001**, *30*, 225–228.
28. Li, G.; Zhang, R.; Xu, X.L.; Zhang, Y.F. CT image reconstruction of coal rock three-dimensional fractures and body fractal dimension under triaxial compression test. *Rock Soil. Mech.* **2015**, *36*, 1633–1642.
29. Yang, Y.C.; Peng, R.D.; Zhou, H.W. Computation of Fractal Dimension for Digital Image in a 3-D Space. *J. China Univ. Min. Technol.* **2009**, *38*, 251–258.
30. Cai, M.F. *Rock Mechanics and Engineering*; Science Press: Beijing, China, 2013.
31. Wang, L.; Liu, H.Q.; Li, S.B.; Chen, L.P.; Liu, H.Q. Fracture evolution characteristics of prefabricated crack coal under different confining pressures. *J. Min. Saf. Eng.* **2023**, *40*, 786–797.

fractal and fractional

Article

Analysis and Warning Prediction of Tunnel Deformation Based on Multifractal Theory

Chengtao Yang [1], Rendong Huang [1], Dunwen Liu [1,*], Weichao Qiu [2], Ruiping Zhang [2] and Yu Tang [1,3,*]

1 School of Resources and Safety Engineering, Central South University, Changsha 410083, China; 225512138@csu.edu.cn (C.Y.); hla@mail.csu.edu.cn (R.H.)
2 Road & Bridge North China Engineering Co., Ltd., Beijing 101100, China; 225511019@csu.edu.cn (W.Q.); 215512141@csu.edu.cn (R.Z.)
3 College of Water Resources and Civil Engineering, Hunan Agricultural University, Changsha 410128, China
* Correspondence: dunwen@csu.edu.cn (D.L.); tangyu12@csu.edu.cn (Y.T.)

Abstract: To better analyze the fluctuation characteristics and development law of tunnel deformation data, multifractal theory is applied to tunnel deformation analysis. That is, the multifractal detrended fluctuation analysis (MF-DFA) model is first utilized to carry out the multifractal characterization of tunnel deformation data. Further, Mann–Kendall (M–K) analysis is utilized to construct the dual criterion ($\Delta\alpha$ indicator criterion and $\Delta f(\alpha)$ indicator criterion) for the tunnel deformation early warning study. In addition, the particle swarm optimization long-short-term memory (PSO-LSTM) prediction model is used for predicting tunnel settlement. The results show that, in reference to the tunnel warning level criteria and based on the Z-value results of the indicator criterion, the warning level of all four sections is class II. At the same time, through the analysis of tunnel settlement predictions, the PSO-LSTM model has a better prediction effect and stability for tunnel settlement. The predicted results show a slow increase in tunnel settlement over the next 5 days. Finally, the tunnel warning level and the predicted results of tunnel settlement are analyzed in a comprehensive manner. The deformation will increase slowly in the future. Therefore, monitoring and measurement should be strengthened, and disaster preparedness plans should be prepared.

Keywords: tunnel; MF-DF: deformation warning; settlement prediction; PSO-LSTM

1. Introduction

With the continuous development of transportation construction in China, the scale and technical level of tunnel construction have reached new heights [1]. By the end of 2022, the national railroad operating mileage reached 155,000 km, boasting 17,873 operational railroad tunnels with a cumulative length of approximately 22,000 km. Of these, 259 extra-long railroad tunnels and 12 tunnels exceeding 20 km in length are in operation [2]. Notably, over the past five years, railroad tunnel development has experienced remarkable acceleration. This surge is attributed to the rapid growth of the high-speed railroad industry and the consistent enhancement of tunnel trimming technology, resulting in the emergence of numerous tunnels traversing diverse types of special strata [3]. However, concurrent with this progress, there exist design and construction irregularities in high-speed railroad tunnels that may lead to deformation, water leakage, cracking, and other issues [4–8].

In tunnel construction, the control of tunnel deformation stands out as pivotal technology [9]. Consequently, the monitoring and stability evaluation of tunnel deformation are paramount. Current methods for deformation monitoring and analysis encompass numerical simulation methods [10–13], time series analysis methods [14–17], fuzzy comprehensive evaluation methods [18–21], and machine learning methods [22–25]. Yet, the complexity of the geological environment and deformation-inducing factors introduce a nonlinear characteristic to the tunnel deformation monitoring time-course curve, causing fluctuations [26]. Therefore, the multifractal detrended fluctuation analysis (MF-DFA) method presents an effective approach for tunnel deformation analysis.

Fractal theory, pioneered by the French mathematicians Mandelbrot et al. [27], describes chaotic, complex, and self-similar systems with regularity. Researchers, such as Zuo et al., have applied fractal theory to analyze and predict tunnel surface settlement, evaluating surface settlement stability [28]. Ye et al. established a model based on fractal theory to study the interaction between microstructure evolution and tunnel leakage behavior, offering insight into the tunnel leakage mechanism [29]. Additionally, MF-DFA, initially proposed by Grassberger [30], finely describes the volatility of tunnel deformation monitoring data at different levels. At present, the multifractal theory has been widely used. Lei et al. used multifractal theory and subcomponent combination prediction to synthesize the warning level of landslides [31]. Mao et al. revealed early warning signals of rock slope deformation based on the multiple fractal time-varying response characteristics of micro-seismic waveforms associated with rock rupture [32]. Zhou et al. applied multifractal theory to the deformation pattern analysis of dams and utilized MF-DFA, multivariate multifractal detrended fluctuation analysis (MV-MFDFA), and asymmetric multifractal detrended fluctuation analysis (A-MFDFA) to resolve the multiple fractal features of deformation states and their asymmetry [33].

Accurate prediction of tunnel deformation is crucial for disaster prevention and early warning. Currently, soil settlement prediction methods include empirical methods [34–36], theoretical methods [37–40], modeling methods [41–44], and machine learning methods [45–48]. Long short-term memory (LSTM) is particularly effective in handing time-related information and is widely used in temporal tunnel settlement prediction [49]. Li et al. employed three machine learning algorithms (LSTM, RF, and GRU) to forecast surface settlement. Their findings indicated that the LSTM algorithm demonstrated superior accuracy in predicting maximum surface settlement and effectively anticipated settlement progression in various strata [50]. Cao et al. proposed the complete ensemble empirical mode decomposition with adaptive noise long short-term memory (CEEMDAN-LSTM) model, combining high prediction accuracy with acceptable computational efficiency [51]. Duan et al. utilized the auto regressive integrated moving average (ARIMA) and LSTM models for predicting structural deformation trends during tunnel operation, determining that LSTM performs better with high data quality and sufficient samples [52].

From the above research results, it can be concluded that research on tunnel deformation analysis and early warning prediction is necessary. Therefore, this paper analyzes tunnel deformation data. Firstly, the deformation rate and the deformation of the tunnel are investigated by multiple fractal characterizations using MF-DFA. Secondly, the Mann–Kendall (M–K) analysis method [53,54] is used to evaluate the early warning classification of tunnel deformation. Furthermore, the prediction of tunnel deformation is realized by the particle swarm optimization LSTM (PSO-LSTM) prediction model. Finally, the results of tunnel deformation early warning classification and the tunnel deformation prediction results are jointly addressed. The tunnel deformation law is evaluated comprehensively to provide theoretical guidance for its monitoring, measurement, and prevention. This study calculates and analyzes the multifractal features of tunnel monitoring data by integrating the M–K method to determine the tunnel's warning level. Furthermore, the PSO-LSTM model is found to exhibit high accuracy in predicting tunnel deformation, offering a novel perspective for advancing research on tunnel early warning prediction.

2. Materials and Methods

The process of tunnel deformation analysis and early warning prediction, established in this paper based on multiple fractal theory, is illustrated in Figure 1. The steps can be summarized as follows:

First, tunnel deformation monitoring and measurement are conducted to obtain tunnel deformation data. Second, the monitoring data are segmented into 12 groups, and multifractal eigenvalues are calculated and analyzed using the M–K method. The warning level is then determined in conjunction with the warning level criteria. Third, the parameters of the LSTM prediction model are optimized using the PSO optimization algorithm, and the PSO-LSTM prediction model is formulated for predicting tunnel settlement. Fourth, the warning level and prediction results are jointly analyzed to obtain the warning outcomes. Finally, corresponding preventive measures are implemented based on the results of the warning level.

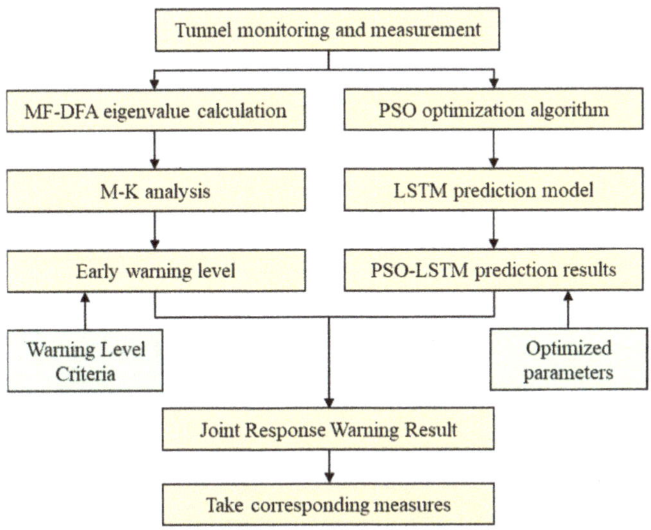

Figure 1. Flow chart of tunnel deformation analysis and early warning prediction.

This comprehensive approach provides a systematic framework for tunnel deformation analysis, early warning prediction, and subsequent preventive action.

2.1. Project Overview and Monitoring Data

The high-speed rail tunnel has a total length of 254.9 m, with the starting and ending mileage marked as DK115 + 133.0~DK115 + 387.9. It is positioned between the Sangujian No. 1 Tunnel and the Xikeng bridge. Geographically, it is situated on the western side of the mountain, in Tamkou Village, Xidi Town, Yixian County, Huangshan City. The surface vegetation on the mountain is characterized by bamboo forests, low shrubs, trees, and low weeds. The terrain is steep, with a natural slope ranging from 30° to approximately 55°. The tunnel is surrounded by dense mountain vegetation, and the entrance and exit are marked by ravine confluence, making traffic inconvenient. The perimeter rock grade is classified as IV, and the excavation employs the three-step method.

The tunnel is situated in the low hills, originating from Huangshan Mountain. It is characterized by undulating terrain with mountain elevations ranging from 250 m to 325 m and a maximum elevation difference of approximately 75 m. The mountain exhibits a monoclinic structure, with steeper south and north slopes. The tunnel primarily traverses carbonaceous mudstone of the Cambrian Hotang formation, with a maximum tunnel depth of about 65 m. (Refer to Figure 2 for a visualization of the tunnel cross-section.)

To ensure the safety of tunnel construction operations, a total station is utilized for manual measurement of tunnel settlement and convergence. This paper involves monitoring the settlement of the tunnel vault and tunnel convergence utilizing the rear rendezvous method and the opposite side measurement method, respectively. For arch settlement monitoring, a single monitoring point is positioned at the arch of each section. In contrast, for tunnel convergence monitoring, two monitoring points are symmetrically arranged on both sides of the tunnel, employing the opposite side measurement method. Consequently, three monitoring points are allocated in each section, with a monitoring frequency of 2 times/day. Further details are depicted in Figure 3.

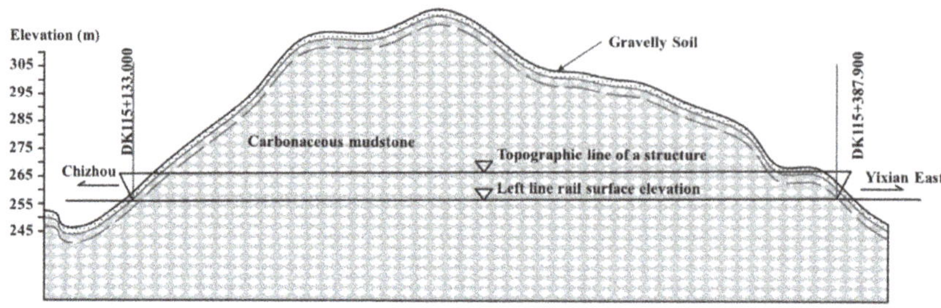

Figure 2. Tunnel section.

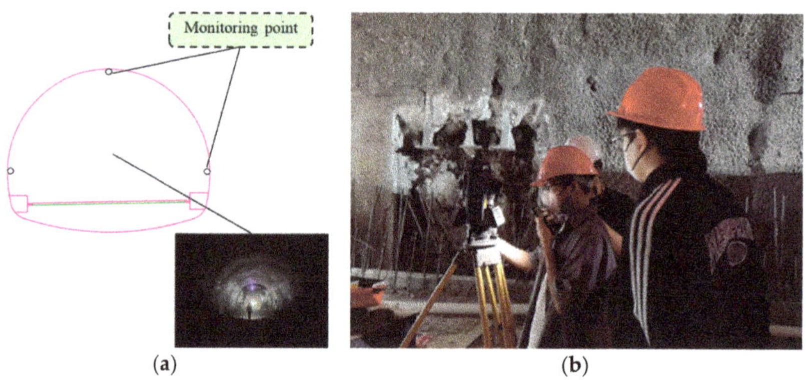

Figure 3. Tunnel deformation monitoring and measurement: (**a**) tunnel monitoring point layout, (**b**) manual monitoring and measurement.

As shown in Figures 4 and 5, this paper examines the monitoring data regarding tunnel settlement and convergence in the section from DK115 + 261~DK115 + 290. Additionally, the data from monitoring points are counted. The cumulative deformation–time diagram and deformation rate–time diagram of tunnel settlement and convergence are obtained, respectively.

In Figure 4, it is evident that both the cumulative tunnel settlement deformation and the cumulative tunnel convergence deformation exhibit an increasing trend over time. Notably, after 1000 h, the growth of these deformation measures reaches a turning point, transitioning into a slower growth trend. Throughout the monitoring period, the data display both significant and minor fluctuations, which can be effectively analyzed using multiple fractal theory.

Looking at Figure 5, it is apparent that both the tunnel settlement deformation rate and the tunnel convergence deformation rate demonstrate significant and minor fluctuations, displaying the characteristic traits of multiple fractals. These fluctuations can be effectively analyzed using multiple fractal theory.

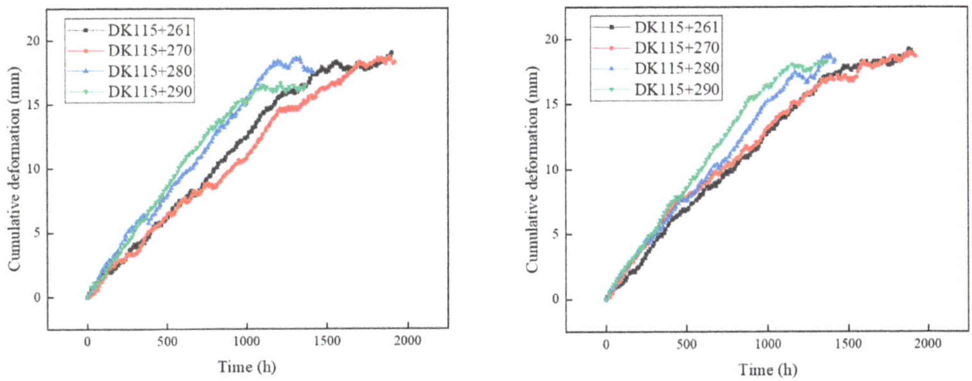

(**a**) (**b**)

Figure 4. Tunnel cumulative deformation–time diagrams: (**a**) tunnel cumulative settlement–time diagram, (**b**) tunnel cumulative convergence–time diagram.

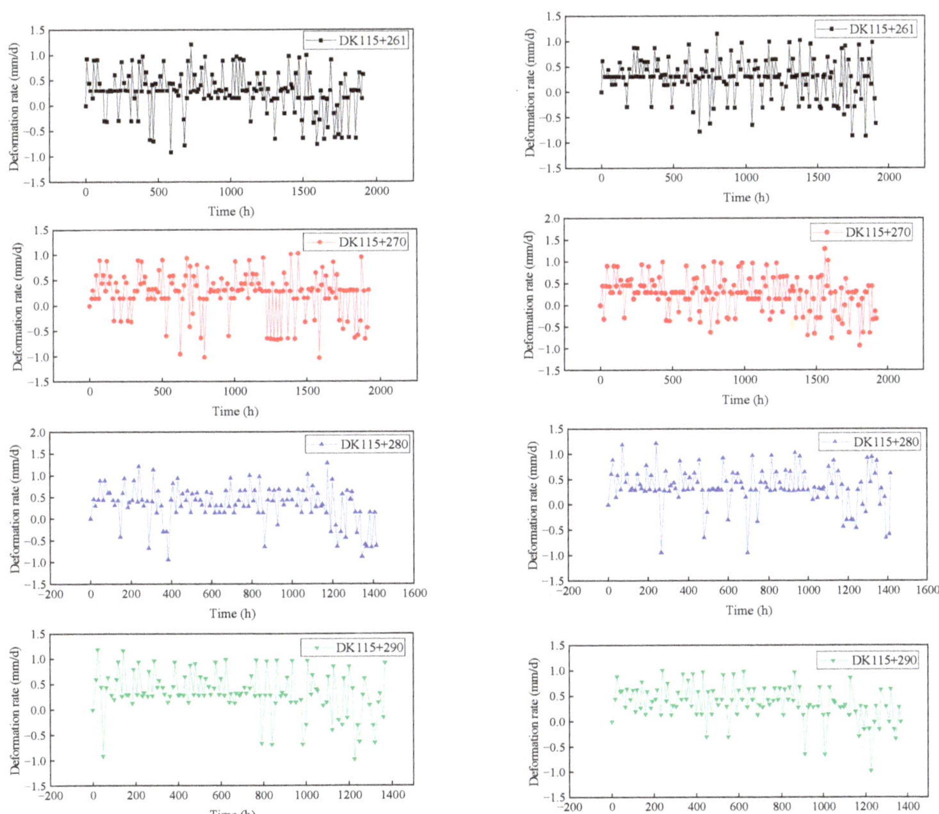

(**a**) (**b**)

Figure 5. Tunnel deformation rate–time diagrams: (**a**) tunnel settlement rate–time diagram, (**b**) tunnel convergence rate–time diagram.

2.2. MF-DFA

The MF-DFA model, which belongs to the multiple non-uniform fractal method, not only reveals the multiple fractal characteristics within the deformation sequence but also effectively evaluates their deformation trends [55]. The calculation steps are as follows:

Step 1: the data presented in Figures 4 and 5, respectively, are set as a time series $x(t)$ with a series length of N, a series mean of $\overline{x}$, and a series of cumulative deviations of $x(t)$ with respect to $\overline{x}$ as $y(t)$:

$$y(t) = \sum_{i=1}^{t}(x(i) - \overline{x}) \tag{1}$$

Step 2: Set a time scale s. Equalize the sequence $y(t)$ in terms of s, dividing it into a total of m equal-length consecutive and non-overlapping sub-intervals, $m = int(N/s)$.

In practical arithmetic, N may not necessarily be divisible, resulting in potential tail data redundancy. Thus, alongside positive-order division, a reverse-order processing method is also employed simultaneously. The division operation is repeated from the end of the sequence to obtain $2m$ subintervals.

Step 3: Fit a trend to each subinterval and subtract the trend portion from the original. Obtain the corresponding residual series denoted as $z_v(t)$:

$$z_v(t) = y_v(t) - p_v^k(t) \tag{2}$$

where $y_v(t)$ represents the subinterval, and $p_v^k(t)$ is a kth order fitting polynomial to the vth subinterval. v ranges from 1 to 2 m, and t ranges from 1 to s.

Step 4: calculate the mean square deviation $F^2(s,v)$ of the residual sequence $z_v(t)$:

$$F^2(s,v) = \frac{1}{s}\sum_{t=1}^{s}(z_v(t))^2 \tag{3}$$

Step 5: Optimize the traditional MF-DFA division by employing a sliding window, which involves moving the values through a window of a specific length along the sequence at a defined step size. This helps to minimize pseudo-fluctuations in the data and maximize the utilization of data information.

The window length is denoted as s, the sequence length as N, and the sliding step is set to 1. The number of subintervals obtained in one run is N − S + 1, and the q-order fluctuation function is calculated according to Equation (4):

$$F_q(s) = \begin{cases} \left\{ \frac{1}{N-s+1} \sum\limits_{v=1}^{N-s+1} \left[F^2(s,v)\right]^{q/2} \right\}^{1/q}, q \neq 0 \\ \exp\left\{ \frac{1}{2(N-s+1)} \sum\limits_{v=1}^{N-s+1} \ln\left[F^2(s,v)\right] \right\}, q = 0 \end{cases} \tag{4}$$

Step 6: Repeat the previous steps to generate a series of point values for s-$F_q(s)$. If this time series exhibits a long-range correction, then $F_q(s)$ has a power-law relationship with s, as shown in Equation (5):

$$F_q(s) \propto s^{h(q)} \tag{5}$$

By taking the logarithms of both sides of the previous equation, we obtain Equation (6):

$$lgF_q(s) = h(q)lgs + lgb \tag{6}$$

where $F_q(s)$ represents the q-order fluctuation function of the series, $h(q)$ is the corresponding generalized Hurst exponent, and b is a constant coefficient.

A plot of $lgF_q(s)$-lgs was created and fitted to determine the generalized Hurst exponent $h(q)$. A fixed $h(q)$ suggests a mono-fractal sequence without multifractal features. When $h(q) < 0.5$, the data sequence behaves as a memory process with inverse persistence.

A value of $h(q) = 0.5$ indicates uncorrelated stochastic behavior. For $h(q) > 0.5$, the data sequence behaves as a memory process with positive persistence. Values of $h(q) > 1$ indicate long-range positively correlated processes with strong non-stationarity.

Step 7: the multifractal spectrum $f(\alpha)$, which characterizes the fractal intensity and singularity of the time series, can typically be determined using Equation (7):

$$\begin{cases} \tau(q) = qh(q) - 1 \\ \alpha = \tau'(q) \\ f(\alpha) = q\alpha - \tau(q) \end{cases} \tag{7}$$

where $\tau(q)$ is the Renyi index, also known as the scalar function. If it is a nonlinear up-convex function of q, the displacement sequence exhibits multifractal characteristics. If it is a linear function of q, the displacement sequence exhibits single fractal characteristics.

Step 8: calculate the multifractal spectral parameter $\Delta\alpha$ and $\Delta f(\alpha)$.

$$\begin{cases} \Delta\alpha = \alpha_{max} - \alpha_{min} \\ \Delta f(\alpha) = \Delta f(\alpha_{max}) - \Delta f(\alpha_{min}) \end{cases} \tag{8}$$

where the $\Delta\alpha$ parameter is mainly used to evaluate the width of the multifractal spectrum of the data deformation sequence. As the value of $\Delta\alpha$ increases, the intensity of multiple fractals becomes stronger, leading to more intense fluctuations. The $\Delta f(\alpha)$ parameter is mainly used to evaluate the proportion of large and small fluctuations in the waveform of the data deformation sequence. As the value of $\Delta f(\alpha)$ decreases, the proportion of large fluctuation waveforms increases.

2.3. M–K Test Method

The M–K test method is a non-parametric method. Non-parametric tests are also known as non-distributional tests. It has the advantage that the sample does not have to follow a particular distribution and is not affected by a few outliers. It is suitable for type and order variables, and it is relatively simple to calculate [56]. The specific calculation steps are as follows:

Step 1: The 12 sets of eigenvalues obtained from each set of data by the formula in 2.2 are set as a time series of 12 sample sizes $\{x_1, x_2, \ldots\ldots, x_{12}\}$. For all $k(j \leq n$ and $k \neq j)$, the distributions of x_k and x_j are different and the difference function $Sgn(x_j - x_k)$ is computed:

$$Sgn(x_j - x_k) = \begin{cases} +1, & (x_j - x_k) > 0 \\ 0, & (x_j - x_k) = 0 \\ -1, & (x_j - x_k) < 0 \end{cases} \tag{9}$$

Step 2: calculate the test statistic S:

$$S = \sum_{k=1}^{n-1} \sum_{j=k+1}^{n} Sgn(x_j - x_k) \tag{10}$$

Step 3: S is normally distributed with mean 0. Calculate the variance $Var(S)$:

$$Var(s) = \frac{n(n-1)(2n+5)}{18} \tag{11}$$

Step 4: calculate the standard normal statistical variable Z:

$$Z = \begin{cases} \frac{S-1}{\sqrt{Var(s)}}, & S > 0 \\ 0, & S = 0 \\ \frac{S+1}{\sqrt{Var(s)}}, & S < 0 \end{cases} \tag{12}$$

Step 5: The trend characteristics of the corresponding evaluation object can be judged by the size of Z. The Z_α value is the critical value under the condition of corresponding significant level α. In this paper, the significance test with 99% confidence level is selected. That is, for a significance level $\alpha = 0.01$, then $Z_{0.01} = 2.32$.

If $Z \geq Z_\alpha$, it means that the evaluation object has an increasing trend, and if it is larger, it indicates a stronger trend.

If $-Z_\alpha < Z < Z_\alpha$, it means that the evaluation object has a smooth trend.

If $Z \leq -Z_\alpha$, it means that the evaluation object has a decreasing trend, and if it is smaller, it indicates a stronger trend.

2.4. PSO-LSTM Prediction Modeling

2.4.1. LSTM

LSTM, a special variant of recurrent neural networks, was introduced by Hochreiter in 1997 [57]. It incorporates temporal memory units capable of learning dependency information across various time periods in a time series. This network is particularly effective at processing and predicting intervals and delayed events within time series data. The specific calculation steps are as follows:

Step 1: tunnel monitoring statistics are normalized and calculated by the formula:

$$S = \sum_{k=1}^{n-1} \sum_{j=k+1}^{n} Sgn(x_j - x_k) \tag{13}$$

where $x(i,j)$ represents the original data, $x'(i,j)$ denotes the normalized data, and $x_{max}(j)$ and $x_{min}(j)$ are the maximum and minimum values of the parameters of the jth model, respectively.

Step 2: LSTM consists of multiple cells, and the formulae for the forgetting gate, input gate, and output gate in each cell are:

$$\begin{cases} f_t = \sigma\left(W_f h_{t-1} + W_f x_t + b_f\right) \\ i_t = \sigma(W_i h_{t-1} + W_i x_t + b_i) \\ O_t = \sigma(W_o h_{t-1} + W_o x_t + b_o) \end{cases} \tag{14}$$

where h_{t-1} represents the hidden state from the previous moment, and σ denotes the sigmoid function. f_t, i_t, and O_t are the outcomes of the state settlements for the oblivious gate, the input gate, and the output gate, respectively. W_f, W_i, and W_o are the weight matrices for the oblivious gate, the input gate, and the output gate, respectively. b_f, b_i, and b_o represent the biases for oblivious gate, the input gate, and the output gate, respectively.

Step 3: C_t is a vector of candidate values, and the product of the input values and the vector of candidate values is used to update the cell state, calculated as:

$$\begin{cases} C_t = \tan h(W_c h_{t-1} + W_c x_t + b_c) \\ C_t = f_t C_{t-1} + iC_t \\ h_t = o_t \tan h(C_i) \\ f(x) = \frac{1}{1+e^{-x}} \\ f(x) = \tan h(x) \end{cases} \tag{15}$$

where W_c represents the weight matrix for the input unit state, while b_c is the bias term for the input unit state. The activation function is $\tan h$. o_t denotes the neuron output value. h_t is the current moment's hidden state.

2.4.2. PSO

PSO is inspired by the behavior of birds foraging in nature. In this algorithm, each optimization problem is referred to as a "particle", and the process of particle swarm optimization is likened to the foraging behavior of a bird flock. These particles are equipped with a memory function to store optimization information, which is then shared within the flock. The optimal position information is subsequently selected as the optimal position information of the entire flock [58].

During each iteration, the position information of each particle is updated. The change in the position information x_{id}^k for the ith evolutionary example consists of a linear summation of the previous position information x_{id}^{k-1} and the previous velocity information V_{id}^{k-1}:

$$x_{id}^k = x_{id}^{k-1} + V_{id}^{k-1} \tag{16}$$

Furthermore, each particle possesses a velocity vector that dynamically changes in real time due to various factors. It includes the example velocity information V_{id}^k of the ith evolution, with the previous velocity WV_{id}^{k-1}, the individual optimal position $C_1r_1\left(pbest_{id} - x_{id}^{k-1}\right)$, and the population optimal position $C_2r_2\left(gbest_{id} - x_{id}^{k-1}\right)$. The formula is as follows:

$$V_{id}^k = WV_{id}^{k-1} + C_1r_1\left(pbest_{id} - x_{id}^{k-1}\right) + C_2r_2\left(gbest_{id} - x_{id}^{k-1}\right) \tag{17}$$

where C_1 and C_2 are acceleration constants, and r_1 and r_2 are random functions.

2.4.3. PSO-LSTM

LSTM demonstrates exceptional feature extraction capabilities from data exhibiting spatial and temporal correlation. This study applies the LSTM model for predicting tunnel settlement data. In order to further reduce the model error, PSO is employed to optimize the LSTM parameters and establish the PSO-LSTM model.

The LSTM model is constructed and trained using the training set. Subsequently, the prediction results and the actual values are compared and analyzed for error in the test set. The accuracy of the model is assessed using three indicators: the coefficient of determination (R^2), the mean absolute error (MAE), and the root mean square error ($RMSE$). The calculation formula is:

$$\begin{cases} R^2 = 1 - \frac{\sum_{i=1}^{n}[y_c(i) - y_0(i)]^2}{\sum_{i=1}^{n}[y_0(i) - \overline{y_c}]^2} \\ MAE = \frac{1}{n}\sum_{i=1}^{n}|y_c(i) - y_0(i)| \\ RMSE = \sqrt{\frac{1}{n}\sum_{i=1}^{n}[y_c(i) - y_0(i)]^2} \end{cases} \tag{18}$$

where n represents the number of predicted outcomes, $y_o(i)$ represents the true outcome, $y_c(i)$ represents the predicted outcome, and $\overline{y_o}$ represents the mean of the true values.

3. Results and Discussion

3.1. Multifractal Characterization of Tunnel Deformation Rates

3.1.1. Characterization of Tunnel Deformation Rate Data

Before investigating the multifractal characteristics of the deformation rate of tunnel settlement and convergence, the basic statistical characteristics of the sample data should be understood. Here, the descriptive statistics of the monitoring data in Figure 2 are shown in Table 1.

According to the K–S criterion [59,60], the four evaluation indexes for average, standard deviation, skewness, and kurtosis are considered:

1. From the average: the average tunnel deformation rate of each cross-section is greater than 0, which shows positive deformation.
2. From the standard deviation: The standard deviation values of the tunnel deformation rate at each cross-section are close. In the deformation rate of tunnel settlement, the standard deviation at DK115 + 280 is the largest, which can be visualized in Figure 2a.

In the deformation rate of tunnel convergence, the standard deviation at DK115 + 270 is the largest, which can be visualized in Figure 2b. The standard deviations are all close to or greater than the mean, indicating a greater degree of dispersion in the data.

3. From the skewness: The skewness value of each cross-section is greater than 0, indicating that the data distribution is right-skewed. That is, the dispersion on the right side of the mean is stronger than that on the left side, showing a certain degree of long tail on the right side. All skewness values are greater than the mean and standard deviation, indicating a greater degree of skewness in the data distribution.
4. From the kurtosis: the kurtosis value for each cross-section is greater than 0, indicating that the peaks of the data are steeper than the peaks of the normal distribution resulting in a spiky state.

Table 1. Descriptive statistics of monitoring data.

Typology	Cross-Section	Average (mm)	Standard Deviation (mm)	Skewness	Kurtosis	J–B Statistic
Tunnel settlement deformation rate	DK115 + 261	0.2611	0.4294	0.5138	0.2869	7.3987
	DK115 + 270	0.2343	0.4292	0.8202	0.6275	20.6918
	DK115 + 280	0.3354	0.4405	0.7564	0.7089	13.7218
	DK115 + 290	0.3454	0.4327	0.7001	0.9317	13.0826
Tunnel convergence deformation rate	DK115 + 261	0.2711	0.3971	0.4901	0.3564	7.0700
	DK115 + 270	0.2509	0.4171	0.3174	0.0704	2.7189
	DK115 + 280	0.3469	0.3950	0.7898	1.5932	24.7456
	DK115 + 290	0.3686	0.3523	0.8295	1.7593	27.0453

In summary, the combined Jarque–Bera (J–B) statistic results are all significantly greater than 0. Under the joint effect of kurtosis and skewness, the deformation rate of each cross-section does not follow a normal distribution. This means that the deformation rate of the tunnel has obvious fractal characteristics, which can be investigated by the relevant methods of fractal theory.

3.1.2. Multifractal Analysis

1. MF-DFA key parameter settings

The values of the parameters in multifractal theory affect the calculation results in different ways. The features of the non-stationary time series differ significantly in various scenarios, including the length of the signal time window and the fluctuation trend. Therefore, the key parameters need to be tried and predicted to estimate more reliable results.

In order to effectively use multiple fractal theory to analyze the fractal characteristics of the tunnel displacement deformation rate, a double logarithmic scatterplot of $lgF_q(s) - lgs$ is plotted by varying the fluctuation order q ($q \in [-10, 10]$). Least squares fitting was used with the slope of the generalized Hurst exponent $h(q)$. Taking a set of multifractal features of tunnel settlement deformation rate and tunnel convergence deformation rate as an example, Figure 6 shows the trend of the q-order fluctuation function $lgF_q(s) - lgs$ through double logarithmic fitting.

The curve-fitting for the tunnel section settlement rate in Figure 6a shows better results when the value of lgs ranges from 1.5 to 1.65. At this time, the multifractal time length s is set as smin = 32 and smax = 45.

The curve-fitting for the tunnel section convergence rate in Figure 6b demonstrates improved performance when the value of lgs ranges from 1.1 to 1.4. At this time, the multifractal time length s is set as smin = 13 and smax = 25.

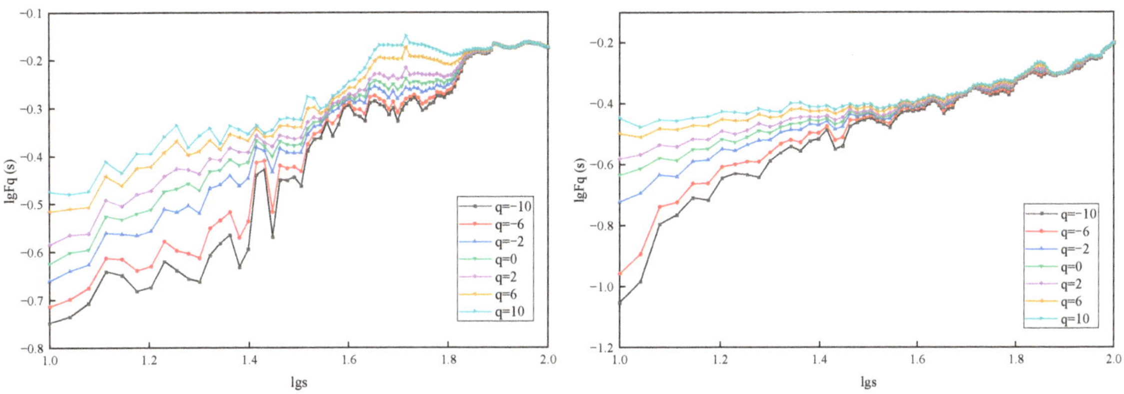

(**a**) (**b**)

Figure 6. Trend plot of q-order fluctuation function $lgF_q(s) - lgs$ double logarithmic fit: (**a**) settlement rate of section DK115 + 261, (**b**) convergence rate of section DK115 + 261.

2. Multifractal characterization of tunnel deformation rates

According to the settings of the above parameters, the data series for tunnel settlement deformation rate and tunnel convergence deformation rate were subjected to multifractals using MATLAB software (R2018b), respectively. The variation of the generalized Hurst index of the tunnel deformation rate data series is shown in Figure 7. The variation of the scalar function $\tau(q)$ of the tunnel deformation rate data series is shown in Figure 8. The multifractal spectrum of the tunnel deformation rate data series is shown in Figure 9.

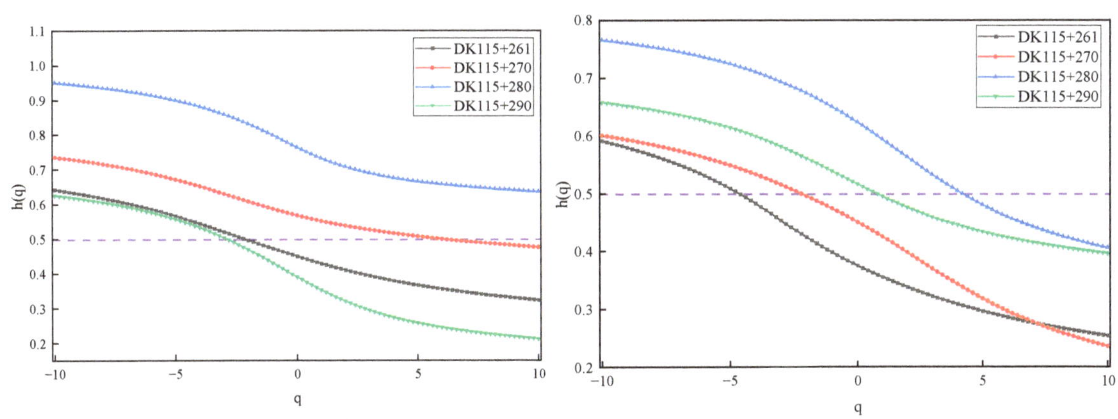

(**a**) (**b**)

Figure 7. Variation of generalized Hurst index: (**a**) tunnel settlement rate; (**b**) tunnel convergence rate.

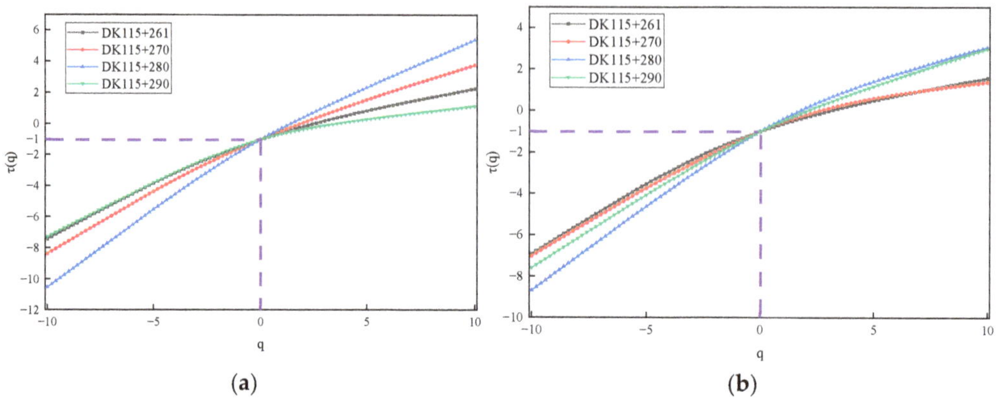

Figure 8. Variation of the scalar function $\tau(q)$: (**a**) tunnel settlement rate; (**b**) tunnel convergence rate.

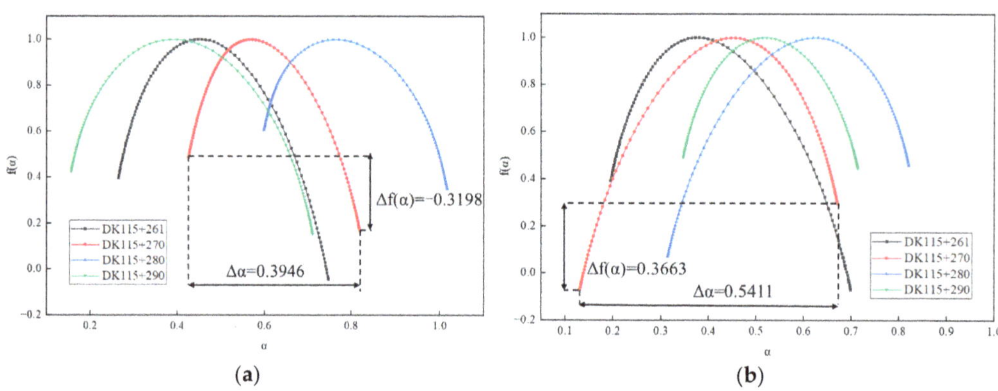

Figure 9. Multifractal spectrum: (**a**) tunnel settling rate; (**b**) tunnel convergence rate.

As can be seen from Figure 7, as q ranges from -10 to 10, the generalized Hurst index of each section measured data is non-constant. Rather, it exhibits a nonlinear decreasing trend with the change of q, suggesting that the measured data of each cross-section display clear multifractal characteristics. When $h(q) < 0.5$, the data series exhibits a memory process with inverse persistence, while $h(q) > 0.5$ indicates a memory process with positive persistence.

As can be seen from Figure 7a, at different fluctuation orders q, the generalized Hurst exponent curves of tunnel settlement rate at DK115 + 290 are concentrated in the lower fluctuations of the other sections. This indicates a weak multifractal nature.

As can be seen from Figure 7b, at different fluctuation orders q, the generalized Hurst exponent curves of the tunnel convergence rate of DK15 + 261 concentrate in the lower fluctuations of the other sections. This indicates a weak multifractal nature.

On the basis of $h(q)$, the Renyi index, i.e., the scaling function $\tau(q)$, is calculated. As can be seen from Figure 8, the consistency of the scalar function of the monitoring data of each cross-section is good, and the central part is up-convex, which satisfies $\tau(0) = -1$. Additionally, there is an overall nonlinear relationship, which further confirms that the monitoring data of each section have multifractal characteristics.

As can be seen from Figure 9, each image of the multifractal spectrum shows a single-peak convex distribution, which resembles a quadratic function curve. The local scales of the multiple fractals of the deformation rate of the tunnel in each section are not constant, reflecting the diversity of local variations at different moments. The singularity intensity α

is mainly concentrated on the two sides of the image, reflecting the uneven distribution of the fractal structure of the data series. The uneven distribution of α also confirms the multifractal characteristics of the measured point series; $f(\alpha)$ characterizes the fractal dimension of the subintervals of the data series with the same singularity index α, which is correlated with the distributional characteristics and fractal intensity of the data series. The statistics of the multifractal features of the tunnel settlement rate and tunnel convergence rate measurement data sequences are calculated from Equation (8), as shown in Table 2.

Table 2. Multifractal characterization statistics.

Feature \ Cross-Section		DK115 + 261	DK115 + 270	DK115 + 280	DK115 + 290
Tunnel settlement rate	$\Delta\alpha$	0.4829	0.3946	0.4187	0.5547
	$\Delta f(\alpha)$	−0.4415	−0.3198	−0.2551	−0.2704
Tunnel convergence rate	$\Delta\alpha$	0.5042	0.5411	0.5069	0.3674
	$\Delta f(\alpha)$	−0.4360	0.3663	0.3860	−0.0455

As can be seen from Figure 9a, the fractal spectrum of the tunnel settlement rate measurement data series for each section shows a clear right hook, indicating that the influence of small fluctuations is dominant. The results indicate that the tunnel settlement rate primarily exhibits minor fluctuations throughout the entire monitoring process. During the tunnel construction process, the settlement of the tunnel vault is minimally impacted by excavation and other external influences, remaining within the normal range of settlement changes.

As can be seen from Figure 9b, the sequence fractal spectrum of the tunnel convergence rate measurement data of DK115 + 290 is basically symmetrical, with good overall synergy and a stable development state. The sequence fractal spectrum of the tunnel convergence rate measurement data of DK115 + 261 shows an obvious right hook, indicating that the influence of small fluctuations is dominant. The fractal spectrum of the sequence of the tunnel convergence rate measurement data of DK115 + 270 and DK115 + 280 shows an obvious left hook, indicating that the influence of large fluctuations is dominant.

The fluctuation of the tunnel convergence rate in section DK115 + 290 remains stable throughout the entire monitoring process. During the tunnel construction process, the tunnel convergence in this section is minimally affected by excavation and other external influences, remaining within the normal range of convergence changes. Similarly, in section DK115 + 261, the tunnel convergence rate primarily exhibits minor fluctuations, with the tunnel convergence being less affected by excavation and other external influences, and remaining within the normal range of convergence changes. However, in sections DK115 + 270 and DK115 + 280, the tunnel convergence rate predominantly displays significant fluctuations. During the tunnel construction process in these sections, tunnel convergence is greatly affected by excavation and other external influences, leading to sudden changes. Consequently, it is imperative to reinforce monitoring in these areas.

Comparison of multifractal spectral width ($\Delta\alpha$). For the tunnel settlement rate measurement data series, the multifractal spectral width $\Delta\alpha = 0.5547$ for DK115 + 290 is the maximum value. It shows that its multifractal intensity is larger, and the fluctuation is more intense and complex. The multifractal spectral width $\Delta\alpha = 0.3946$ of DK115 + 270 is the minimum value. It shows that its multifractal intensity is smaller, and the fluctuation is smoother. For the tunnel convergence rate measurement data series, the multifractal spectrum width $\Delta\alpha = 0.5411$ for DK115 + 270 is the maximum value. It shows that its multifractal intensity is larger, and the fluctuation is more intense and complex. The multifractal spectral width $\Delta\alpha = 0.3674$ of DK115 + 290 is the minimum value. It shows that its multiple fractal intensity is smaller, and the fluctuation is smoother. The results suggest that a larger $\Delta\alpha$ corresponds to more dramatic and intricate fluctuations, while smaller data changes indicate that tunnel deformation is less affected by external influences such as excavation. Overall, the multiple fractal spectral widths of the measured data series of the tunnel settlement rate and tunnel convergence rate of the same section show an inverse change relationship, and the sum of the two is stable between 0.9 and 1, as shown in Figure 10a.

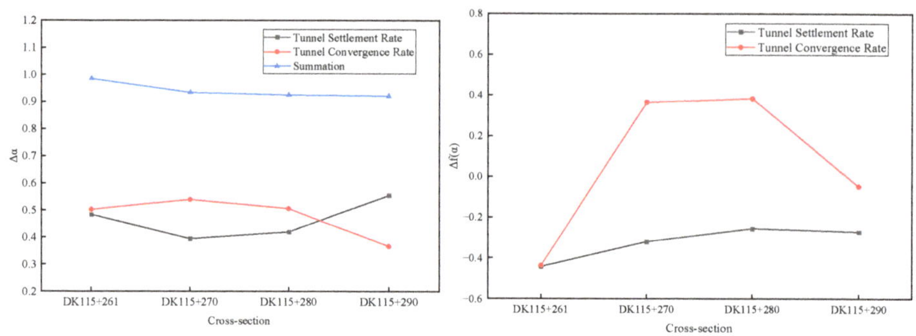

(a) **(b)**

Figure 10. Plot of multifractal feature statistics: (**a**) multifractal spectral width $\Delta\alpha$; (**b**) proportion of size fluctuations $\Delta f(\alpha)$.

Compare the proportions of large and small fluctuations ($\Delta f(\alpha)$). For the tunnel settlement rate measurement data series, the $\Delta f(\alpha)$ in DK115 + 280 is $\Delta f(\alpha) = -0.2251$, which is the maximum value. This indicates that the proportion of small fluctuations is larger. The $\Delta f(\alpha)$ for DK115 + 261 is $\Delta f(\alpha) = -0.4415$, which is the minimum value. This indicates that the proportion of small fluctuations is small. For the tunnel convergence rate measurement data series, the $\Delta f(\alpha)$ for DK115 + 280 is $\Delta f(\alpha) = 0.3663$, which is the maximum value. This indicates that the proportion of small fluctuations is larger. The $\Delta f(\alpha)$ for DK115 + 261 is $\Delta f(\alpha) = -0.4360$, which is the minimum value. This indicates that the proportion of small fluctuations is small. The findings suggest that a larger $\Delta f(\alpha)$ corresponds to small data changes, indicating that tunnel deformation is less affected by external influences such as excavation. Overall, the $\Delta f(\alpha)$ in the measured data series of the tunnel settlement rate and tunnel convergence rate of the same section shows the same directional change relationship, as shown in Figure 10b.

3.2. Tunnel Deformation Warning Classification Study

3.2.1. Tunnel Deformation Warning Level Classification Criteria

Based on the research results in the literature [31,61,62], the tunnel displacement monitoring criterion is constructed through the $\Delta\alpha$ and $\Delta f(\alpha)$ parameters derived from the tunnel displacement monitoring data in order to realize the tunnel displacement warning level classification. The specific criteria are set as shown in Table 3.

Table 3. Tunnel warning level classification criteria.

Warning Level	$\Delta\alpha$ Indicator Criterion	$\Delta f(\alpha)$ Indicator Criterion	Treatment Measures
I	Decreasing trend	Increasing trend	Suspend construction. Use additional temporary support, grouting, and other measures to reinforce the deformed section.
II	Other trend combinations		Conduct a comprehensive evaluation of design and construction measures, reinforce monitoring and measurement protocols, develop disaster prevention plans, and implement appropriate engineering countermeasures when necessary.
III	Steady trend	Steady trend	Normal construction.

The trends of the two indicators need to be satisfied at the same time. In case of inconsistency in warning levels, the final warning level is determined according to the most unfavorable principle.

3.2.2. Warning Classification of Tunnel Deformation

The tunnel displacement monitoring data in Figure 4 are used to obtain the required sets of $\Delta\alpha$ and $\Delta f(\alpha)$ parameters. In order to realize the trend judgment of the two discriminant indicators, 100 groups of tunnel displacement monitoring data are divided into one group. Synthesizing the actual situation and monitoring data, a total of 12 groups are divided. Additionally, calculate $\Delta\alpha$ and $\Delta f(\alpha)$ parameters for each group of tunnel displacement monitoring data, as shown in Figure 11.

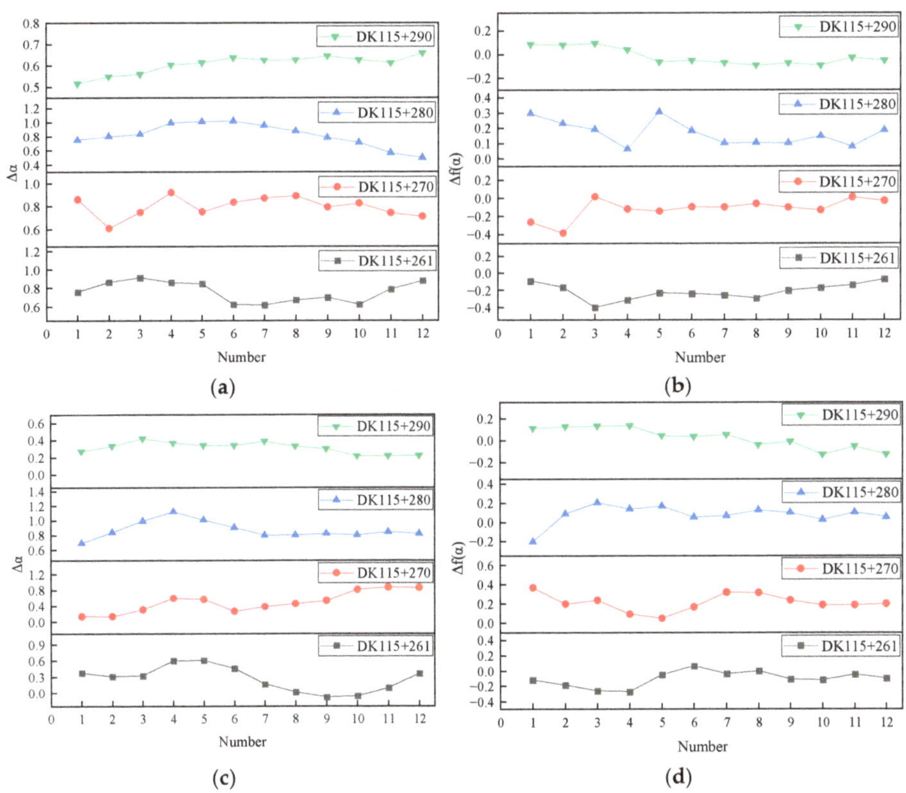

Figure 11. Calculated values of $\Delta\alpha$ and $\Delta f(\alpha)$ parameters: (**a**) $\Delta\alpha$ parameter values for tunnel settlement, (**b**) $\Delta f(\alpha)$ parameter values for tunnel settlement, (**c**) $\Delta\alpha$ parameter values for tunnel convergence, (**d**) $\Delta f(\alpha)$ parameter values for tunnel convergence.

The trends of the two discriminant indicators were judged using the M–K test to achieve the early warning grading of tunnel displacement. The results are analyzed as follows:

- The analysis of $\Delta\alpha$ index criterion results: Through calculations and statistical analysis, the results of the $\Delta\alpha$ index criterion are obtained (see Table 4). In the cross-section settlement monitoring data, the Z values for DK115 + 261 and DK115 + 270 fall within the range $[-2.32, 2.32]$, which is a steady trend. The Z value for DK115 + 280 is less than -2.32, which is a decreasing trend. The Z value for DK115 + 290 is more than 2.32, which is an increasing trend. In the cross-section convergence monitoring data, the Z value for DK115 + 280 falls within the range [−2.32, 2.32], which is a steady trend. The Z values for DK115 + 261 and DK115 + 290 are less than −2.32, which is a decreasing trend. The Z value for DK115 + 270 is more than 2.32, which is an increasing trend.

- The analysis of $\Delta f(\alpha)$ index criterion results: Through statistical calculations, the results of $\Delta f(\alpha)$ index criterion are obtained (see Table 5). In the section settlement monitoring data, the Z value for DK115 + 261 falls between the ranges [−2.32, 2.32], which is a steady trend. The Z value for DK115 + 270 is greater than 2.32, which is an increasing trend. The Z values for DK115 + 280 and DK115 + 290 are less than −2.32, which is a decreasing trend. In the section convergence monitoring data, the Z values for DK115 + 261, DK115 + 270, and DK115 + 280 fall within the range [−2.32, 2.32], which is a steady trend. The Z value for DK115 + 290 is less than −2.32, which is a decreasing trend.
- The analysis of final warning results: On the basis of the results for the $\Delta\alpha$ indicator criterion and $\Delta f(\alpha)$ indicator criterion, the final warning results of the four monitoring cross-sections are analyzed (see Table 6). In the section settlement, the warning level for DK115 + 261 is level III, and all other sections are at level II. In section convergence, the warning level of DK115 + 280 is grade III, and all other sections are grade II. Therefore, according to the most unfavorable principle of synthesis, the final warning level of the four sections are all level II. That is, monitoring and measurement should be strengthened, and corresponding engineering countermeasures should be taken if necessary.

Table 4. Results of the $\Delta\alpha$ indicator criterion.

Cross-Section	Section Settlement		Section Convergence	
	Z-Value	**Growing Trend**	**Z-Value**	**Growing Trend**
DK115 + 261	−1.3540	Steady trend	−2.5849	Decreasing trend
DK115 + 270	−0.8616	Steady trend	5.0468	Increasing trend
DK115 + 280	−2.5849	Decreasing trend	−0.8616	Steady trend
DK115 + 290	5.5391	Increasing trend	−2.8311	Decreasing trend

Table 5. Results for the $\Delta f(\alpha)$ indicator criterion.

Cross-Section	Section Settlement		Section Convergence	
	Z-Value	**Growing Trend**	**Z-Value**	**Growing Trend**
DK115 + 261	2.0926	Steady trend	1.6002	Steady trend
DK115 + 270	2.8311	Increasing trend	−0.3693	Steady trend
DK115 + 280	−2.5849	Decreasing trend	−0.8616	Steady trend
DK115 + 290	−4.3082	Decreasing trend	−5.2929	Decreasing trend

Table 6. Final warning results of tunnel displacement.

Cross-Section	DK115 + 261	DK115 + 270	DK115 + 280	DK115 + 290
Section settlement	III	II	II	II
Section convergence	II	II	III	II
Combined warning levels	II	II	II	II

3.3. *Tunnel Settlement Prediction*

3.3.1. Prediction Model Selection

In order to select the tunnel settlement prediction model, the performance of the PSO-LSTM and Back Propagation Neural Network (BP-ANN) prediction models is compared and analyzed. The monitoring data of section DK115 + 261 is used as the sample data to establish the settlement prediction model of PSO-LSTM and BP-ANN. 80% of the data is used as the sample training set, and 20% of the data is allocated to the test set. The delay step of prediction model is set to 6, and the prediction is carried out across 1 time point. The adaptation change curve of PSO-LSTM and BP-ANN are shown in Figure 12.

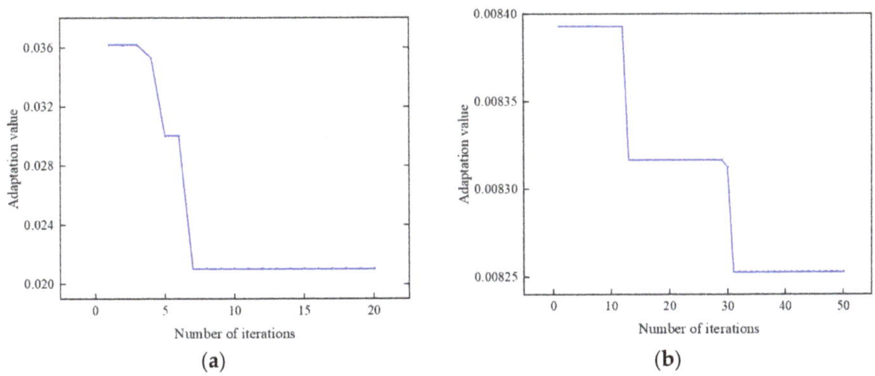

Figure 12. Adaptation change curves for different prediction models: (**a**) PSO-LSTM adaptation change curve; (**b**) BP-ANN adaptation change curve.

The tunnel section settlement is predicted, and the prediction results of the two models are compared. The tunnel settlement prediction results for different models are shown in Figure 13. Further, the performance of each model is evaluated using the model evaluation index. The calculation results of the evaluation indexes for the model test set are shown in Table 7. PSO-LSTM predicts significantly better than BP-ANN. Consequently, PSO-LSTM was selected for predicting tunnel settlement data.

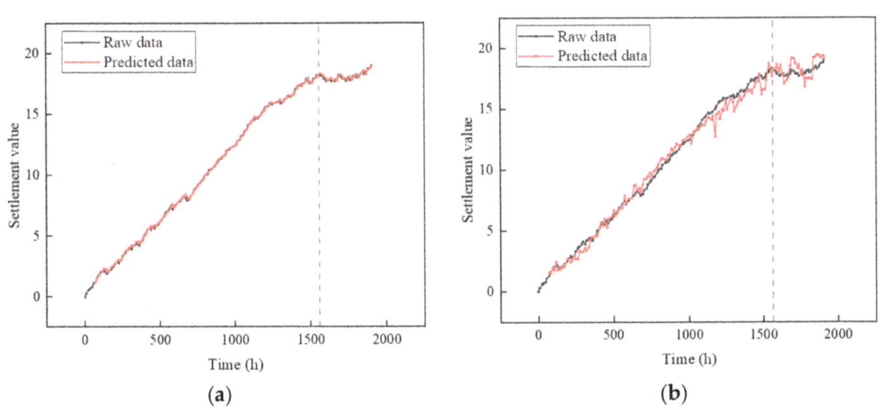

Figure 13. Tunnel settlement prediction results for different prediction models: (**a**) PSO-LSTM prediction results; (**b**) BP-ANN prediction results.

Table 7. Results of calculated performance evaluation for the model.

Prediction Model	R^2	*MAE*	*RMSE*
PSO-LSTM	0.98	0.05	0.06
BP-ANN	0.71	0.16	0.18

3.3.2. Optimization of PSO-LSTM Parameters

To make the forecast results more concise, subsequent forecasts will be conducted daily. A program was written using Python software (3.9.5) to optimize the parameters of the LSTM prediction model using the PSO algorithm program. The parameters in PSO are set and a range is delineated for the optimization parameters in LSTM: the number of neurons ranges from 10 to 200, the number of training rounds ranges from 10 to 100, and the batch size ranges from 1 to 10. Initialize each parameter of the PSO-LSTM prediction model. Since the total amount of monitoring data is different for each section, after each set of data is input into the PSO program and finishes running, the optimal number of neurons, the number of training rounds, and the batch size of the model are obtained for the 4 groups. The details are shown in Table 8.

Table 8. PSO-LSTM optimal parameters.

Optimal Parameter	Number of Neurons	Training Rounds	Batch Size
DK115 + 261	77	72	1
DK115 + 270	82	72	1
DK115 + 280	11	99	6
DK115 + 290	20	81	2

3.3.3. PSO-LSTM Prediction Results

In this study, the PSO-LATM prediction model was constructed using the Python software platform to complete the prediction analysis of cumulative tunnel settlement. The parameters for the LSTM neural network were calculated by PSO. Eighty percent of the data from each section was used as a sample training set and 20% as a test set. R^2, MAE, and $RMSE$ were used as evaluation indexes for predicting accuracy. Tunnel settlement was predicted for the next 5 days, and the prediction results are shown in Table 9.

Table 9. Predicted cumulative tunnel settlement values.

Cross-Section	Test Set Evaluation Metrics			Predicted Results				
	R^2	MAE	$RMSE$	Day 1	Day 2	Day 3	Day 4	Day 5
DK115 + 261	0.96	0.28	0.06	18.80	18.81	18.81	18.82	18.82
DK115 + 270	0.97	0.10	0.01	18.31	18.32	18.34	18.35	18.36
DK115 + 280	0.98	0.04	0.04	17.50	17.51	17.51	17.51	17.52
DK115 + 290	0.98	0.02	0.02	16.52	16.54	16.55	16.57	16.59

For the test set results of settlement prediction for four cross-sections, the test set R^2 values for the prediction model are 0.96, 0.97, 0.98, and 0.98. The MAE values are 0.28, 0.10, 0.04, and 0.02. The $RMSE$ values are 0.06, 0.01, 0.04, and 0.02. Among these, the closer the R^2 values for prediction result converge to 1, the smaller the MAE and $RMSE$ become, and the higher the model prediction accuracy is. Therefore, the settlement prediction results of the PSO-LSTM settlement prediction model for the four sections are analyzed. The model's predictions for the tunnel section settlement are in good agreement with the measured data.

Analyzing the prediction results, the future settlements of the four sections generally show a slowly increasing trend. The accuracy of tunnel warning classification is further verified, and, based on the evaluation indexes, the PSO-LSTM model demonstrates high accuracy in predicting tunnel settlement.

4. Conclusions

In this study, the multifractal theory was utilized to the analyze tunnel deformation, and the sliding time window was employed to enhance the segmentation of traditional multifractal subintervals. Furthermore, the M–K analysis method was utilized to ascertain the multifractal feature trends and establish the warning level for the tunnel section. Subsequently, the accuracy of the warning level was validated by predicting tunnel settlement using the PSO-LSTM prediction model. The following conclusions were drawn:

1. The tunnel settlement and convergence rates of the four sections exhibit distinct fractal sequence characteristics. The width of the multifractal spectra in the measured data series of tunnel settlement rate and tunnel convergence rate within the same section shows an inverse relationship, with the sum of the two remaining stable between 0.9 and 1. Additionally, the proportion of size fluctuations in the measured data series of the tunnel settlement rate and the tunnel convergence rate within the same section demonstrate a consistent trend.
2. The analysis of tunnel settlement prediction indicates that the PSO-LSTM prediction model delivers superior predictive performance and stability in tunnel settlement forecasts.
3. A comprehensive analysis of the tunnel warning level and tunnel settlement prediction results reveals a class II tunnel deformation warning level, which aligns with the actual tunnel conditions. This approach, leveraging quantitative data as a reference, enables a more precise determination of the tunnel warning level.

Author Contributions: Conceptualization, C.Y. and Y.T.; methodology, C.Y.; software, C.Y. and R.H.; validation, R.H., W.Q. and R.Z.; formal analysis, D.L.; investigation, R.H.; resources, D.L.; data curation, W.Q. and R.Z.; writing—original draft preparation, C.Y.; writing—review and editing, Y.T. and D.L.; visualization, W.Q. and R.Z.; supervision, W.Q. and R.Z.; project administration, D.L.; funding acquisition, Y.T. All authors have read and agreed to the published version of the manuscript.

Funding: This project was supported by the Scientific Research Fund of Hunan Provincial Education Department, grant number 23C0091.

Data Availability Statement: The data that support the findings of this study are available upon request from the authors.

Acknowledgments: The authors would like to thank Road and Bridge North China Engineering Co., Ltd. for their assistance in conducting the monitoring of data.

Conflicts of Interest: Authors Weichao Qiu and Ruiping Zhang were employed by the company Road & Bridge North China Engineering Co., Ltd. The remaining authors declare that the research was conducted in the absence of any commercial or financial relationships that could be construed as a potential conflict of interest.

References

1. Liu, C.; Liu, Y.; Chen, Y.; Zhao, C.; Qiu, J.; Wu, D.; Liu, T.; Fan, H.; Qin, Y.; Tang, K. A State-of-the-Practice Review of Three-Dimensional Laser Scanning Technology for Tunnel Distress Monitoring. *J. Perform. Constr. Facil.* **2023**, *37*, 03123001. [CrossRef]
2. Cao, Y.; Zhang, Z.; Cheng, F.; Su, S. Trajectory Optimization for High-Speed Trains via a Mixed Integer Linear Programming Approach. *IEEE Trans. Intell. Transp. Syst.* **2022**, *23*, 17666–17676. [CrossRef]
3. Zhao, T. Research on Invert Heave Mechanism and Control Technique of High-speed Railway Tunnel in Mudstone. Ph.D. Thesis, Lanzhou Jiaotong University, Lanzhou, China, 2022.
4. Chen, H.; Lai, H.; Qiu, Y.; Chen, R. Reinforcing Distressed Lining Structure of Highway Tunnel with Bonded Steel Plates: Case Study. *J. Perform. Constr. Facil.* **2020**, *34*, 04019082. [CrossRef]
5. Li, Z.; Lai, J.; Ren, Z.; Shi, Y.; Kong, X. Failure mechanical behaviors and prevention methods of shaft lining in China. *Eng. Fail. Anal.* **2023**, *143*, 106904. [CrossRef]
6. Weng, X.; Li, H.; Hu, J.; Li, L.; Xu, L. Behavior of Saturated Remolded Loess Subjected to Coupled Change of the Magnitude and Direction of Principal Stress. *Int. J. Géoméch.* **2023**, *23*, 04022244. [CrossRef]
7. Wang, Z.; Cai, Y.; Fang, Y.; Lai, J.; Han, H.; Liu, J.; Lei, H.; Kong, X. Local buckling characteristic of hollow π-type steel-concrete composite support in hilly-gully region of loess tunnel. *Eng. Fail. Anal.* **2023**, *143*, 106828. [CrossRef]
8. Zhou, M.; Fang, Q.; Peng, C. A mortar segment-to-segment contact method for stabilized total-Lagrangian smoothed particle hydrodynamics. *Appl. Math. Model.* **2022**, *107*, 20–38. [CrossRef]
9. Zhen, Y.; Guo, P.; Wang, L.; Chen, X.; Duan, X.; Wang, A. Key Technologies for Treating High Ground Stress and Large Deformation of Soft Rock in Daliangshan Tunnel of Yunlin Expressway. *Tunn. Constr.* Available online: https://link.cnki.net/urlid/44.1745.U.20231107.1430.003 (accessed on 8 November 2023).

10. Zhang, C.; Han, K.; Zhang, D. Face stability analysis of shallow circular tunnels in cohesive-frictional soils. *Tunn. Undergr. Space Technol.* **2015**, *50*, 345–357. [CrossRef]
11. Li, W.; Zhang, C.; Zhang, D.; Ye, Z.; Tan, Z. Face stability of shield tunnels considering a kinematically admissible velocity field of soil arching. *J. Rock Mech. Geotech. Eng.* **2021**, *14*, 505–526. [CrossRef]
12. Li, G.; Hu, Y.; Tian, S.-M.; Weibin, M.; Huang, H.-L. Analysis of deformation control mechanism of prestressed anchor on jointed soft rock in large cross-section tunnel. *Bull. Eng. Geol. Environ.* **2021**, *80*, 9089–9103. [CrossRef]
13. Wu, H.-N.; Shen, S.-L.; Chen, R.-P.; Zhou, A. Three-dimensional numerical modelling on localised leakage in segmental lining of shield tunnels. *Comput. Geotech.* **2020**, *122*, 103549. [CrossRef]
14. Shen, S.-L.; Wu, H.-N.; Cui, Y.-J.; Yin, Z.-Y. Long-term settlement behaviour of metro tunnels in the soft deposits of Shanghai. *Tunn. Undergr. Space Technol.* **2014**, *40*, 309–323. [CrossRef]
15. Lai, H.; Zhao, X.; Kang, Z.; Chen, R. A new method for predicting ground settlement caused by twin-tunneling under-crossing an existing tunnel. *Environ. Earth Sci.* **2017**, *76*, 726. [CrossRef]
16. Suwansawat, S.; Einstein, H.H. Describing settlement troughs over twin tunnels using a superposition technique. *Geotech. Geoenviron. Eng.* **2007**, *133*, 445–468. [CrossRef]
17. Chou, W.I.; Bobet, A. Predictions of ground deformations in shallow tunnels in clay. *Tunn. Undergr. Space Technol.* **2002**, *17*, 3–19. [CrossRef]
18. Zhang, J.-Z.; Huang, H.-W.; Zhang, D.-M.; Phoon, K.K.; Liu, Z.-Q.; Tang, C. Quantitative evaluation of geological uncertainty and its influence on tunnel structural performance using improved coupled Markov chain. *Acta Geotech.* **2021**, *16*, 3709–3724. [CrossRef]
19. Li, Y.; Li, J.; Zhao, J.; Zhao, T.; Guo, D. Research on a Safety Evaluation System for Railway-Tunnel Structures by Fuzzy Comprehensive Evaluation Theory. *Civ. Eng. J.-Staveb. Obz.* **2023**, *32*, 122–136. [CrossRef]
20. Li, Z.; Meng, X.; Liu, D.; Tang, Y.; Chen, T. Disaster Risk Evaluation of Superlong Highways Tunnel Based on the Cloud and AHP Model. *Adv. Civ. Eng.* **2022**, *2022*, 8785030. [CrossRef]
21. Yan, X.; Li, H.; Liu, F.; Liu, Y. Structural Safety Evaluation of Tunnel Based on the Dynamic Monitoring Data during Construction. *Shock Vib.* **2021**, *2021*, 6680675. [CrossRef]
22. Tan, X.; Chen, W.; Zou, T.; Yang, J.; Du, B. Real-time prediction of mechanical behaviors of underwater shield tunnel structure using machine learning method based on structural health monitoring data. *J. Rock Mech. Geotech. Eng.* **2023**, *15*, 886–895. [CrossRef]
23. Tang, L.; Na, S. Comparison of machine learning methods for ground settlement prediction with different tunneling datasets. *J. Rock Mech. Geotech. Eng.* **2021**, *13*, 1274–1289. [CrossRef]
24. Zhang, W.; Li, Y.; Wu, C.; Li, H.; Goh, A.; Liu, H. Prediction of lining response for twin tunnels constructed in anisotropic clay using machine learning techniques. *Undergr. Space* **2022**, *7*, 122–133. [CrossRef]
25. Suwansawat, S.; Einstein, H.H. Artificial neural networks for predicting the maximum surface settlement caused by EPB shield tunneling. *Tunn. Undergr. Space Technol.* **2006**, *21*, 133–150. [CrossRef]
26. Yang, H.; Song, K.; Zhou, J. Automated Recognition Model of Geomechanical Information Based on Operational Data of Tunneling Boring Machines. *Rock Mech. Rock Eng.* **2022**, *55*, 1499–1516. [CrossRef]
27. Schappacher, N. Mandelbrot, Benoît B. The Fractal Geometry of Nature. In *Kindlers Literatur Lexikon (KLL)*; WH Freeman: New York, NY, USA, 2020; pp. 1–2.
28. Zuo, C.; Liu, D.; Ding, S.; Li, L. Analysis and Prediction of Tunnel Surface Subsidence Based on Fractal Theory. *J. Yangtze River Sci. Res. Inst.* **2016**, *33*, 51–56. [CrossRef]
29. Ye, D.; Liu, G.; Tian, Y.; Sun, Z.; Yu, B. A Fractal Model for the Micro–Macro Interactions on Tunnel Leakage. *Fractals* **2022**, *30*, 2250142. [CrossRef]
30. Grassberger, P. Generalized dimensions of strange attractors. *Phys. Lett. A* **1983**, *97*, 227–230. [CrossRef]
31. Lei, H.; Zhou, X.; Wang, Y. Research on Landslide Early Warning and Prediction Based on Combined Response of Multifractal Characteristics and Sub Item Prediction. *J. Geod. Geodyn.* **2022**, *42*, 885–891.
32. Mao, H.; Zhang, M.; Jiang, R.; Li, B.; Xu, J.; Xu, N. Study on deformation pre-warning of rock slopes based on multi-fractal characteristics of microseismic signals. *Chin. J. Rock Mech. Eng.* **2020**, *39*, 560–571.
33. Zhou, L.; Liu, Z. Multifractal feature analysis method for measured data of dam deformation. *Adv. Sci. Technol. Water Resour.* **2021**, *41*, 18–24. [CrossRef]
34. Hughes, H.M. The relative cuttability of coal-measures stone. *Min. Sci. Technol.* **1986**, *3*, 95–109. [CrossRef]
35. Hamidi, J.K.; Shahriar, K.; Rezai, B.; Rostami, J. Performance prediction of hard rock TBM using Rock Mass Rating (RMR) system. *Tunn. Undergr. Space Technol.* **2010**, *25*, 333–345. [CrossRef]
36. Goh, A.T.C.; Zhang, W.; Zhang, Y.; Xiao, Y.; Xiang, Y. Determination of earth pressure balance tunnel-related maximum surface settlement: A multivariate adaptive regression splines approach. *Bull. Eng. Geol. Environ.* **2018**, *77*, 489–500. [CrossRef]
37. Ozdemir, L. Development of Theoretical Equations for Predicting Tunnel Boreability. Ph.D. Thesis, Colorado School of Mines, Golden, CO, USA, 1977.
38. Resendiz, D.; Romo, M.P. Settlements upon soft-ground tunneling: Theoretical solution. *Int. J. Rock Mech. Min. Sci. Geomech. Abstr.* **1983**, *151*, 65–74.

39. Bai, B. Fluctuation responses of saturated porous media subjected to cyclic thermal loading. *Comput. Geotech.* **2006**, *33*, 396–403. [CrossRef]
40. Bai, B.; Nie, Q.; Zhang, Y.; Wang, X.; Hu, W. Cotransport of heavy metals and SiO_2 particles at different temperatures by seepage. *J. Hydrol.* **2020**, *597*, 125771. [CrossRef]
41. Rowe, R.K.; Lee, K.M. Subsidence owing to tunnelling. II. Evaluation of a prediction technique. *Can. Geotech. J.* **1992**, *29*, 941–954. [CrossRef]
42. Yuan, B.; Li, Z.; Su, Z.; Luo, Q.; Chen, M.; Zhao, Z. Sensitivity of Multistage Fill Slope Based on Finite Element Model. *Adv. Civ. Eng.* **2021**, *2021*, 6622936. [CrossRef]
43. Yuan, B.; Li, Z.; Zhao, Z.; Ni, H.; Su, Z.; Li, Z. Experimental study of displacement field of layered soils surrounding laterally loaded pile based on transparent soil. *J. Soils Sediments* **2021**, *21*, 3072–3083. [CrossRef]
44. Yuan, B.; Li, Z.; Chen, Y.; Ni, H.; Zhao, Z.; Chen, W.; Zhao, J. Mechanical and microstructural properties of recycling granite residual soil reinforced with glass fiber and liquid-modified polyvinyl alcohol polymer. *Chemosphere* **2021**, *286*, 131652. [CrossRef]
45. Chou, J.; Lin, C. Predicting Disputes in Public-Private Partnership Projects: Classification and Ensemble Models. *J. Comput. Civ. Eng.* **2013**, *27*, 51–60. [CrossRef]
46. Bouayad, D.; Emeriault, F. Modeling the relationship between ground surface settlements induced by shield tunneling and the operational and geological parameters based on the hybrid PCA/ANFIS method. *Tunn. Undergr. Space Technol.* **2017**, *68*, 142–152. [CrossRef]
47. Santos, O.J.; Celestino, T.B. Artificial neural networks analysis of Sao Paulo subway tunnel settlement data. *Tunn. Undergr. Space Technol.* **2008**, *23*, 481–491. [CrossRef]
48. Hasanipanah, M.; Noorian-Bidgoli, M.; Armaghani, D.J.; Khamesi, H. Feasibility of PSO-ANN model for predicting surface settlement caused by tunneling. *Eng. Comput.* **2016**, *32*, 705–715. [CrossRef]
49. Chen, Y.; Zhao, B.; Wang, H.; Zheng, J.; Gao, Y. Time-Series InSAR Ground Deformation Prediction Based an LSTM Model. *Yangtze River*. Available online: https://link.cnki.net/urlid/42.1202.TV.20231117.0949.002 (accessed on 17 November 2023).
50. Li, C.; Li, J.; Shi, Z.; Li, L.; Li, M.; Jin, D.; Dong, G. Prediction of Surface Settlement Induced by Large-Diameter Shield Tunneling Based on Machine-Learning Algorithms. *Geofluids* **2022**, *2022*, 4174768. [CrossRef]
51. Cao, Y.; Zhou, X.; Yan, K. Deep Learning Neural Network Model for Tunnel Ground Surface Settlement Prediction Based on Sensor Data. *Math. Probl. Eng.* **2021**, *2021*, 9488892. [CrossRef]
52. Duan, C.; Hu, M.; Zhang, H. Comparison of ARIMA and LSTM in Predicting Structural Deformation of Tunnels during Operation Period. *Data* **2023**, *8*, 104. [CrossRef]
53. Mann, H.B. Nonparametric Tests Against Trend. *Econometrica* **1945**, *13*, 245–259. [CrossRef]
54. Kendall, M.G.; Gibbons, J.D. Rank Correlation Method. *Biometrika*. 1990. Available online: http://www.jstor.org/stable/2333282 (accessed on 18 June 2014).
55. Wang, L.I.; Gao, X.; Zhou, W. Testing for Intrinsic Multifractality in the Global Grain Spot Market Indices: A Multifractal Detrended Fluctuation Analysis. *Fractals* **2023**, *31*, 2350090. [CrossRef]
56. Wang, F.; Shao, W.; Yu, H.; Kan, G.; He, X.; Zhang, D.; Ren, M.; Wang, G. Re-evaluation of the Power of the Mann-Kendall Test for Detecting Monotonic Trends in Hydrometeorological Time Series. *Front. Earth Sci.* **2020**, *8*, 14. [CrossRef]
57. Hochreiter, S.T.U.M.; Schmidhuber, J. Long short-term memory. *Neural Comput.* **2010**, *9*, 1735–1780. [CrossRef] [PubMed]
58. Kennedy, J.; Eberhart, R. Particle Swarm Optimization. In Proceedings of the ICNN'95-International Conference on Neural Networks, Perth, WA, Australia, 27 November–1 December 1995.
59. Wu, Z. Time-Varying Risk Aversion and Crude Oil Futures Price Volatility. Master's Thesis, Anhui University of Finance & Economics, Bengbu, China, 2023.
60. Xie, W. Study on the Interdependence Structure and Risk Spillover Effect between Cryptocurrency and Chinese financial Assets. Ph.D. Thesis, Nanjing University of Information Science & Technology, Nanjing, China, 2023.
61. Guo, Y.; Zheng, J.; Liu, H. An early warning method for tunneling-induced ground surface settlement considering accident precursors and consequences. *Tunn. Undergr. Space Technol.* **2023**, *140*, 105214. [CrossRef]
62. Wang, J.; Wang, X. Early Warning and Prediction of Side Displacement and Deformation of Soft Soil Foundation Pit. *J. Yangtze River Sci. Res. Inst.* **2021**, *38*, 91–96.

fractal and fractional

MDPI

Article

Multifractal Characteristics of Smooth Blasting Overbreak in Extra-Long Hard Rock Tunnel

Wanmao Zhang [1], Dunwen Liu [1,*], Yu Tang [1,2,*], Weichao Qiu [3] and Ruiping Zhang [3]

[1] School of Resources and Safety Engineering, Central South University, Changsha 410083, China; zhangwm@csu.edu.cn
[2] College of Water Resources and Civil Engineering, Hunan Agricultural University, Changsha 410128, China
[3] Road & Bridge North China Engineering Co., Ltd., Beijing 101100, China; 225512138@csu.edu.cn (W.Q.); 215512141@csu.edu.cn (R.Z.)
* Correspondence: dunwen@csu.edu.cn (D.L.); tangyu12@csu.edu.cn (Y.T.)

Abstract: With the development of infrastructure construction in mountainous areas, the number of new extra-long tunnels is increasing. However, these tunnels often face the challenge of complex and variable surrounding rock grades, resulting in a large number of overbreak and underbreak due to the untimely adjustment of smooth blasting parameters. This study focuses on the optimization of the peripheral hole charging structure and blasting parameters for extra-long hard rock tunnels, aiming to improve the effectiveness of smooth blasting technology. The results of this study demonstrate a significant improvement in the effect of smooth blasting after implementing bidirectional polymerization blasting in the tunnel. A comparison between the bidirectional shaped charge and spaced decoupled charge blasting reveals that the former yields better results. To obtain accurate data on the tunnel section profile during excavation, a laser cross-section meter is used for measurement. Furthermore, this study quantitatively compares the optimization effect of smooth blasting parameters. The multifractal characteristics of the tunnel profile overbreak point sequences are analyzed under different smooth blasting schemes using the multifractal detrended fluctuation analysis (MF-DFA) method. It is found that both the spaced decoupled charge and the bidirectional shaped charge blasting exhibit multifractal features in the overbreak measurement point sequences. The calculation results of the multifractal features of the tunnel profile under different smooth blasting plans are in line with the actual situation.

Keywords: extra-long tunnel; smooth blasting; laser profiler; overbreak and underbreak; multifractal detrending fluctuation analysis

1. Introduction

With the rapid development of economic globalization, the scale and quantity of railroad tunnel construction in China have significantly increased. China has the largest number of tunnels, the fastest development speed, and the most complex geological and structural forms in the world [1]. By the end of 2022, the total mileage of China's railroads reached 155,000 km, with over 42,000 km of high-speed railroads in operation. Among them, 4178 high-speed railroad tunnels, totaling 7032 km, have been constructed, including 105 tunnels longer than 10 km, with a total length of approximately 1339 km. In recent years, the construction of extra-long railroad tunnels in China has rapidly advanced. With the gradual expansion of the construction scale of railroad tunnels, tunnel construction technology is also in constant development [2].

In the process of constructing extra-long tunnels, excavation is the most critical and time-consuming process, which significantly impacts the construction period. Long tunnels have limited working faces, large project volumes, complex hydrogeological conditions, and numerous uncontrollable factors [3,4]. As a result, the construction period of extra-long tunnels is typically the determining factor for the overall project. With the development of infrastructure construction in mountainous areas, the number of new extra-long tunnels is on the rise. This increase in tunnel construction complexity, longer construction periods, and higher construction risks pose challenges for dealing with complex geological conditions [5]. Blasting is the first and most crucial process in tunnel excavation, and subsequent processes such as mucking and support are based on successful blasting operations [6]. The effectiveness of blasting has a significant impact on project progress, quality, and cost [7], particularly in long hard rock tunnels where smooth blasting plays a critical role in maintaining the construction schedule [8,9]. Therefore, how to ensure the effects of smooth blasting in extra-long tunnels will be a difficult problem in tunnel construction.

The effectiveness of smooth blasting in extra-long tunnels is influenced by various complex factors. In addition to the blasting parameters (perimeter hole spacing, charge, decoupling coefficient, and minimum burden), the surrounding rock conditions (degree of fissure development, rock properties, and rock strength) and drilling accuracy also play important roles [10–13]. Therefore, in tunnels with complex geological conditions and variable surrounding rock grades, the untimely adjustment of blasting parameters often leads to a large number of overbreak and underbreak, resulting in project delays [7,14–17]. Relevant project data indicate that the time spent on dealing with construction quality issues such as overbreak accounts for 30% of the total construction period of the tunnel [18]. Overbreak not only increases the amount of concrete required for the initial support or even the secondary lining, but also affects the safety of the tunnel cavity [19]. Addressing underbreak is also a critical aspect of tunnel construction. If underbreak is treated using drill and blast methods, it can be time-consuming, and dynamite blasting to address undercutting may lead to overbreak.

It is well-known that new blasting methods such as supercritical carbon dioxide, soundless chemical demolition agents, and high-pressure gas expansion have the advantages of safety and environmental protection [20,21]. These new blasting methods can effectively control the overbreak and underbreak problems of tunnel blasting. However, due to the high cost of new blasting, it cannot be widely used in extra-long hard rock tunnels. Traditional blasting is widely used in extra-long tunnels due to its mature technology and economic benefits. Currently, the optimization of blasting parameters and improvement of blasting equipment are the main focus in the research of extra-long hard rock tunnel blasting technology [10,22]. The efficiency of blasting construction is a crucial factor that determines the progress of the entire project. To optimize the drilling and blasting technology for rapid tunnel boring in hard rock tunnels, scholars have conducted research on explosive selection, drilling accuracy, trenching program, blasting footage, detonation network, and other aspects [3,10,23]. Through the improvement of blasting parameters and blasting equipment, the efficiency of extra-long hard rock tunnel boring projects has significantly improved. Some scholars have optimized the blasting parameters using evaluation and prediction models [24–27]. For instance, Jang et al. [28] predicted the distribution of the fragmentation size of the debris after blasting using a neural network model, obtained the relationship between the rock fragmentation size and blasting parameters, and optimized the blasting parameters accordingly. Ma et al. [29] conducted a series of bursting tests in the tunnel excavation face to improve the construction efficiency of large section tunnels and reduce production costs. They determined the critical distance of emulsified explosives under the bursting hole constraint and proposed the bare surface blasting technology without a detonating cord. Pan et al. [30] determined the peripheral hole parameters by an eccentric charging structure and studied the blasting effect of different charging structures based on the Riedel–Hiermaier–Thoma (RHT) model. They concluded that the eccentric charging structure has an obvious eccentric pressure and optimized the parameters of surface blasting to control the phenomenon of under-excavation. Numerous scholars at home and abroad have conducted extensive research on the quality control of tunnel smooth blasting, which mainly involves repeated tests and the optimization of blasting parameters [22,31], computer simulation [32–36], three-dimensional laser scanning [37,38], and other techniques [39,40], thus proposing reasonable control techniques.

Multifractals were introduced by Grassberger in the 1980s [41]. The multifractal theory uses geometric probability to describe the local singularities of measures and functions. It utilizes generalized information dimensions and multifractal spectra to describe fractal objects and is an important development in fractal theory [42,43]. Compared with single fractal methods, multifractal methods describe the fractal structure through spectral functions, which can more finely characterize the volatility of fractal objects at different levels [44]. In recent years, some researchers have applied multifractals to the field of geotechnical engineering [45–47]. There are a large number of studies in the characterization of structural surface features of rock bodies and the characterization of rock acoustic emission signals [48–50]. In the field of tunnel blasting, Yin et al. [51] calculated the fractal dimension of tunnel blasting contour lines. The indicator levels and weights were obtained by cluster analysis and the combined assignment method. An unconfirmed metric model was established and applied to the tunnel blasting evaluation. Li et al. [52] carried out explosion tests on granite specimens under different stress states and analyzed the morphology and fractal characteristics of radial fracturing on the rock surface after blasting. In addition, a large number of scholars have studied the multifractal characteristics of undesirable geological bodies such as rock bursts, faults, and karsts that may exist in tunnel construction [53–55]. Therefore, the multifractal theory can be used to analyze the characteristics of overbreak and underbreak of tunnel blasting contour lines and evaluate the construction effect of tunnel smooth blasting.

In previous studies on the drilling and blasting method for extra-long tunnels, the focus has primarily been on optimizing the blasting parameters and improving the drilling accuracy. The objective of this approach is to reduce the problem of overbreak and underbreak during tunnel excavation and enhance the effectiveness of smooth blasting. However, for long tunnels with complex and variable surrounding rock grades, the untimely adjustment of blasting parameters often results in significant overbreak and underbreak issues. To address this challenge, this study employs a combination of theoretical analysis, field tests, and multifractal analysis to conduct an in-depth investigation into the technology of smooth blasting in extra-long hard rock tunnels. The main areas of research encompass the following aspects: (1) Optimizing the peripheral hole charging structure and blasting parameters to enhance the effectiveness of smooth blasting in tunnels. (2) Measuring the actual excavation contour line of the tunnel section using a laser section meter. This enables the acquisition of specific data on the overbreak and underbreak of the tunnel profile line, facilitating a quantitative comparison of the optimization effect of smooth blasting parameters. (3) Utilizing the multifractal detrending fluctuation analysis method to analyze the multifractal characteristics of the sequence of measurement points for overbreak and underbreak in the tunnel profile under different smooth blasting schemes. The calculated results are subsequently compared with the actual situation.

2. Materials and Methods

2.1. Overview of the Tunnel Project

2.1.1. Tunnel Project Introduction

A high-speed railway tunnel, classified as an extra-long tunnel, spans a total length of 10.8 km. The tunnel's inlet mileage is DK102+727.765, while the exit mileage is DK113+573.89, resulting in a total length of 10,846.125 m. The elevation of the inlet track stands at 460.2 m, whereas the exit track's elevation is 264.0302 m. Positioned within the middle mountainous region of the tectonic structure, the tunnel traverses the mountain range, characterized by significant terrain undulations. The mountain's elevation ranges from 250 m to 885 m, with a maximum difference of approximately 635 m. The mountain slopes are steep, with a natural gradient of 40° to 55°, and local areas reaching a steepness of around 500 m. The tunnel itself reaches a maximum depth of approximately 500 m. The surface vegetation along the tunnel route is dense, primarily consisting of tall trees with weeds and shrubs interspersed. The tunnel project roadmap and layout plan can be observed in Figure 1.

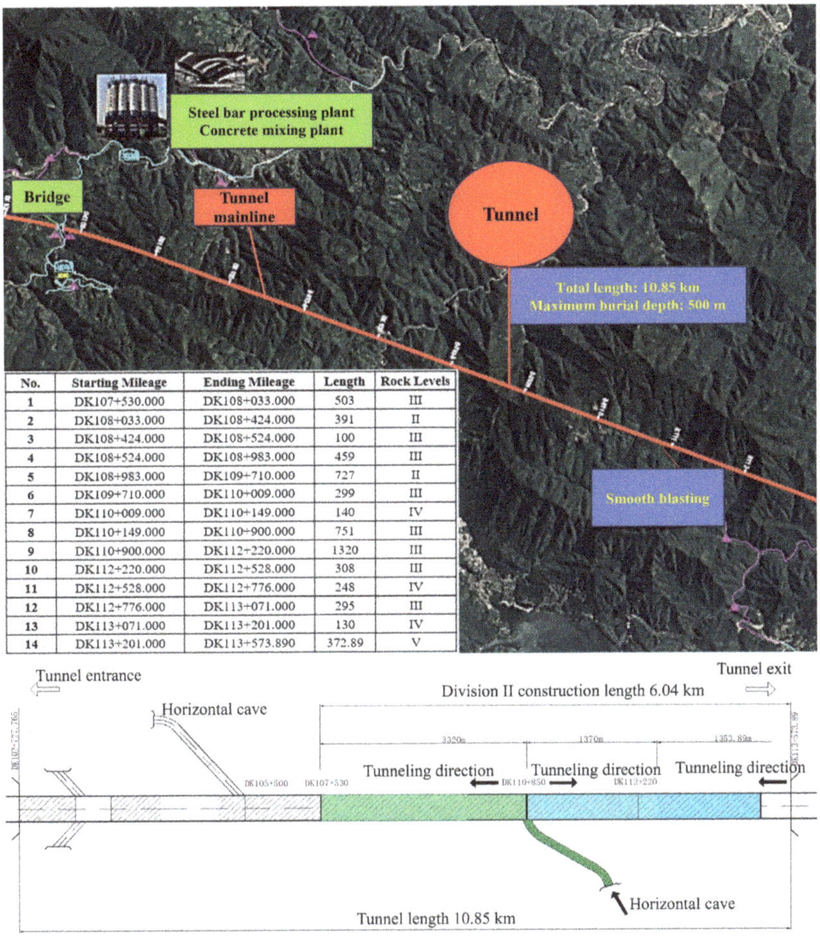

No.	Starting Mileage	Ending Mileage	Length	Rock Levels
1	DK107+530.000	DK108+033.000	503	III
2	DK108+033.000	DK108+424.000	391	II
3	DK108+424.000	DK108+524.000	100	III
4	DK108+524.000	DK108+983.000	459	III
5	DK108+983.000	DK109+710.000	727	II
6	DK109+710.000	DK110+009.000	299	III
7	DK110+009.000	DK110+149.000	140	IV
8	DK110+149.000	DK110+900.000	751	III
9	DK110+900.000	DK112+220.000	1320	III
10	DK112+220.000	DK112+528.000	308	III
11	DK112+528.000	DK112+776.000	248	IV
12	DK112+776.000	DK113+071.000	295	III
13	DK113+071.000	DK113+201.000	130	IV
14	DK113+201.000	DK113+573.890	372.89	V

Figure 1. The tunnel project roadmap and layout plan.

The tunnel incorporates two transverse holes, with this study focusing on the section between tunnel DK107+530 and exit DK113+574. The cross-hole is positioned on the right side of the line's forward direction and intersects with the main tunnel at DK110+850. The angle between the cross-hole and the line mileage direction is 60°, and the overall slope is 1.3%. The cross-hole's mileage is PKHDK0+945, 0utilizing double lanes for trackless transportation, with a length of 945 m.

The mountain structure of the tunnel area predominantly consists of a backward sloping structure from north to south, primarily exhibiting monoclinic terrain. The slopes on both sides of the tunnel are steeper. The rock formation at the tunnel entrance is relatively fragmented, posing risks of dangerous rocks and rockfall incidents. The surrounding rocks of the mountain consist mainly of metasedimentary siltstone, sandstone, kyanite siltstone, and Aurignacian siltstone. The tunnel site area contains a fracture zone and several joints. Three types of parent materials were selected and sent to the Testing Center of Geological Engineering Survey Institute for rock and mineral identification. The analysis revealed that the surrounding rock lithology composition is primarily quartz, followed by rock chips, feldspar, and black mica. Laboratory testing of exploration core samples and cave slag samples yielded the compressive strength values of the perimeter rock, as shown in Table 1, with an average value of 118.6 MPa. The study focuses on the work area of the tunnel, which has a total length of 3320 m, and the perimeter rock is classified as grade III, necessitating the use of the full cross-section construction method.

Table 1. Perimeter rock compressive strength test results.

Specimen Number	Specimen 1	Specimen 2	Specimen 3	Specimen 4	Average Value
Compressive strength	106.0 MPa	119.1 MPa	129.1 MPa	120.3 MPa	118.6 MPa

2.1.2. Original Tunnel Excavation Design

The study focuses on an extra-long tunnel with a length exceeding 10 km. Smooth blasting method has been employed to enhance efficiency. Firstly, auxiliary cross-holes have been installed to increase the excavation working surface, facilitating the construction of a long tunnel with shorter strikes. Additionally, large-scale machinery such as three-arm rock drilling carts, hydraulic trestle bridges, and intelligent lining carts have been utilized for support operations, enabling mechanized and rapid construction of the extra-long tunnel. However, certain engineering challenges have arisen during construction, significantly impacting efficiency. For instance, the drilling and blasting method employed exhibits poor blasting effects, and difficulties arise in the construction of auxiliary cross-holes, resulting in low efficiency. Furthermore, issues regarding the adaptability of large-scale machinery to construction conditions are prevalent. Figure 2 illustrates the blasting effect of the drilling and blasting method employed in construction.

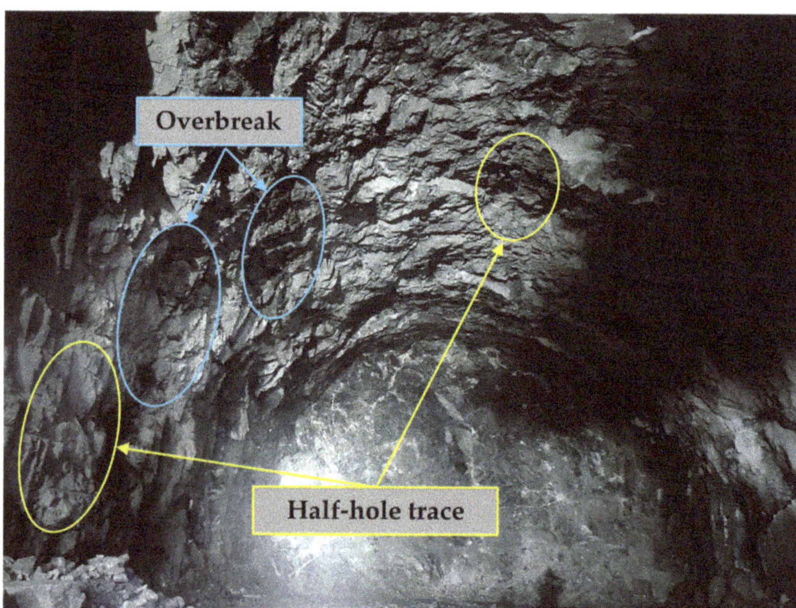

Figure 2. Blasting effect of the original design drill and blast method of construction.

The original design utilizes conventional blasting parameters, with a peripheral hole distance of approximately 0.6 m and a distance of 0.8 m between the auxiliary hole and the tunnel excavation contour line. The hole spacing ranges from 1 m to 1.4 m. However, during the initial stages of construction, these parameters prove to be poorly adaptable to the peripheral rock conditions of the long and hard rock tunnel. Insufficient precision in the drilling and charging operations results in significant overbreak and underbreak in the tunnel blasting peripheral rock cross-section profile. The charging structure of the peripheral holes is located at the bottom, with no utilization of smooth blasting for construction. As depicted in Figure 2, the half-hole trace rate on the tunnel wall after blasting is low, resulting in an uneven wall surface and evident overbreak and underbreak. The maximum value of overbreak at the measuring point of the tunnel section is close to 0.8 m. A significant portion of the explosive energy during blasting is concentrated at the bottom of the hole, leading to pronounced local crushing after blasting. This particularly damages the surrounding rock, posing potential risks to the safety, stability, and quality control of the surrounding rock.

Consequently, it is crucial to promptly adjust and optimize the parameters of smooth blasting to effectively control the overbreak and underbreak during tunnel blasting. Additionally, the fractal characteristics of the tunnel profile before and after the optimization of smooth blasting are analyzed using multifractal theory. The flowchart of the study is presented in Figure 3.

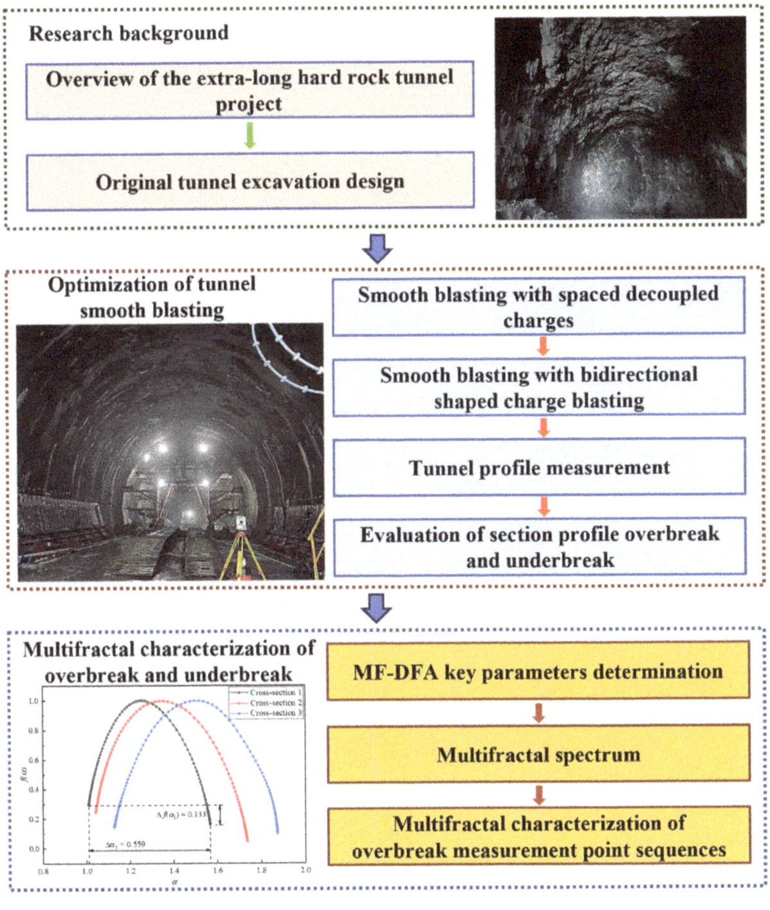

Figure 3. Flowchart of this study.

2.2. Optimization of Tunnel Smooth Blasting

To address the issue of overbreak and underbreak, the smooth blasting coefficients were preliminarily optimized and determined. This involved considering factors such as the number of holes, hole spacing, charge amount, and detonation method, using smooth blasting theory. The design of the smooth blasting parameters took into account the physical and mechanical properties of the surrounding rock, as well as the development of internal joints and fissures in the rock mass.

2.2.1. Spaced Decoupled Charge Blasting

The peripheral hole blasting parameters and charging structure are crucial factors influencing the smooth blasting effect. The original peripheral hole blasting parameters for the tunnel are as follows.

1 Blasting equipment

The explosives are No. 2 emulsified explosives. The length and diameter of the cartridge are 200 mm and 32 mm, respectively, and the weight of a single cartridge is 200 g. The detonator is a millisecond differential time-delay detonator with a detonating cord.

2. Parameters of hole arrangement

The layout of smooth blasting holes with spaced decoupled charging is depicted in Figure 4. The excavation method employed at the site is one-time blasting in full section with invert arch. A total of 65 peripheral holes were set up, with a spacing of 50–60 cm and a hole depth of 420 cm. The auxiliary holes adjacent to the peripheral holes were positioned 80 cm away from the tunnel excavation contour line, resulting in a light blasting layer thickness of 76 cm. The parameters of the blast holes are presented in Table 2.

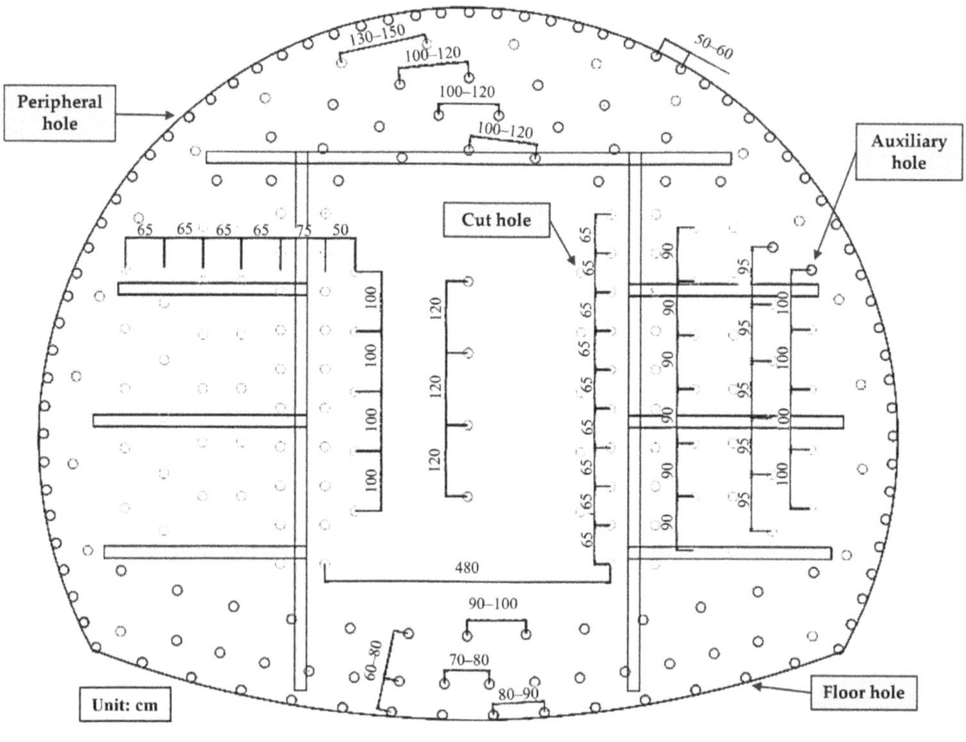

Figure 4. Arrangement of blast holes.

Table 2. Parameter settings of the shell hole.

Event	Blasthole Depth (m)	Number of Holes	Number of Rolls per Hole (Rolls/Hole)	Charge of Single Hole (kg/Hole)	Subtotal Dosage (kg)	Total Number of Holes	Total Amount of Explosives (kg)
Cut hole	4.4	34	14.5	2.9	98.6	252	437.1
Auxiliary hole	4.2	139	10	2	278		
Peripheral hole	4.2	65	2.5	0.5	32.5		
Floor hole	4.2	14	10	2	28		

3. Charge structure

The decoupled charging structure is employed for the peripheral hole blasting. Each hole has a loading capacity of 0.5 kg, with a loading concentration of 0.2 kg/m and an interval of 55 cm. The uncoupling coefficient is 1.31. The plugging material used is a pre-mixed flexible mortar, with a plugging length of 60 cm. The spacing and uncoupling loading structure are depicted in Figure 5. The other holes are loaded continuously.

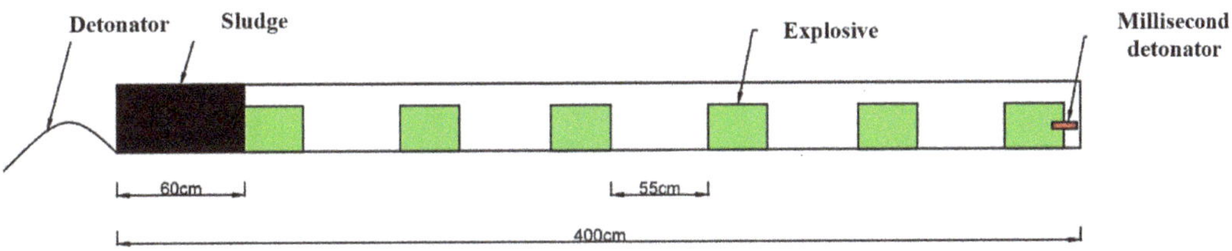

Figure 5. Spaced decoupled charge structure.

For the peripheral holes of the decoupled charging method, smooth blasting is conducted based on the characteristics of the surrounding rock. Detonating cord connection is utilized, and the spacing of the peripheral holes is strictly controlled. The holes are detonated using multi-stage detonators, with a detonation order of hollowing holes, auxiliary holes, base plate holes, and smooth blasting perimeter holes. The detonation time difference between adjacent holes is not less than 50 ms. The detonation time difference control between the holes in a row and the surrounding hole is between 100 to 150 ms. The quality of hole plugging is an important factor that affects the smooth blasting effect. To maximize the explosive energy and extend the effective action time of the blast gas, a mixture of clay and fine sand with good viscosity is used for plugging the holes. This aims to improve rock fragmentation and the effect of smooth blasting on peripheral holes.

2.2.2. Bidirectional Shaped Charge Blasting

In the 1970s, shaped charge blasting was applied in geotechnical engineering. Shaped charge blasting fully utilizes the high compressive and low tensile properties of rocks to form tangential tensile stress. The coupling effect between the shaped charge tank and the explosive charge is perpendicular to the direction of the shaped charge [56,57]. Tangential tensile stress causes initial cracks on the wall of the blast hole. When the radial tensile stress generated by the explosion stress wave acts on the rock between two adjacent main cracks, the rock can be pulled apart to form a circumferential crack, which can connect with each other, thus forming rock fracture. The remaining gas pressure ejects the rock mass, separating it from the parent rock and maximizing the conversion of detonation pressure into tensile action on the surrounding rock.

Due to the complex geological conditions of the extra-long hard rock tunnel project, the smooth blasting effect is not ideal in many instances. To achieve the best smooth blasting effect, the bidirectional shaped charge blasting technology is introduced into the smooth blasting construction of the tunnel. The peripheral hole blasting parameters are reasonably adjusted. The bidirectional shaped charge blasting hole program and the order of detonation are similar to the spaced decoupled charging smooth blasting technology. This study focuses on adjusting the peripheral hole parameters and charge structure.

1. Peripheral hole parameters

Peripheral hole depth is 420 cm, the number of peripheral holes is adjusted to 51, the hole spacing is adjusted to 60~70 cm, and the thickness of the light explosion layer is adjusted to 60 cm.

2. Charge structure

The structure of the glossy blasting charge for bidirectional shaped charge blasting is shown in Figure 6. The peripheral hole is continuously loaded with bidirectional shaped tube charge. Considering that the bottom of the hole is subject to greater rock entrapment, the charge is reinforced at the bottom of the hole. In order to improve the utilization rate of blasting energy, water bags were added at the bottom of the hole and the hole opening. The incompressibility of water is utilized to increase the time of action of the blasting load. The length of the poly energy pipe is 250 cm, its equivalent diameter is 23.5 cm, the charge of a single hole is 0.6 kg, the concentration of the charge is 0.15 kg/m, the non-coupling coefficient is 1.79, and the length of the blast mud blockage is 40 cm. The other holes are loaded in the same structure as the conventional smooth blasting.

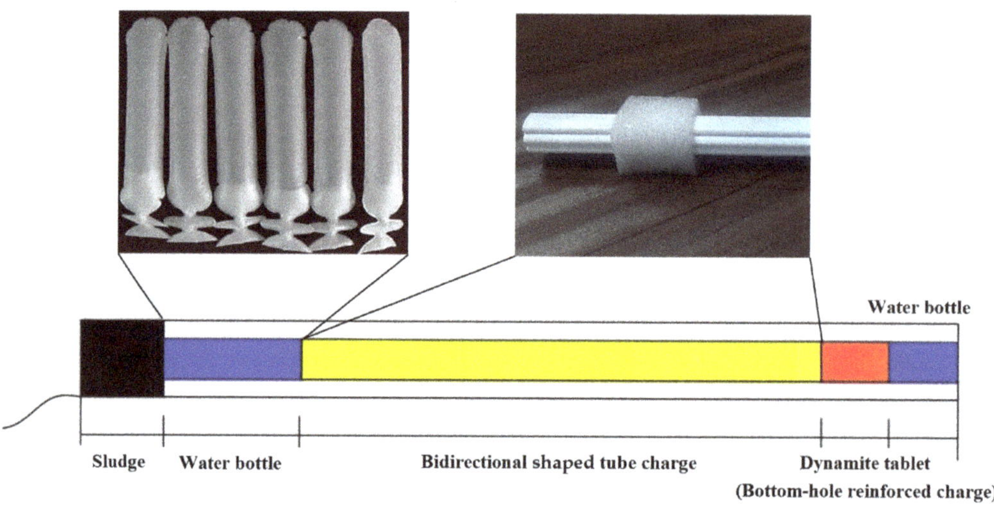

Figure 6. Bidirectional shaped charge structure (physical drawing of shaped charge tube and water bag).

The blasting test utilized a specialized PVC pipe with symmetrical grooves on both sides, resulting in an equivalent diameter of 23.5 cm. For tunnel blasting, conventional water bags were employed, filled with water, with a diameter of 3.5 cm and a length of 25 cm.

2.3. Tunnel Profile Measurement

The laser profilometer, also known as the laser tunnel limiter, is widely used in tunnel section measurement due to its simplicity, high accuracy, and intuitive image. It operates based on the polar co-ordinate method, which combines polar co-ordinate measurement with computer technology and specialized graphic post-processing software. This allows for the convenient acquisition of the actual excavation contour line of the measured tunnel section and a comparison with the design contour line to obtain overbreak and underbreak data. The laser profiler can be used to measure both the current section and the forward section, with the current section primarily used for measuring overbreak and underbreak.

The distance between each measuring site should be selected based on the specific conditions of the tunnel site. To facilitate subsequent statistical analysis of the overbreak and underbreak data, 120 uniformly distributed measurement points are typically used for each measurement section. Figure 7 illustrates the design and field implementation of the tunnel profile overbreak and underbreak measurement.

Figure 7. Measurement of overbreak and underbreak of the tunnel section: (**a**) measurement design drawing; and (**b**) field implementation drawing.

2.4. Multifractal Detrended Fluctuation Analysis Methodology

Fractal theory can be categorized into geometric self-similarity or uniform fractals, and statistical self-similarity or non-uniform fractals, also known as multifractals. Geometric self-similarity is often described using a simple fractal dimension (D), while statistical self-similarity requires the multifractal spectrum $f(\alpha) - \alpha$ to characterize it. The multifractal spectrum, also known as the singularity spectrum, is a mathematical tool used to describe the nature of multifractals [58]. It is utilized to analyze data with a multifractal structure, such as images and time series. The multifractal property refers to the existence of several different fractal dimensions in a system, where physical quantities exhibit different fractal characteristics at different scales. The multifractal spectrum measures the fractal dimensions at these different scales by dividing the fractal body into several small intervals. The singularity index α represents the fractal dimension of each subinterval, and the corresponding $f(\alpha)$ values represent the fractal dimension of each subinterval. Intervals with the same α value form a subset of the fractal, resulting in an infinite sequence of different α values. The multifractal spectral function, $f(\alpha)$, is obtained from this infinite sequence.

The MF-DFA is an extension of detrended fluctuation analysis (DFA) that effectively reveals the dynamic behaviors in nonlinear and nonsmooth signals. Compared to traditional multifractal computation methods, MF-DFA utilizes the length of the sequence data and divides the sequence into equal time lengths in both directions [59]. The polynomials are fitted to each segment using the least squares method to eliminate the influence of the non-stationary trend of the time series. MF-DFA analyzes the scalar behavior of the series at different levels using different orders of fluctuation functions, allowing for a fine characterization of the fractal features and revealing the multifractal features hidden in non-stationary time series [60].

The calculation of MF-DFA involves five steps. The detailed calculation process is shown in Figure 8.

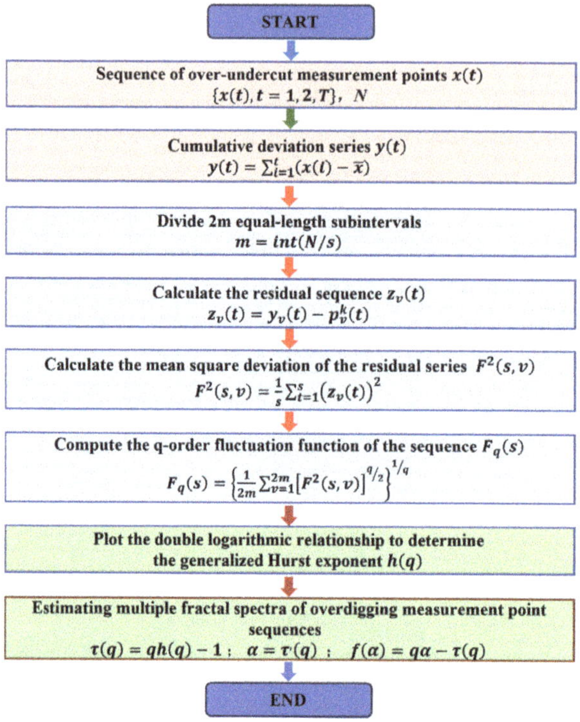

Figure 8. Flowchart of MF-DFA calculation.

Step 1: Given a nonlinear, nonsmooth time series $x(t)$ with a sequence length of N and a sequence mean of $\overline{x}$, calculate the cumulative deviation series $y(t)$ with respect to the mean $\overline{x}$:

$$y(t) = \sum_{i=1}^{t}(x(i) - \overline{x}) \tag{1}$$

Step 2: The sequence is $y(t)$, equalized in terms of time scale s and divided into m equal-length continuous and non-overlapping subintervals:

$$m = int(N/s) \tag{2}$$

Since the length of the time series, N, may not be an integer multiple of s, there will be residual values in the division process. To fully utilize the data information, a reverse-order processing method is applied. This means that, after the initial positive-order division, the same operation is repeated starting from the end of the sequence. As a result, 2 m equal-length subintervals are obtained.

Step 3: A trend is fitted to each subinterval and subtracted from the original time series. The residual series obtained is denoted as $z_v(t)$:

$$z_v(t) = y_v(t) - p_v^k(t) \tag{3}$$

Specifically, $y_v(t)$ represents the v-th subinterval, while $p_v^k(t)$ represents a k-th-order fitting polynomial for the v-th subinterval. Here, v takes values in the range [1, 2 m] and t takes values in the range [1, s].

Step 4: The residual sequence, $z_v(t)$, is calculated for each subinterval v. This forms the $F^2(s,v)$ dataset:

$$F^2(s,v) = \frac{1}{s}\sum_{t=1}^{s}(z_v(t))^2 \tag{4}$$

Step 5: The mean value of the $F^2(s,v)$ dataset is calculated, and the q-order volatility function of the series, $F_q(s)$, is obtained using Equation (4):

$$F_q(s) = \left\{\frac{1}{2m}\sum_{v=1}^{2m}\left[F^2(s,v)\right]^{q/2}\right\}^{1/q} \tag{5}$$

The value of q can be any non-zero real number and is related to the degree of exposure to fluctuations in $F_q(s)$. It is worth noting that, when q = 2, the MF-DFA degenerates to the standard DFA. Additionally, when q = 0, there is a limiting form of Equation (4):

$$F_0(s) = exp\left\{\frac{1}{4m}\sum_{v=1}^{2m}\ln\left[F^2(s,v)\right]\right\} \tag{6}$$

Conventional MF-DFA is prone to pseudo-fluctuations in the division of intervals, causing interference in the subsequent analysis. In addition, the timing length may not be able to rectify the sub-interval length, resulting in redundancy of data. If the redundancy is ignored, it will cause the loss of data information. If the reverse-order processing method is adopted, the order of the original data will be disturbed, affecting the acquisition of information. In view of this, considering the special characteristics of the over-undercut data of the tunnel profile, a sliding window is used here to optimize the way of dividing subintervals of the traditional MF-DFA. A window of a certain length is used to slide the values on the sequence according to a certain step size to reduce the pseudo-fluctuation of the data and make full use of the data information. Let the length of the window be s, the length of the sequence be N, and take the sliding step to be 1; then, the number of sub-intervals obtained in one run is $N - s + 1$, and we replace Equations (5) and (6) with Equation (7):

$$F_q(s) = \begin{cases} \left\{\frac{1}{N-s+1}\sum\limits_{v=1}^{N-s+1}\left[F^2(s,v)\right]^{q/2}\right\}^{1/q}, q \neq 0 \\ exp\left\{\frac{1}{2(N-s+1)}\sum\limits_{v=1}^{N-s+1}\ln\left[F^2(s,v)\right]\right\}, q = 0 \end{cases} \tag{7}$$

The q-order fluctuation function corresponding to a certain scale s can be obtained by the above steps. By varying the values of s and repeating the above steps, a series of $s - F_q(s)$ point values are obtained. If there is long-range correlation in this time series, there will be a power law relationship between s and $F_q(s)$ as shown in Equation (8):

$$F_q(s) \propto s^{h(q)} \tag{8}$$

Taking logarithms on both sides of the above equation gives the form of Equation (9):

$$\lg F_q(s) = h(q)\lg s + \lg b \tag{9}$$

where $F_q(s)$ is the q-order volatility function of the series, $h(q)$ is the corresponding generalized Hurst exponent, and b is a constant coefficient.

By creating a double logarithmic scatterplot of $\lg F_q(s) - \lg s$ and fitting it, the slope can be determined as the generalized Hurst exponent, $h(q)$. If $h(q)$ is a constant, this indicates that the sequence is unifractal without multifractal features. However, if $h(q)$ is a nonlinear subtractive function of q, this suggests that the sequence exhibits multifractal characteristics. Specifically, when $h(q) < 0.5$, the sequence demonstrates a memory process with inverse persistence. When $h(q) = 0.5$, the sequence behaves as an uncorrelated stochastic process. When $h(q) > 0.5$, the sequence exhibits a memory process with positive persistence. Moreover, when $h(q) > 1$, the sequence behaves as a long-range positively correlated process with strong non-stationarity.

The fractal intensity and fractal singularity of a time series can usually be characterized by the multifractal spectrum $f(\alpha)$ to be characterized by solving Equations (10)–(12) as follows:

$$\tau(q) = qh(q) - 1 \tag{10}$$

$$\alpha = \tau'(q) \tag{11}$$

$$f(\alpha) = q\alpha - \tau(q) \tag{12}$$

The Renyi index $\tau(q)$ is a scalar function that can be used to determine whether a sequence exhibits multifractal characteristics. If $\tau(q)$ is a nonlinear up-convex function of q, this suggests that the sequence is multifractal. On the other hand, if $\tau(q)$ is a linear function of q, this indicates that the sequence has a single fractal feature. Therefore, the Renyi index is often used as a criterion to determine the multifractality of a sequence. The singular intensity α, and the multifractal spectrum $f(\alpha)$, are also important in characterizing the multifractal properties of a sequence. When the plot of $\alpha - f(\alpha)$ is convex with a single peak and resembles a quadratic function, this suggests that the sequence has a multifractal feature. On the other hand, when $\alpha - f(\alpha)$ clusters around a point, this indicates that the sequence is unifractal.

In addition, the fractal spectral width $\Delta\alpha$ and the fractal intensity Δh are commonly used to quantitatively characterize the multifractal properties of a sequence. $\Delta\alpha$ mainly reflects the singularity of the over-undercut data of the tunnel profile and the spatial variability of the sequence. On the other hand, Δh mainly reflects the fractal intensity of the sequence and is proportional to the parameter values:

$$\Delta h = \max(h(q)) - \min(h(q)) \tag{13}$$

$$\Delta\alpha = \max(\alpha(q)) - \min(\alpha(q)) \tag{14}$$

3. Results and Discussion

3.1. Analysis of the Optimized Effect of Tunnel Smooth Blasting

3.1.1. Smooth Blasting with Spaced Decoupled Charge

Based on the theory of spaced decoupled charge smooth blasting technology, the aforementioned decoupled charge blasting design was applied in the test section of tunnel excavation. The resulting effect of the smooth blasting is depicted in Figure 9.

As can be seen from Figure 9, the tunnel profile was basically shaped after blasting. However, the tunnel wall exhibits unevenness, and the traces of half-hole marks on the tunnel wall are not clearly visible. The rate of half-hole marks on the tunnel wall is 56%, and the maximum unevenness exceeds 30 cm, indicating a noticeable occurrence of overbreak and underbreak. Upon analysis, it was found that, although the compressive strength of the tunnel rock reached 118.6 MPa, the integrity of the surrounding rock was poor, with certain joints and fissures. Additionally, the deep surrounding rock of the peripheral holes experienced a significant clamping force, resulting in insufficient blasting effectiveness. Consequently, the incomplete separation of the light blasting layer of rock led to overbreak and underbreak.

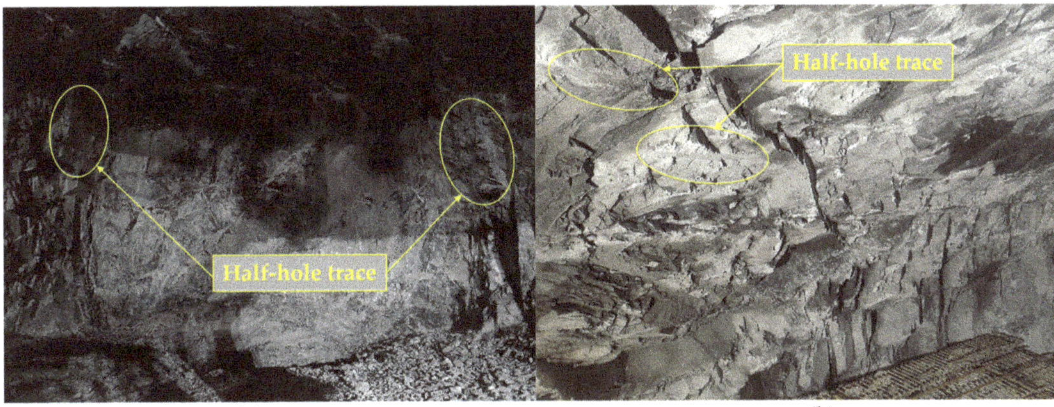

(a) (b)

Figure 9. The effect of spaced decoupled charge of smooth blasting: (**a**) overall diagram of half-hole trace; and (**b**) detail diagram of half-hole trace.

3.1.2. Smooth Blasting with Bidirectional Shaped Charge

To address this issue, the design scheme of bidirectional shaped charge blasting, based on the theory of smooth blasting technology, was implemented in the test section of tunnel excavation. After blasting the peripheral hole with a bidirectional shaped charge, the improved smooth blasting effect is shown in Figure 10.

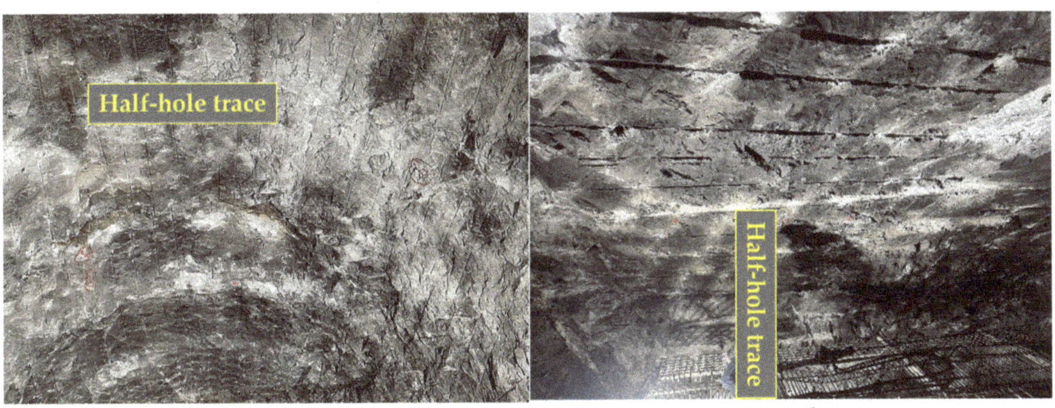

(a) (b)

Figure 10. Bidirectional shaped charge blasting effect of smooth blasting diagram: (**a**) overall diagram of half-hole trace; and (**b**) detail diagram of half-hole trace.

As can be seen in Figure 10, the glossy smooth effect was significantly enhanced after the tunnel was loaded with the bidirectional shaped charge. The excavation contour line of the tunnel appeared flat after blasting, and the traces of half-hole marks on the tunnel wall were clearly visible. The rate of half-hole marks on the tunnel wall reached 89%, and the maximum unevenness was less than 10 cm. Laser sectional meter measurements confirmed that the tunnel section profile was nearly free of overbreak and underbreak phenomena.

3.1.3. Overbreak and Underbreak of Tunnel Profile of Smooth Blasting

For the long hard rock tunnel perimeter rock section, two different smooth blasting technologies, namely, spaced decoupled charging and bidirectional shaped charge, were tested. Multiple groups of field tests were conducted using distinct smooth blasting programs. A laser profiler was utilized to scan the tunnel profile after blasting, and the scanning results are presented in Figure 11. Based on the scanning results, data on tunnel overbreak and underbreak were extracted. The distribution curves of the overbreak and underbreak of the tunnel profile are depicted in Figure 12.

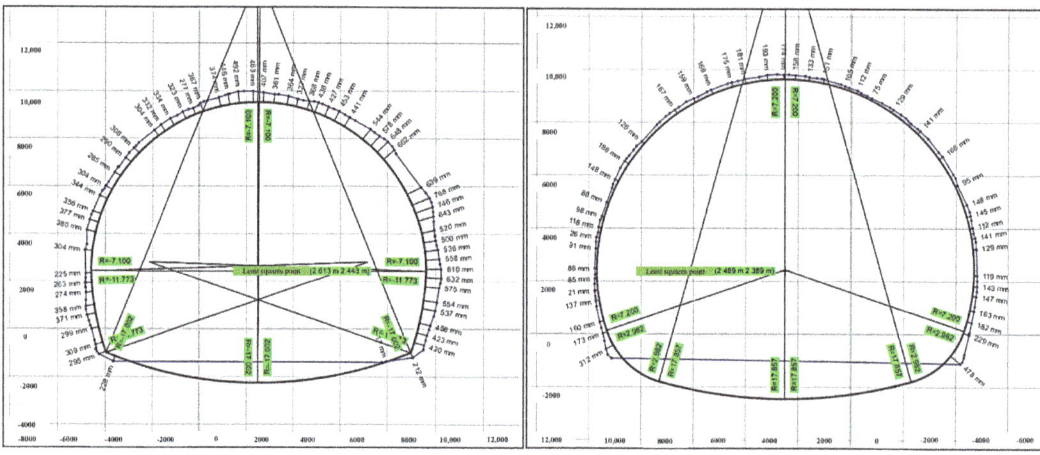

(**a**) (**b**)

Figure 11. Scanning results of tunnel profile after blasting with two charge methods: (**a**) spaced decoupled charge, and (**b**) bidirectional shaped charge.

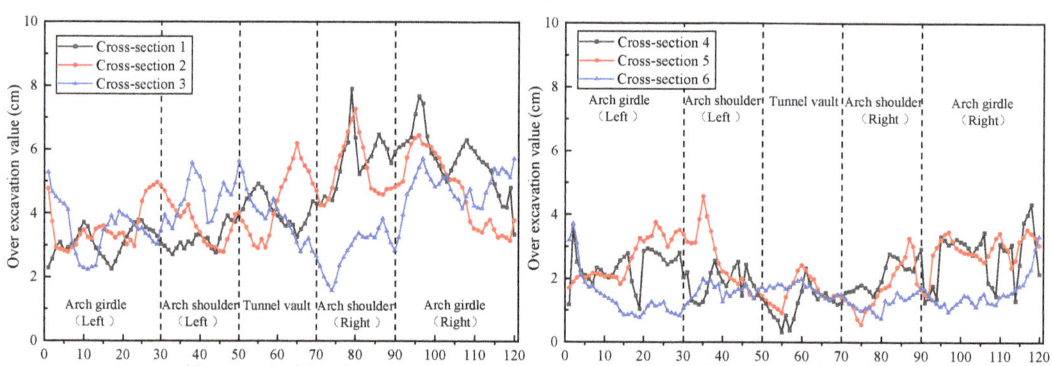

(**a**) (**b**)

Figure 12. Distribution curve of overbreak and underbreak of the profile of the blasting tunnel section for two charge methods: (**a**) spaced decoupled charge, and (**b**) bidirectional shaped charge.

According to the scanning results shown in Figure 11, the overbreak and underbreak area of the interval decoupled charge blasting section is calculated to be 13.01 m^2, while the bidirectional shaped charge blasting only resulted in an overbreak and underbreak area of 4.61 m^2. This indicates that the bidirectional shaped charge blasting has a significantly better smooth blasting effect compared to the interval decoupled charge blasting. As depicted in Figure 12, the average overbreak value for each measurement point in the interval decoupled charge blasting is 4.21 cm, with a maximum overbreak value of 7.91 cm occurring at the right arch shoulder. The overbreak at the right arch shoulder and arch waist is more pronounced in the interval decoupled charge blasting. On the other hand, for the bidirectional shaped charge blasting, the average overbreak value at each measurement point is 1.99 cm, with a maximum overbreak value of 4.58 cm occurring at the left arch shoulder. The overbreak values of each area of the tunnel section after bidirectional shaped charge blasting are not significantly different, indicating that the smooth blasting effect is better than that of the interval decoupled charge blasting.

To further compare the blasting effects of the two charging methods in extra-long hard rock tunnels, statistical analysis was conducted, as shown in Table 3.

Table 3. Comparison of the effects of spaced decoupled faceted blasting and bidirectional shaped charge blasting technology.

Sports Event	Spaced Decoupled Charge Smooth Blasting	Bidirectional Shaped Charge BSmooth Blasting	Efficiencies
Number of smooth holes	65	51	Decrease 21.5%
Semi-porous trace rate/%	56	89	Increase 59.9%
Peripheral hole explosives/kg	32.5	30.6	Decrease 5.8%
Maximum unevenness/cm	30	10	Decrease 66.7%
Amount of concrete m^3/m	13.01	4.61	Decrease 64.6%

It can be observed from Table 3 that the bidirectional shaped charge blasting resulted in a decrease of 14 smooth blasting holes compared to the interval decoupled charge blasting, with a 21.5% decrease in the number of holes around the optimized blasting scheme. Additionally, the half-hole trace rate on the tunnel wall increased from 56% to 89%, improving the utilization rate of the shell hole. The explosive dosage of peripheral holes was reduced by 5.8%, leading to an improved utilization rate of blasting energy. These improvements effectively addressed the issue of overbreak and underbreak in the tunnel and saved 8.4 m^3 of concrete cubic meters for each meter of advance. As a result, the optimized smooth blasting scheme ensured the safe and efficient completion of the extra-long hard rock tunnel.

3.2. *Multifractal Characteristics of Tunnel Profile Overbreak*

3.2.1. MF-DFA Key Parameters Determination

The characteristics of the non-stationary time series obtained under different computational parameters vary significantly, including the signal time window length and fluctuation trend. Therefore, it is necessary to try different key parameters and preset them to obtain more reliable results. The parameter values have different effects on the calculation results. In order to analyze the fractal characteristics of the overbreak and underbreak of the tunnel profile before and after the optimization of smooth blasting more efficiently using multifractal theory, multifractal analysis is conducted on the overbreak characteristics of the tunnel profile. The fluctuation order q is taken in the range of $[-10, 10]$, and a double logarithmic scatter plot of $\lg F_q(s) - \lg s$ is plotted. Least squares fitting is used, and the slope of the fitting line is the generalized Hurst exponent $h(q)$. Taking a set of overbreak characteristics of the tunnel profile after spaced decoupled faceted blasting and bidirectional shaped charge blasting as an example, the q-order fluctuation function $F_q(s) - s$ is fitted using double logarithmic scale, as shown in Figure 13.

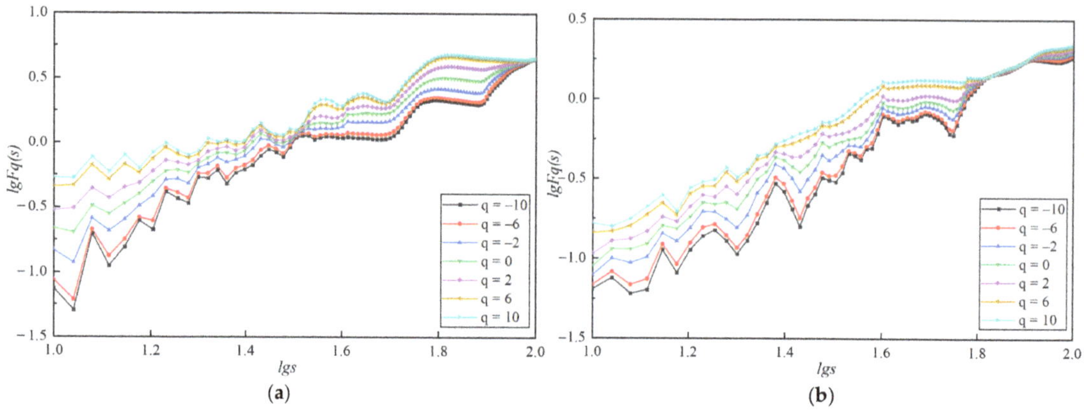

Figure 13. q-order wave function $F_q(s) - s$ double logarithmic fitting trend image: (**a**) spaced decoupled charge, and (**b**) bidirectional shaped charge.

3.2.2. Multifractal Characterization of Overbreak Section Profile

A sliding time window optimization MF-DFA is then used to perform a multifractal analysis on the sequence of overbreak points of the tunnel profile. The fluctuation order q is in the range of $[-10, 10]$, the scale s is in the range of $[10, 100]$, and the sliding window step is 1. The changes of the generalized Hurst exponent of the sequence of overbreak points are shown in Figure 14. The variation of the scalar function $\tau(q)$ of the overbreak point sequence is shown in Figure 15. The multifractal spectrum of the overbreak measurement point sequence is shown in Figure 16.

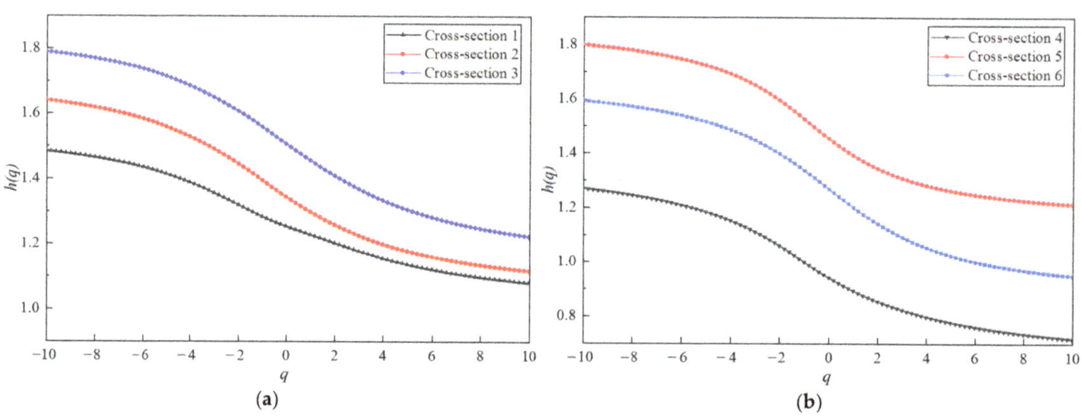

Figure 14. Variation of the generalized Hurst index: (**a**) spaced decoupled charge, and (**b**) bidirectional shaped charge.

It can be observed from Figure 14 that the generalized Hurst exponent of each overbreak measurement point sequence is not constant for both spaced decoupled charging and bidirectional shaped charge blasting when q varies between $[-10, 10]$. Instead, it shows a nonlinear decreasing trend with q, indicating that each sequence of overbreak measurement points exhibits obvious multifractal characteristics. It is not sufficient to describe them with a single fractal theory. Under different fluctuation orders q, the generalized Hurst exponent curves of partially spaced decoupled charge blasting are concentrated in the lower fluctuation range compared to bidirectional shaped charge blasting, indicating a weaker multifractal nature. However, the values of the spaced decoupled charge and bidirectional shaped charge blasting are significantly larger than 0.5. These $h(q)$ values indicate that the sequence of overbreak measurement points exhibits good memory and long-range correlation from the whole to the local components, combining non-stationarity and randomness.

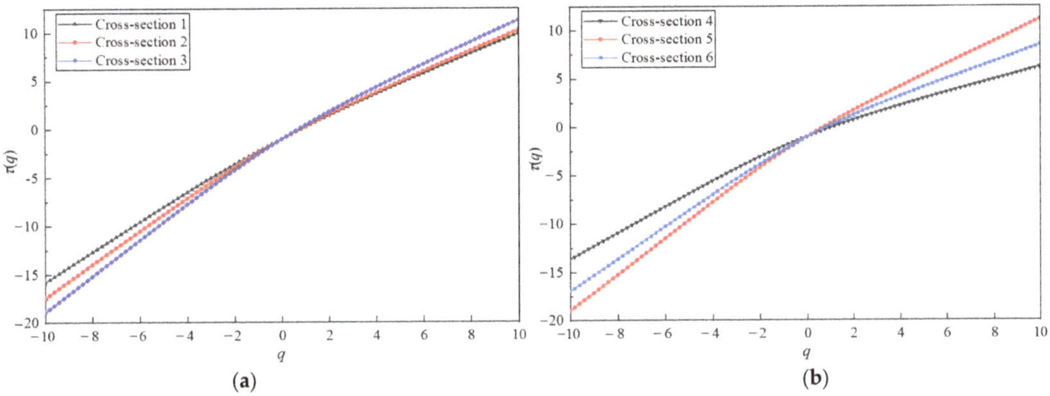

Figure 15. Variation of the scalar function $\tau(q)$: (**a**) spaced decoupled charge, and (**b**) bidirectional shaped charge.

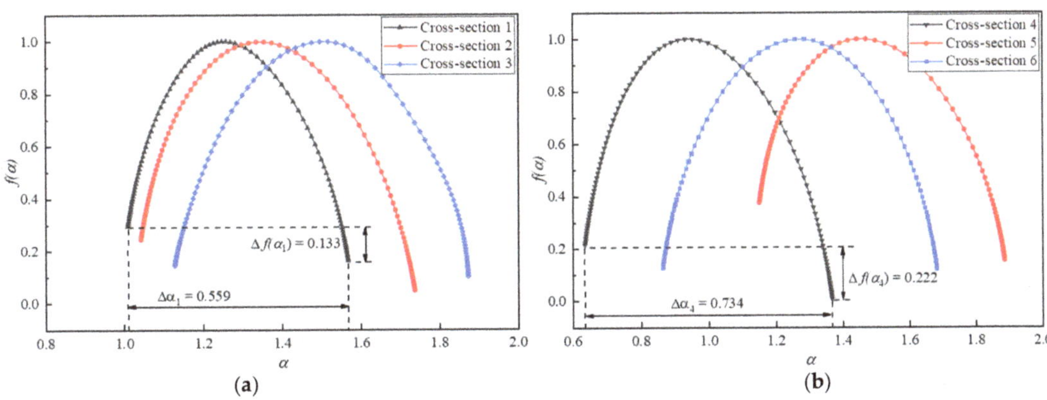

Figure 16. Multifractal spectra: (**a**) spaced decoupled charge, and (b) bidirectional shaped charge.

Based on the $h(q)$, the scaling function $\tau(q)$ is calculated. From Figure 15, it can be observed that the scalar function of each overbreak measurement point sequence shows good consistency for both spaced decoupled charging and bidirectional shaped charge blasting. The function is up-convex, satisfying $\tau(q) = -1$ and exhibiting an overall non-linear relationship. This further confirms the multifractal characteristics of the overbreak measurement point sequence.

Figure 16 shows the multifractal spectra of the overbreak point sequence of the tunnel profile under different smooth blasting plans. The spectra display a single convex distribution, similar to a quadratic function curve. The local scales of the multifractals in the overbreak point sequences vary, indicating the diversity of local changes at different moments. The singular intensities α are mainly concentrated on the two sides of the image, reflecting the uneven distribution of the fractal structure in the overbreak point sequence. This uneven distribution further confirms the multifractal characteristics of the measurement point sequence. The multifractal spectra of the spaced decoupled charge blasting and bidirectional shaped charge blasting show good overall synergy and stable development status. The spectra of the overbreak point sequences in cross-section 4 and cross-section 5 exhibit obvious right hooks, indicating that small fluctuations have a slightly dominant influence in the overbreak point sequences of these sections. The fractal dimension of the sub-intervals that characterize the overbreak point sequence with the same singularity index α is the same. The fractal dimension of the sub-intervals in the overbreak point sequence is related to the distribution characteristics and fractal strength of the overbreak point sequence.

In Figure 16, $\Delta\alpha$ represents the multifractal spectral width, which characterizes the multifractal strength of the sequence at the measurement point and the complexity of the fluctuations. A larger $\Delta\alpha$ indicates a stronger multifractal strength and more intense and complex fluctuations. The opposite is also true. $\Delta f(\alpha)$ represents the proportion of large and small fluctuations in the sequence. A larger proportion of small and medium fluctuations in the sequence leads to a larger $\Delta f(\alpha)$. The calculation method for α and $\Delta f(\alpha)$ can be expressed as follows:

$$\Delta\alpha = \alpha_{max-}\alpha_{min} \tag{15}$$

$$\Delta f(\alpha) = \Delta f(\alpha_{max}) - \Delta f(\alpha_{min}) \tag{16}$$

Combining Figures 14 and 16, it can be observed that the fractal intensity Δh and the width of the multifractal spectrum $\Delta\alpha$ of the fractal spectrum of the overbreak measurement point sequence in cross-section 1 are minimized. This is because the overbreak measurement point sequences in the same section are not independent of each other but are related to each other to some extent. Overall, the values of the six cross-section overbreak point sequences are close to each other, and the fluctuations have similar odd values. The $\Delta\alpha$ values are close to each other, indicating the similar singularity and spatial variability of the fluctuations. The probability distribution ranges of the fluctuations at each measurement point are relatively close to each other, but the fractal intensity Δh values show slight differences. The fractal strength of the overbreak sequence in the tunnel section after the bidirectional shaped charge blasting is slightly stronger than that of the spaced decoupled charge blasting. Based on the actual situation, it can be initially inferred that the optimization of the tunnel smooth blasting significantly reduces the phenomenon of overbreak in the tunnel profile, leading to differences in the multifractal characteristics of the measurement points in each cross-section.

The statistics of multifractal features for the over-undercut sequence of the tunnel profile under different smooth blasting plans are calculated, and the results are shown in Table 4.

Table 4. Multifractal characterization statistics.

Index \ Cross-Section	Cross-Section 1	Cross-Section 2	Cross-Section 3	Cross-Section 4	Cross-Section 5	Cross-Section 6
$\Delta\alpha$	0.559	0.695	0.746	0.734	0.737	0.818
$\Delta f(\alpha)$	0.133	0.196	0.137	0.222	0.292	0.644

Comparing the widths of the multifractal spectra $\Delta\alpha$ of the cross-section profile overbreak point sequences, it can be observed that the width of the multifractal spectrum for the spaced decoupled charge blasting is slightly smaller than that for the bidirectional shaped charge blasting. This indicates that the bidirectional shaped charge blasting results in a larger multifractal intensity in the overbreak point sequence and slightly more complex fluctuations. Furthermore, comparing the proportion of large and small fluctuations $\Delta f(\alpha)$ in the profile undercut point sequences, it can be seen that the bidirectional shaped charge blasting results in a larger proportion of small and medium fluctuations.

These findings align more closely with the actual situation and indicate that the optimization of tunnel smooth blasting through bidirectional shaped charge blasting significantly reduces the phenomenon of overbreak in the tunnel profile. The intensity of the multifractals in the measurement point sequence is larger, and the fluctuation of the cross-section profile overbreak data is slightly more complex, with a larger proportion of small fluctuations.

4. Conclusions

For extra-long tunnels with complex and variable perimeter rock grades, the untimely adjustment of smooth blasting parameters often leads to a significant number of overbreak and underbreak, resulting in project delays. Therefore, this paper focuses on conducting in-depth research on smooth blasting technology for extra-long hard rock tunnels in the context of a high-speed rail project. The research approach combines theoretical analysis, field tests, and multifractal analysis to provide comprehensive insights. The main research conclusions are as follows:

1. The peripheral hole charging structure and blasting parameters are improved and optimized to enhance the effect of tunnel smooth blasting. The implementation of the bidirectional shaped charge significantly improves the smooth blasting effect. After blasting, the tunnel excavation contour line becomes flat, and the half-hole trace marks on the tunnel wall become more distinct. In comparison to spaced decoupled charging blasting, the smooth blasting effect with a bidirectional shaped charge is notably superior.
2. The laser profiler is utilized to measure the actual excavation contour line of the tunnel section, obtaining specific data on overbreak and underbreak. A further quantitative comparison of the optimization effect of smooth blasting parameters is conducted. It is observed that the overbreak and underbreak value of each area of the tunnel section after bidirectional shaped charge blasting does not differ significantly, and the smooth blasting effect is evidently better than that of spaced decoupled charge blasting.
3. MF-DFA is employed to analyze the multifractal features of the overbreak point sequences of the tunnel profile under different smooth blasting plans. It is concluded that both spaced decoupled charge and bidirectional shaped charge blasting result in measurement point sequences with multifractal features in the overbreak area. The statistical calculation results of the multifractal features of the tunnel profile under different smooth blasting plans align more closely with the actual situation.

The smooth blasting tests conducted in this study were carried out in a single surrounding rock grade. Future studies can focus on optimizing the smooth blasting parameters and conducting the multifractal characterization of the sequence of overbreak and underbreak points for different surrounding rock grades of tunnels.

Author Contributions: Methodology, D.L. and Y.T.; software, W.Z.; validation, W.Z. and W.Q.; formal analysis, W.Z. and W.Q.; data curation, D.L. and R.Z.; writing—original draft preparation, W.Z.; writing—review and editing, D.L. and Y.T.; project administration, W.Q. and R.Z.; funding acquisition, D.L. and Y.T. All authors have read and agreed to the published version of the manuscript.

Funding: The research was funded by the Initiation fund for postdoctoral research of Central South University, grant number 228697.

Data Availability Statement: The data that support the findings of this study are available upon request from the authors. The data are not publicly available due to privacy.

Acknowledgments: The authors would like to thank Road and Bridge North China Engineering Co., Ltd. for their assistance with conducting the field experiments.

Conflicts of Interest: The authors declare no conflict of interest. Authors Weichao Qiu and Ruiping Zhang were employed by the company Road & Bridge North China Engineering Co., Ltd., Beijing 101100, China. The remaining authors declare that the research was conducted in the absence of any commercial or financial relationships that could be construed as a potential conflict of interest.

References

1. Cui, P.; Ge, Y.; Li, S.; Li, Z.; Xu, X.; Zhou, G.G.; Chen, H.; Wang, H.; Lei, Y.; Zhou, L.; et al. Scientific challenges in disaster risk reduction for the Sichuan-Tibet Railway. *Eng. Geol.* **2022**, *309*, 106837. [CrossRef]
2. Lv, Y.; Jiang, Y.; Hu, W.; Cao, M.; Mao, Y. A review of the effects of tunnel excavation on the hydrology, ecology, and environment in karst areas: Current status, challenges, and perspectives. *J. Hydrol.* **2020**, *586*, 124891. [CrossRef]
3. Liu, X.; Tao, T.; Tian, X.; Luo, Q.; Xie, C. Layout method and numerical simulation study of reduced-hole blasting in large-section tunnels. *Front. Earth Sci.* **2022**, *10*, 976419. [CrossRef]
4. Zhou, P.; Li, J.; Jiang, Y.; Zhou, F.; Lin, M.; Lin, J.; Wang, Z. Damage mechanism of tunnels in the high-content salt rock stratum. *Bull. Eng. Geol. Environ.* **2021**, *80*, 7633–7652. [CrossRef]
5. Yan, P.; Lu, W.; Chen, M.; Hu, Y.; Zhou, C.; Wu, X. Contributions of In-Situ Stress Transient Redistribution to Blasting Excavation Damage Zone of Deep Tunnels. *Rock Mech. Rock Eng.* **2015**, *48*, 715–726. [CrossRef]
6. Wang, S.; Tang, Y.; Li, X.; Du, K. Analyses and predictions of rock cuttabilities under different confining stresses and rock properties based on rock indentation tests by conical pick. *Trans. Nonferrous Met. Soc.* **2021**, *31*, 1766–1783. [CrossRef]
7. Hu, Y.; Lu, W.; Chen, M.; Yan, P.; Yang, J. Comparison of Blast-Induced Damage between Presplit and Smooth Blasting of High Rock Slope. *Rock Mech. Rock Eng.* **2014**, *47*, 1307–1320. [CrossRef]
8. Wang, S.; Li, X.; Yao, J.; Gong, F.; Li, X.; Du, K.; Tao, M.; Huang, L.; Du, S. Experimental investigation of rock breakage by a conical pick and its application to non-explosive mechanized mining in deep hard rock. *Int. J. Rock Mech. Min. Sci.* **2019**, *122*, 104063. [CrossRef]
9. Li, C.; Yang, R.; Wang, Y.; Kang, Y.; Zhang, Y. Theory and numerical simulation of deep hole cut blasting based on dispersed charge and staged detonation. *Int. J. Rock Mech. Min. Sci.* **2023**, *169*, 105453. [CrossRef]
10. Yilmaz, O.; Unlu, T. An application of the modified Holmberg-Persson approach for tunnel blasting design. *Tunn. Undergr. Space Technol.* **2014**, *43*, 113–122. [CrossRef]
11. Monjezi, M.; Dehghani, H. Evaluation of effect of blasting pattern parameters on back break using neural networks. *Int. J. Rock Mech. Min. Sci.* **2008**, *45*, 1446–1453. [CrossRef]
12. Mahdevari, S.; Haghighat, H.S.; Torabi, S.R. A dynamically approach based on SVM algorithm for prediction of tunnel convergence during excavation. *Tunn. Undergr. Space Technol.* **2013**, *38*, 59–68. [CrossRef]
13. Oggeri, C.; Ova, G. Quality in tunnelling: ITA-AITES Working Group 16 Final Report. *Tunn. Undergr. Space Technol.* **2004**, *19*, 239–272. [CrossRef]
14. Hamdi, E.; Romdhane, N.B.; Le Cléac'h, J.M. A tensile damage model for rocks: Application to blast induced damage assessment. *Comput. Geotech.* **2011**, *38*, 133–141. [CrossRef]
15. Garcia Bastante, F.; Alejano, L.; Gonzalez-Cao, J. Predicting the extent of blast-induced damage in rock masses. *Int. J. Rock Mech. Min. Sci.* **2012**, *56*, 44–53. [CrossRef]
16. Xia, X.; Li, H.B.; Li, J.C.; Liu, B.; Yu, C. A case study on rock damage prediction and control method for underground tunnels subjected to adjacent excavation blasting. *Tunn. Undergr. Space Technol.* **2013**, *35*, 1–7. [CrossRef]
17. Costamagna, E.; Oggeri, C.; Segarra, P.; Castedo, R.; Navarro, J. Assessment of contour profile quality in D&B tunnelling. *Tunn. Undergr. Space Technol.* **2018**, *75*, 67–80. [CrossRef]
18. Fodera, G.M.; Voza, A.; Barovero, G.; Tinti, F.; Boldini, D. Factors influencing overbreak volumes in drill-and-blast tunnel excavation. A statistical analysis applied to the case study of the Brenner Base Tunnel—BBT. *Tunn. Undergr. Space Technol.* **2020**, *105*, 103475. [CrossRef]
19. Singh, S.P.; Xavier, P. Causes, impact and control of overbreak in underground excavations. *Tunn. Undergr. Space Technol.* **2005**, *20*, 63–71. [CrossRef]
20. Habib, K.; Shnorhokian, S.; Mitri, H. Evaluating the Application of Rock Breakage without Explosives in Underground Construction-A Critical Review of Chemical Demolition Agents. *Minerals* **2022**, *12*, 220. [CrossRef]
21. Liu, D.; Tang, Y.; Cai, C.; Jian, Y. A Rock Fracturing Method Using High-Pressure Gas Expansion: Case Study on Its Application in Hangzhou-Lin'an Intercity Railway. *Adv. Civ. Eng.* **2021**, *2021*, 6654471. [CrossRef]
22. Zare, S.; Bruland, A. Comparison of tunnel blast design models. *Tunn. Undergr. Space Technol.* **2006**, *21*, 533–541. [CrossRef]
23. Kim, Y.; Moon, H. Application of the guideline for overbreak control in granitic rock masses in Korean tunnels. *Tunn. Undergr. Space Technol.* **2013**, *35*, 67–77. [CrossRef]
24. Chen, Y.; Chen, S.; Wu, Z.; Dai, B.; Xv, L.; Wu, G. Optimization of Genetic Algorithm through Use of Back Propagation Neural Network in Forecasting Smooth Wall Blasting Parameters. *Mathematics* **2022**, *10*, 1271. [CrossRef]
25. Liu, K.; Liu, B. Optimization of smooth blasting parameters for mountain tunnel construction with specified control indices based on a GA and ISVR coupling algorithm. *Tunn. Undergr. Space Technol.* **2017**, *70*, 363–374. [CrossRef]
26. Khandelwal, M.; Monjezi, M. Prediction of Backbreak in Open-Pit Blasting Operations Using the Machine Learning Method. *Rock Mech. Rock Eng.* **2013**, *46*, 389–396. [CrossRef]
27. Zou, B.; Wang, J.; Luo, Z.; Hu, L. Intelligent Control of Smooth Blasting Quality in Rock Tunnels Using BP-ANN, ENN, and ANFIS. *Geofluids* **2021**, *2021*, 6612824. [CrossRef]
28. Jang, H.; Topal, E. Optimizing overbreak prediction based on geological parameters comparing multiple regression analysis and artificial neural network. *Tunn. Undergr. Space Technol.* **2013**, *38*, 161–169. [CrossRef]

29. Ma, C.; Xie, W.; Liu, Z.; Li, Q.; Xu, J.; Tan, G. A New Technology for Smooth Blasting without Detonating Cord for Rock Tunnel Excavation. *Appl. Sci.* **2020**, *10*, 6764. [CrossRef]
30. Pan, C.; Xie, L.; Li, X.; Liu, K.; Gao, P.; Tian, L. Numerical investigation of effect of eccentric decoupled charge structure on blasting-induced rock damage. *J. Cent. South Univ.* **2022**, *29*, 663–679. [CrossRef]
31. Huo, X.; Shi, X.; Qiu, X.; Zhou, J.; Gou, Y.; Yu, Z.; Ke, W. Rock damage control for large-diameter-hole lateral blasting excavation based on charge structure optimization. *Tunn. Undergr. Space Technol.* **2020**, *106*, 103569. [CrossRef]
32. Ji, L.; Zhou, C.; Lu, S.; Jiang, N.; Gutierrez, M. Numerical Studies on the Cumulative Damage Effects and Safety Criterion of a Large Cross-section Tunnel Induced by Single and Multiple Full-Scale Blasting. *Rock Mech. Rock Eng.* **2021**, *54*, 6393–6411. [CrossRef]
33. Zou, B.; Xu, Z.; Wang, J.; Luo, Z.; Hu, L. Numerical Investigation on Influential Factors for Quality of Smooth Blasting in Rock Tunnels. *Adv. Civ. Eng.* **2020**, *2020*, 9854313. [CrossRef]
34. Cho, S.H.; Nakamura, Y.; Mohanty, B.; Yang, H.; Kaneko, K. Numerical study of fracture plane control in laboratory-scale blasting. *Eng. Fract. Mech.* **2008**, *75*, 3966–3984. [CrossRef]
35. Zhu, Z. Numerical prediction of crater blasting and bench blasting. *Int. J. Rock Mech. Min. Sci.* **2009**, *46*, 1088–1096. [CrossRef]
36. Wang, Z.; Wang, H.; Wang, J.; Tian, N. Finite element analyses of constitutive models performance in the simulation of blast-induced rock cracks. *Comput. Geotech.* **2021**, *135*, 104172. [CrossRef]
37. Nasseri, M.H.B.; Rezanezhad, F.; Young, R.P. Analysis of fracture damage zone in anisotropic granitic rock using 3D X-ray CT scanning techniques. *Int. J. Fract.* **2011**, *168*, 1–13. [CrossRef]
38. Fekete, S.; Diederichs, M.; Lato, M. Geotechnical and operational applications for 3-dimensional laser scanning in drill and blast tunnels. *Tunn. Undergr. Space Technol.* **2010**, *25*, 614–628. [CrossRef]
39. Sun, S.; Li, S.; Li, L.; Shi, S.; Liu, H.; Hu, J.; Zhou, S. Structural planes surveying and fractal dimension characteristics of tunnel face based on digital photogrammetry. *Arab. J. Geosci.* **2018**, *11*, 622. [CrossRef]
40. Kim, Y.; Bruland, A. A study on the establishment of Tunnel Contour Quality Index considering construction cost. *Tunn. Undergr. Space Technol.* **2015**, *50*, 218–225. [CrossRef]
41. Grassberger, P. Generalized dimensions of strange attractors. *Phys. Lett. A* **1983**, *97*, 227–230. [CrossRef]
42. Kantelhardt, J.W.; Zschiegner, S.A.; Koscielny-Bunde, E.; Havlin, S.; Bunde, A.; Stanley, H.E. Multifractal detrended fluctuation analysis of nonstationary time series. *Phys. A* **2002**, *316*, 87–114. [CrossRef]
43. Liu, J.; Li, Q.; Wang, X.; Wang, Z.; Lu, S.; Sa, Z.; Wang, H. Dynamic multifractal characteristics of acoustic emission about composite coal-rock samples with different strength rock. *Chaos Solitons Fract.* **2022**, *164*, 112725. [CrossRef]
44. Lopes, R.; Betrouni, N. Fractal and multifractal analysis: A review. *Med. Image Anal.* **2009**, *13*, 634–649. [CrossRef] [PubMed]
45. Ehlen, J. Fractal analysis of joint patterns in granite. *Int. J. Rock Mech. Min. Sci.* **2000**, *37*, 909–922. [CrossRef]
46. Nayak, S.R.; Mishra, J.; Palai, G. Analysing roughness of surface through fractal dimension: A review. *Image Vis. Comput.* **2019**, *89*, 21–34. [CrossRef]
47. Li, W.; Zhao, H.; Wu, H.; Wang, L.; Sun, W.; Ling, X. A novel approach of two-dimensional representation of rock fracture network characterization and connectivity analysis. *J. Pet. Sci. Eng.* **2020**, *184*, 106507. [CrossRef]
48. Xie, X.; Li, S.; Guo, J. Study on Multiple Fractal Analysis and Response Characteristics of Acoustic Emission Signals from Goaf Rock Bodies. *Sensors* **2022**, *22*, 2746. [CrossRef]
49. Zhang, Q.; Shen, Y.; Pei, Y.; Wang, X.; Wang, M.; Lai, J. Determination of Integrity Index *Kv.* in CHN-BQ Method by BP Neural Network Based on Fractal Dimension *D. Fractal Fract.* **2023**, *7*, 546. [CrossRef]
50. Zhao, Y.; Feng, Z.; Yang, D.; Liang, W.; Feng, Z. Three-dimensional fractal distribution of the number of rock-mass fracture surfaces and its simulation technology. *Comput. Geotech.* **2015**, *65*, 136–146. [CrossRef]
51. Yin, Y.; Wang, J.; Zou, B.; Zhang, J.; Su, Y.; Sun, Q. Evaluation of Controlled Blasting Quality for Rock-Mass Tunneling Based on Multiple Indices. *J. Constr. Eng. Manag.* **2023**, *149*, 04022155. [CrossRef]
52. Li, X.; Liu, K.; Sha, Y.; Yang, J.; Ma, S.; Hong, Z. Investigation on radial fracturing around borehole under combined static stress and blasting. *Theor. Appl. Fract. Mech.* **2023**, *127*, 104038. [CrossRef]
53. Ren, F.; Zhu, C.; He, M.; Shang, J.; Feng, G.; Bai, J. Characteristics and Precursor of Static and Dynamic Triggered Rockburst: Insight from Multifractal. *Rock Mech. Rock Eng.* **2023**, *56*, 1945–1967. [CrossRef]
54. Dimitrakopoulos, R.; Scott, J.; Li, S. Multi-fractal conditional simulation of fault populations in coal seams using analogues: Method and application. *Int. J. Min. Reclam. Environ.* **2019**, *33*, 340–352. [CrossRef]
55. Ma, L.H.; Jiang, X.; Chen, J.; Zhao, Y.; Liu, R.; Ren, S. Analysis of Damages in Layered Surrounding Rocks Induced by Blasting During Tunnel Construction. *Int. J. Struct. Stab. Dyn.* **2021**, *21*, 2150089. [CrossRef]
56. Liu, D.; Zhang, J.; Tang, Y.; Jian, Y.; Cai, C. Damage Analysis of Concrete Structure under Multidirectional Shaped Charge Blasting Using Model Experiment and Ultrasonic Testing. *Adv. Civ. Eng.* **2021**, *2021*, 6677041. [CrossRef]
57. Chen, W.; Ma, H.; Shen, Z.; Wang, D. Experiment research on the rock blasting effect with radial jet cracker. *Tunn. Undergr. Space Technol.* **2015**, *49*, 249–252. [CrossRef]
58. Kong, X.; Wang, E.; Hu, S.; Shen, R.; Li, X.; Zhan, T. Fractal characteristics and acoustic emission of coal containing methane in triaxial compression failure. *J. Appl. Geophys.* **2016**, *124*, 139–147. [CrossRef]

59. Kong, X.; Wang, E.; He, X.; Li, Z.; Li, D.; Liu, Q. Multifractal characteristics and acoustic emission of coal with joints under uniaxial loading. *Fractals* **2017**, *25*, 1750045. [CrossRef]
60. Kong, B.; Wang, E.; Li, Z.; Lu, W. Study on the feature of electromagnetic radiation under coal oxidation and temperature rise based on multifractal theory. *Fractals* **2019**, *27*, 1950038. [CrossRef]

fractal and fractional

MDPI

Article

The Formative Factors of a Rock Burst Based on Energy Calculations and the Experimental Verification of Butterfly-Shaped Plastic Zones

Wenlong Zhang [1], Jicheng Feng [2,*], Jianju Ren [3], Ji Ma [4], Jianjun Shi [2] and Junfeng Zhang [5]

1 School of Civil Engineering and Architecture, Qingdao Huanghai University, Qingdao 266427, China; wenlong0523@163.com
2 School of Safety Engineering, North China Institute of Science & Technology, Langfang 065201, China
3 School of Resource Engineering, Xi'an University of Architecture and Technology, Xi'an 710055, China; jianjuren5995@163.com
4 College of Safety Science and Engineering, Henan Polytechnic University, Jiaozuo 454003, China; showsma@hpu.edu.cn
5 Shandong Lansheng Mining Technology Co., Ltd., Weifang 262400, China
* Correspondence: sky_fjc@163.com

Abstract: The research on the formation factors of rock burst is one of the main research directions of rock mechanics in recent years, which is helpful to solve the problem of rock burst accidents. So, in this study, the calculation method of energy released during rock burst is first obtained by using different medium models, and then, the formation factors of rock bursts are obtained by comparing the calculation energy with the actual accident energy. The method of energy calculation utilizes the difference between elastoplastic and pure elastic models to innovatively quantify the specific values of energy released before and after the occurrence of the rock burst. It is considered that the stress and plastic zone state before the occurrence of rock burst have an important influence on the occurrence of the accident and are one of the formation factors, while the deviatoric stress field and butterfly-shaped plastic zone create conditions for greater energy release. In addition, the trigger stress constitutes another formation factor. The plastic zone state before rock failure is verified by the experimental test; the location distribution shape of acoustic emission (AE) events during the later stage of compression failure is approximately the same as theoretical result. The results also preliminarily indicated the fractal characteristics of acoustic emission events distribution before sample failure. The study obtained the formative factors of rock burst accident, which provides a new ideas and references for the research on the formation of rock bursts.

Keywords: formative factors of rock burst; released energy calculation method; rock mass; experimental verification; rock burst

1. Introduction

The dynamic disasters of rock engineering perplex safe construction in many engineering fields. The so-called dynamic disasters mainly include rock burst occurring in tunnels [1], rock burst occurring in non-coal mines [2], rock burst occurring in coal mines (different from the disasters occurring in non-coal mines, the medium is coal, and the mining impact is greater) [3], coal and gas outbursts occurring in coal mines [4], slope collapses occurring during metro engineering [5], etc. The common point of these dynamic disasters is the instantaneous dynamic failure of coal or rock, which is often related to the medium state and human construction activities [6]. In terms of dynamic disasters, coal mine rock burst accidents have occurred more frequently in China in recent years, bringing huge pressure to safety production. The study of the mechanism of geotechnical engineering disasters is a fundamental work that can provide direction for subsequent monitoring [7,8], warning [9], and evaluation [10].

In view of the frequent occurrence of rock burst accidents in coal mines, scholars have also carried out a lot of research, mainly focusing on their occurrence mechanism [11], monitoring and early warning indicators [12] and methods [12,13], and risk treatment measures [11,14]. The research on the mechanism of rock burst dynamic disaster is the most important, which has attracted more scholars to study. In recent decades, various mechanisms have been formed, such as early energy theory [15], early strength theory [16], early stiffness theory [17], early rock burst tendency theory [18], intermediate instability theory [19], intermediate three-factor theory [20], intermediate three-criteria theory [21], later dynamic and static load theory [20], later rock burst initiation theory [22], later butterfly rock burst theory [23,24], etc. Some of these mechanisms or theories have put forward some concepts and descriptions, while others have established models and carried out strict mechanical derivation. which may be more scientific.

For the research methods of rock bursts, some studies focus on a specific condition, such as fault influence [25,26], high-strength mining faces [27,28], stress anomaly particularity [29], large vibration interference [30], periodic movement of hard roofs [31], etc. Other studies are conducted only from a certain angle, for example rock strength [32], stress analysis [33], energy analysis [34], special rock properties [20], gas coupling [35]. Some use theoretical reasoning and calculation [36,37], some use laboratory research [38,39], some use numerical simulation [40,41], and some only use a conceptual expression. Rigorous mathematical and mechanical reasoning is highly recommended for scientific and engineering problems, and numerical simulation can be used to reflect or verify regular problems. If the simplest calculation method and the simplest model can be established to explain the common problems for geotechnical engineering, the rock burst mechanism will be easier to reveal (like the famous Oakham criterion [42]: if there are two or more different hypotheses about the same phenomenon, the simpler or falsifiable one should be adopted).

As for the stress state when rock burst occurs, it is clear that it must be a situation with certain characteristics, and then the smaller triggering stress leads to the accident [43]. How the energy changes before and after the occurrence is directly related to the root cause of the occurrence mechanism. In order to study the mechanism of rock burst more scientifically, the energy conversion before, after, and during its occurrence is calculated from an energy perspective, which is an important research content in the field of rock dynamics. This study first proposes a method to calculate the system energy, and then obtains which mechanical state is more likely to reach the accident energy through this method. The results show that the butterfly-shaped plastic zone state caused by the bias stress field is more dangerous. Finally, the whole process AE monitoring of a loaded sample is carried out from the laboratory scale to verify the butterfly mode before failure. The experiment fully considered the specific situation of the original waveform, and adopted a more appropriate event location method, and finally obtained the butterfly-shaped plastic zone. The energy calculation method, rock burst mechanism, and its verification obtained in this study provide a new idea for the study of the rock burst mechanism.

2. Numerical Methodology

Rock burst accidents occur in rock masses, but rock masses are often relatively large and difficult to study via mechanical models and analysis. So, this study narrowed down the scope of the study and reflected big problems through small models. Due to the fact that regular knowledge is only obtained through model establishment and energy calculation, and the stress and situation of small model are less different from the actual environment in which the tunnel is located, the research method of the small model is considered feasible. In addition, considering the feasibility of the calculation, the method proposed in this study is completed by means of FLAC^{3D} numerical simulation and theoretical calculation, the rock mass (total volume is Ω) involved in the rock burst accident is picked out separately as the research object in the numerical simulation, and a hole is set in the middle of it to respect mining space. The rock mass is subjected to an isotropic force, which can be simplified as a three-dimensional stress of (P_1, P_2, P_3), where P_1, P_2, and P_3 are the main force. The rock mass is composed of many units (f(x, y, z)), and the force of these units can also be simplified as (σ_{1i}, σ_{2i}, σ_{3i}), where σ_{1i}, σ_{2i}, and σ_{3i} are the main stress. The energy of each element in this mechanical state can be calculated by Equation (1) [32].

$$f(x,\ y,\ z) = \frac{1}{2E_i}\left[\sigma_{1i}{}^2+\sigma_{2i}{}^2+\sigma_{3i}{}^2 - 2\mu_i(\sigma_{1i}\sigma_{2i} + \sigma_{2i}\sigma_{3i} + \sigma_{1i}\sigma_{3i})\right] \tag{1}$$

where E_i is the elastic modulus of the corresponding element and μ_i is the Poisson ratio of the corresponding element.

The stress state before the rock burst accident is named as pre-state stress field (PSSF), after being affected by a trigger stress field (named TSF; a high probability arises from roof fracture or coal rock fracture [44]), the stress state after being applied is named as the later state stress field (LSSF). During FLAC3D numerical simulation, in the model, the loaded media can be pure elastic or elastoplastic, and the energy of the two media under the named PSSF and LSSF states are U_{PSSF}, U_{PSSF}', U_{LSSF}, and U_{LSSF}', respectively. When the model is purely elastic, whether it is PSSF or LSSF, all elements are purely elastic, and the formula can be expressed as Equation (2), However, when the model is elastoplastic, under the action of PSSF, some elements will become elastoplastic (Ω_e). Under LSSF, a part of elastoplastic elements ($\Delta\Omega_p$) will be added compared with PSSF, so the energy under U_{PSSF}' and U_{LSSF}' states can be expressed as Equations (3) and (4), respectively.

$$U_{PSSF/LSSF} = \iiint_{\Omega(PSSF/LSSF)} f(x,y,z)dV \tag{2}$$

$$U_{PSSF}' = \iiint_{\Omega_e(PSSF)} f(x,y,z)dV_e + \iiint_{\Omega_p(PSSF)} f(x,y,z)dV_p \tag{3}$$

$$U_{LSSF}' = \iiint_{(\Omega_e-\Delta\Omega_p)(LSSF)} f(x,y,z)d(V_e-\Delta V_P) + \iiint_{(\Omega_p+\Delta\Omega_p)(LSSF)} f(x,y,z)d(V_p+\Delta V_P) \tag{4}$$

where V represents all units of the rock mass. V_e and V_p represent the units of pure elastic and elastoplastic, respectively.

Under the action of PSSF and LSSF, the energy difference ($D_{PSSF/LSSF}$) between the pure elastic model and elastic–plastic model rock mass is expressed as Equation (5). The difference of energy difference between PSSF and LSSF is the total energy in the process of mechanical state change. However, the elastic wave energy (W) [45] needs to be multiplied by β (elastic wave energy conversion coefficient, the value is 1~10% [27,46]), as in Equation (6).

$$D_{PSSF/LSSF} = U_{PSSF/LSSF} - U_{PSSF/LSSF}' \tag{5}$$

$$W = \beta(D_{LSSF} - D_{PSSF}) \tag{6}$$

The calculation chart process of energy release during the mechanical state change process of rock mass from PSSF to LSSF is denoted in Figure 1. Firstly, calculate the energy difference between the pure elastic and elastic-plastic models under the mechanical state of PSSF (D_{PSSF}), and then calculate the energy difference under the mechanical state of LSSF (D_{LSSF}). The difference between D_{PSSF} and D_{LSSF} is the released energy from PSSF to LSSF. When calculating the release energy, it is not simply the difference between U_{PSSF} and U_{LSSF} when it is the elastic–plastic model, but the stored energy under the pure elastic model also needs to be considered in order to calculate the accurate release energy value. This study comprehensively considered the differences between pure elastic and elastic–plastic models under two mechanical states, and eliminated the influence of model size. The obtained energy calculation process and results can be compared with the actual energy release of on-site accidents, which can be used to infer the most likely occurrence condition of rock burst, and is of great significance to reveal the mechanism of rock burst.

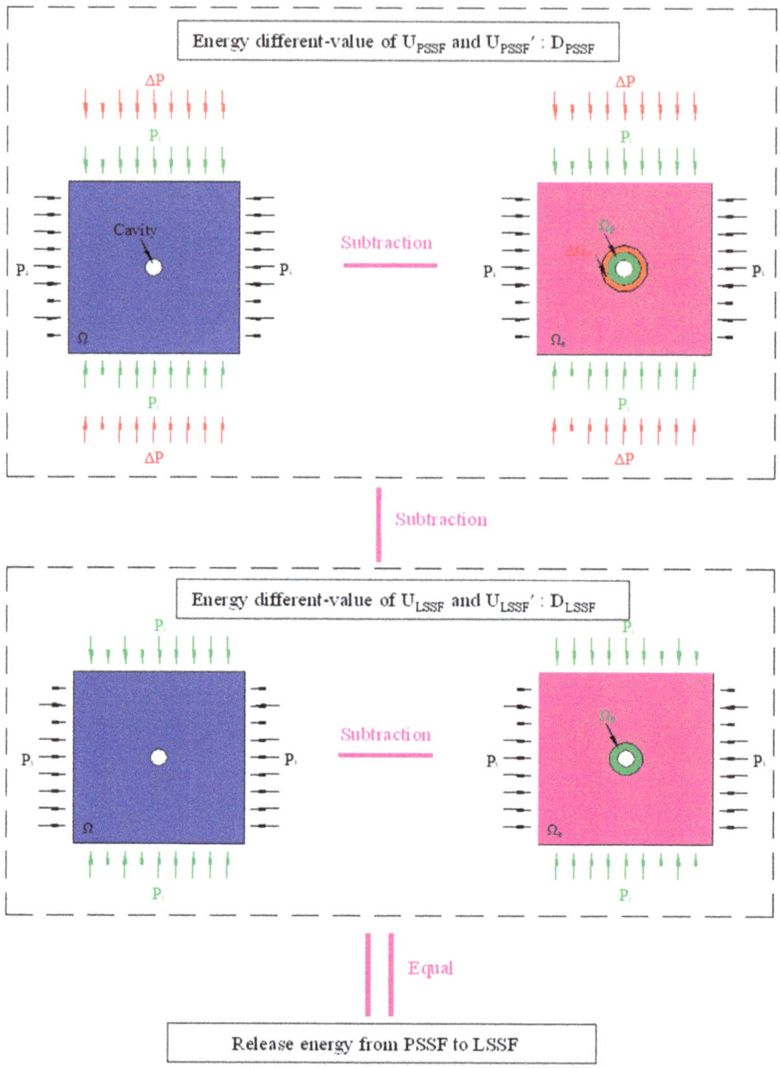

Figure 1. Calculation chart process of energy release during the mechanical state change process of rock mass from PSSF to LSSF.

3. Results

An example model of 200 × 200 × 1 m (the model and force application direction are shown in Figure 2, the diameter of the hole is 5.6 m, and the application direction of P_2 is front and rear) was taken to calculate the rock mass energy under different mechanical states (PSSF as (P_1 = 20 MPa, P_2 = 20 MPa, P_3 = 20 MPa), ΔP as 1 MPa, and only added to P_1) and different model conditions, according to the proposed method and the final energy release result. The shear strength, cohesion, friction angle, and tensile strength of the used medium are 1.3 GPa, 3 MPa, 25°, and 1.77 MPa, respectively.

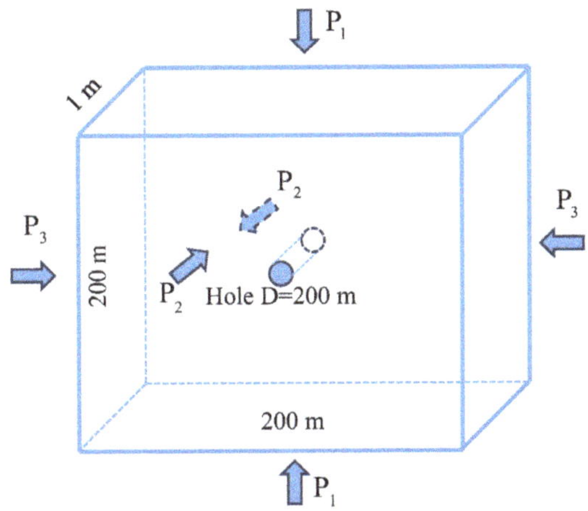

Figure 2. Example model and its external force state.

From the numerical simulation results, it can be seen that under the conditions of the elastoplastic model, the shape of the plastic zone presents a butterfly shape when P_1 is bigger than 50 MPa (η bigger than 2.5, similar as the results in [47]); a more pronounced butterfly shape (extending further) has emerged when P_1 = 55 MPa (at this time, η = 2.75). The concentration phenomenon of plastic zone and stored energy distribution around the hole is closely related to the existence of hole. At the same time, the energy difference also continues to increase and expand with the increase in drilling P_1, as in Figure 3 (D_{LSSF} distribution when P_1 = 40 MPa, 50 MPa, 55 MPa, 58.6 MPa). Another interesting point is that some unit bodies do not release energy when subjected to changes in force, but instead absorb energy (it may be due to the fact that certain unit bodies tend to be subjected to more uniform forces, or the occurrence of tensile phenomena leads to a decrease in the calculated energy value). As the vast majority of unit bodies release energy, the final energy result of the entire sample is releasing state.

The change curve of the storage and different-value of energy are denoted in Figure 4; it can be seen intuitively from the figure that both storage energy of pure elastic medium and elastoplastic medium are increasing with the increase in P_1, but the increasing speed is different (only describing the change in P_1 is because the values of P_2 and P_3 have not changed in all mechanical states). The different-value of energy increases and acceleration are various with the increase in P_1, and mainly divided into three stages. The obvious critical points of the three stages are P_1 = 50 MPa, P_1 = 55 MPa, and P_1 = 58.6 MPa, respectively. Interestingly, P_1 = 50 MPa is the starting point of the "early butterfly", P_1 = 55 MPa is the starting point of the "late butterfly", and P_1 = 58.6 MPa is the value of the "final butterfly". The η values corresponding to the three critical points are 2.5, 2.75 and 2.93, respectively. The above fact shows that the shape of plastic zone is closely related to the different-value of energy, the butterfly shape is strengthened with the increase in P_1 and, meanwhile, the different-value of energy is also increased.

Divide the energy of the unit body by the volume of the unit body to obtain the energy release density map of the rock mass, and the release energy density map of butterfly related mechanical states are shown in Figure 5. The results show that the distribution of energy release density and plastic zone has a strong management, and has experienced a process from small butterfly to large butterfly. From the value of energy distribution, the greater the force, the greater the density value, and the more concentrated the distribution.

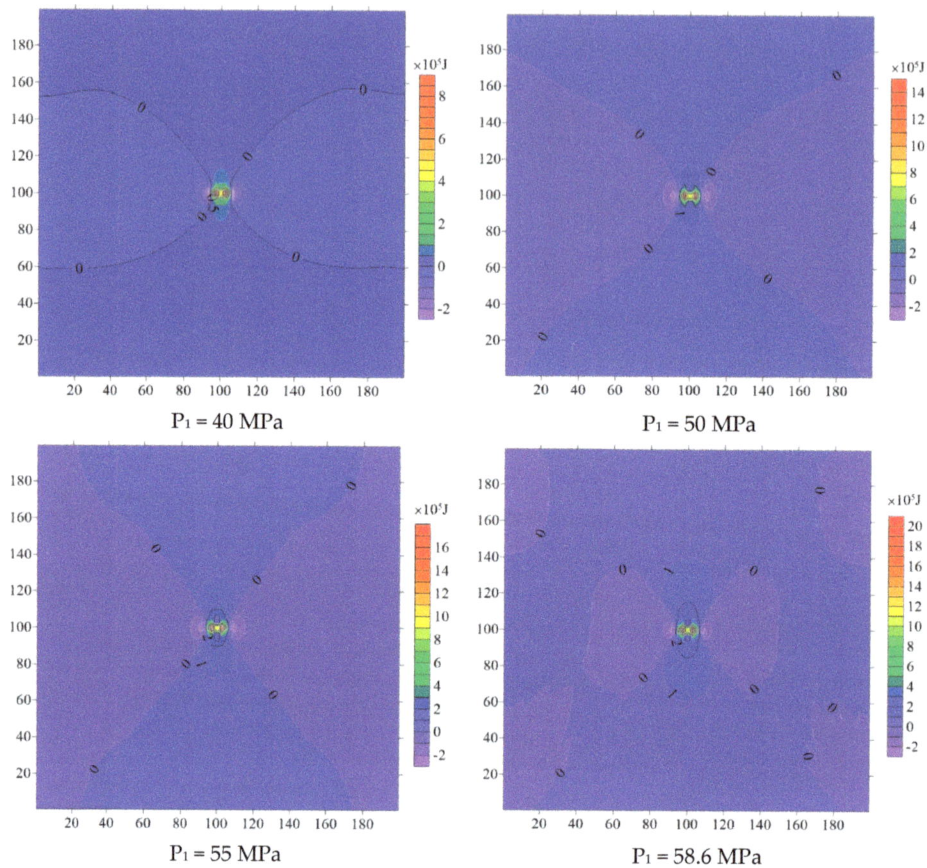

Figure 3. D_{LSSF} distribution when P_1 = 40 MPa, 50 MPa, 55 MPa, 58.6 MPa (the horizontal and vertical coordinates represent the width and height of the model, corresponding to Figure 1).

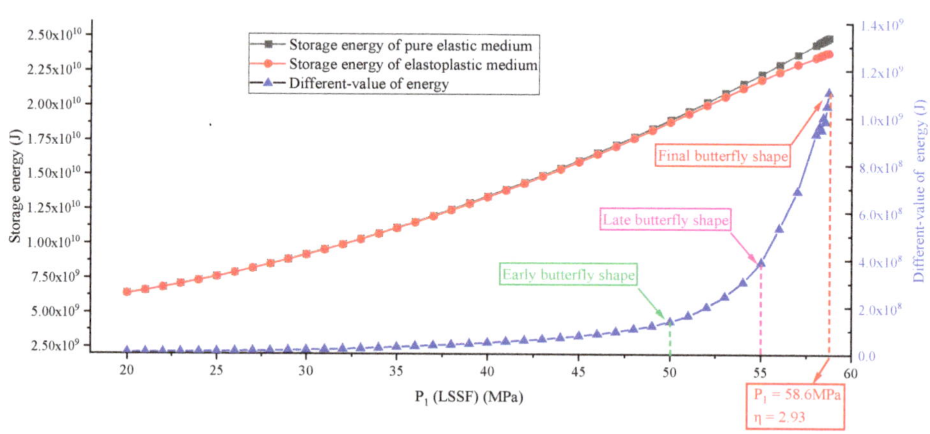

Figure 4. Change curve of storage and different-value of energy.

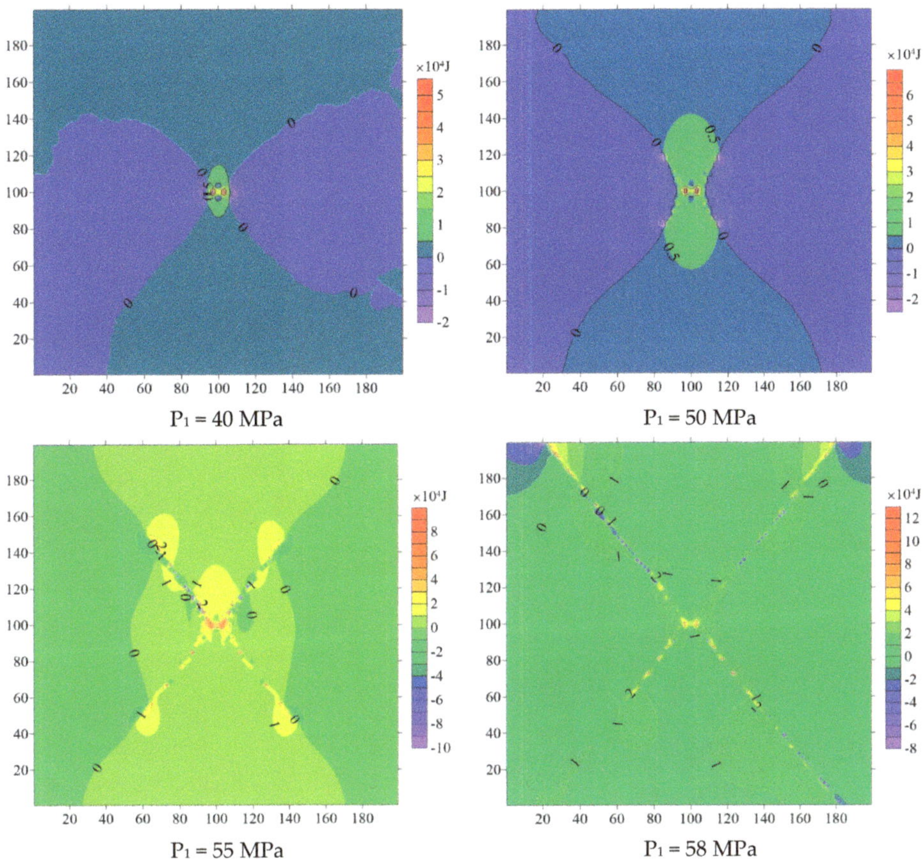

Figure 5. Release energy density map of butterfly related mechanical states (the horizontal and vertical coordinates represent the width and height of the model, corresponding to Figure 1).

The calculation results of the released energy of rock mass are shown in Figure 6, from which it can be seen that the phased characteristics are obvious. The entire process shown in the figure can be divided into three stages: the pregnant period, growth period, and upheaval period, representing the phenomenon of releasing energy from small to large, respectively, and then rapidly increasing. Results show that even with the same TSF, different PSSFs have a great impact on the results of the energy release. When the shape of the plastic zone caused by the PSSF (when there is deviatoric loading [48], and P_1/P_3 is mostly >2.5) of rock mass is butterfly, the change in the released amount forms an inflection point, and the corresponding η values are 2.5 MPa, 2.75 MPa, and 2.93, respectively, the released energy is increased by 12.5 times form η = 2.5 to η = 2.93.

The calculated energy of the rock burst can be compared with the actual energy. It can be seen that the released energy of the accident is matched with the calculated energy from the above energy change results, and the formation factors of rock burst are the butterfly plastic zone caused by PSSF and appropriate TSF in addition to the basic conditions (mining space and force relation are easily satisfied). The PSSFs are related to the protolith stress, mining stress, roadway layout, surrounding rock, coal properties, etc. The TSFs are related to the roof breakage event, coal fracture event, blasting vibration, etc. Further, these factors are finally reflected in PSSF and TSF, otherwise it is difficult to meet the energy conditions.

When the energy release reaches the critical value, the energy conditions for the accident can be met, as in Figure 7.

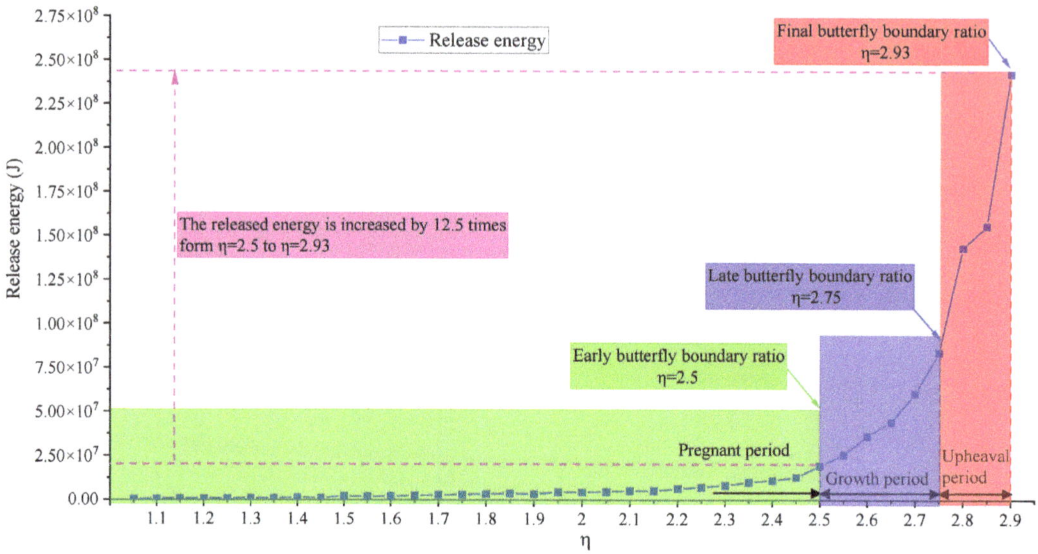

Figure 6. Release energy variation curve with different P_1 (LSSF).

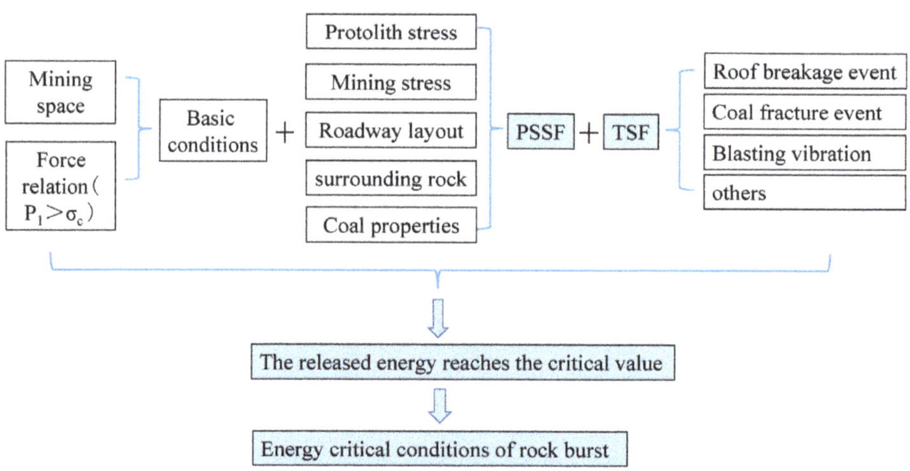

Figure 7. The formation factors of rock burst based on energy calculation.

4. Discussion

The formation factors of rock burst have been obtained in the Section 3. In fact, the conclusive key factor of large ratio PSSF is based on less requirements of TSF under "late butterfly shape" or "final butterfly shape" of the plastic zone, so the required minimum TSF to reach the critical energy value under different stage based on a total energy of 10^7 J is indicated in Figure 8. The results show that in the non-butterfly stage (P_1 less than 50 MPa), the minimum TSF required is 1~14 MPa, while in the butterfly stage (P_1 more than 50 MPa and less than 55 MPa), the value is reduced to 0.2 MPa, and in the late butterfly stage (P_1 more than 55 MPa), the value is even reduced to 0.1 MPa, which is easy to achieve in realistic data. The above facts confirm the key factor role of a large ratio PSSF for a rock burst, which is similar to the result of rock bursts caused by larger difference existing between horizontal and vertical stresses in [49].

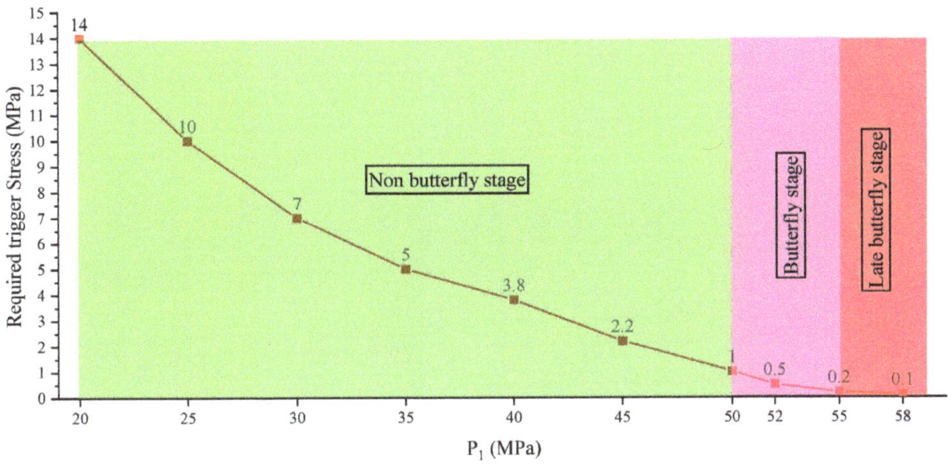

Figure 8. Required minimum TSF to reach the critical energy value under different stage.

In fact, the implicit equation of the plastic zone boundary under the condition of the uniform medium model is obtained in a previous study [24]. The plastic zone distributions of typical mechanical states obtained by Equation (7). The results show that when $P_1 = 40$ MPa, the shape of plastic zone is not butterfly, but when $P_1 = 50$ MPa, the butterfly state is more obvious. When $P_1 = 55$ MPa, the maximum radius of the plastic zone R_{max} extends to 15 m, and when $P_1 = 58.6$ MPa, the R_{max} extends to 90 m. The plastic zone results of previous theoretical calculations are the same as those in this study, while this study focuses more on the variation law of energy corresponding to plastic zone.

$$
\begin{aligned}
&9\left(1-\eta^2\right)\left(\frac{a}{r}\right)^8+\left[-12(1-\eta)^2+6(1-\eta)^2\cos 2\theta\right]\left(\frac{a}{r}\right)^6 \\
&+\left[10\left(1-\eta^2\right)\cos^2 2\theta-4\left(1-\eta^2\right)\sin^2\varphi\cos^2 2\theta-2\left(1-\eta^2\right)\sin^2 2\theta-4\left(1-\eta^2\right)\cos 2\theta+(1+\eta)^2\right]\left(\frac{a}{r}\right)^4 \\
&+\left[-4(1-\eta)^2\cos 4\theta+2(1-\eta^2)\cos 2\theta-4(1-\eta^2)\sin^2\varphi\cos 2\theta-\frac{4C(1-\eta)\sin 2\varphi\cos 2\theta}{P_3}\right]\left(\frac{a}{r}\right)^2 \\
&+\left[\left(1-\eta^2\right)-\sin^2\varphi\left(1+\eta+\frac{2C\cos\varphi}{P_3\sin\varphi}\right)^2\right]=0
\end{aligned}
\tag{7}
$$

Another index of the plastic zone is the area of plastic zone, which not only represents the depth of the plastic zone, but also comprehensively reflects the volume of the plastic zone. Therefore, the variation curves of plastic zone R_{max}, area S, and released energy with the increase in P_1 are compared in Figure 9. From the comparison results, the three indicators all formed a certain inflection point when $\eta = 2.5$, 2.75, and 2.93, but the difference is the rate and degree of change. The most drastic change is the released energy index, followed by R_{max} index, and finally S. Although the change degree of the three indexes is different, they all reflect the mutation phenomenon, which shows that the rock burst mechanism relying on large ratio PSSF and butterfly shape plastic zone is reasonable.

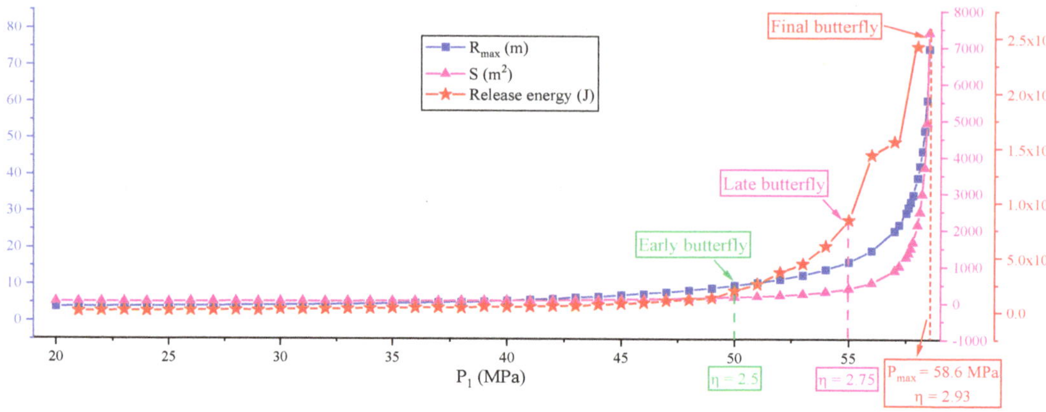

Figure 9. Variation curves of plastic zone R_{max}, area S, and released energy with the increase in P_1.

The above analysis obtained the important influence of butterfly plastic zone based on theoretical calculation and numerical simulation. An experimental verification is more reliable; therefore, the whole process of AE monitoring of sample loading process is implemented. The equipment is denoted in Figure 10a, which mainly includes a universal tester and its control system, an AE acquisition system, and a rock sample. The sample is a kind of sandstone with a rock burst tendency (shown in Figure 10b), and the size is 200 cubic millimeters (a 20 mm diameter central hole is set in the axial direction to represent the roadway, and the direction is assumed to be in the Y direction). The uniaxial loading method is adopted for the experiment, until the sample is completely crushed. The sampling frequency of AE system is 6 MHz, and a total of eight sensors are used for signal acquisition, as in Figure 10c. The study focuses more on the positioning of AE events within the few seconds before rock failure, so as to verify the shape of plastic zone area before failure.

The study should focus on the failure distribution form in the XZ plane (perpendicular to the Y direction) during 1990~1995s, and the influence of boundary conditions should be removed as far as possible. Therefore, the statistical AE event scatter plot and heat map during 1990~1992s and 1993~1995s are stated in Figure 11. The number of AE events during 1990~1992s is less than that during 1993~1995s. The distribution pattern of event points is not obvious, but the heat map results can basically show the butterfly shape. The distribution of each butterfly leaf is not very regular in the heat map results, which may be caused by the heterogeneity of the sample or the influence of force transmission. The AE event scatter plot and heat map during the whole period of 1990~1995s are shown in Figure 12, which shows a more obvious butterfly shape, and is basically consistent with the results of theoretical calculation and numerical simulation. The formation of butterfly-shaped plastic zones not only reflects the morphology before failure, but also indicates that events are concentrated in certain specific areas (fractal characteristics) and may cause large-scale damage after reaching a certain level.

This study reveals the shape of butterfly-shaped plastic zone of the sample in a specific state, which is caused by a large ratio of the stress field (like the state described in [50,51], the influence of some large ratio of foundation ground stress or geological structure often results in deviatoric stress field, which increases the probability of rock bursts), and resulting in a large amount of concentrated energy release. Based on the initial stress state of the rock, the study considers that a small TSF may lead to a huge release under specific stress conditions, and gives the minimum TSF required to form disasters under different states. It considers that the probability of the occurrence of rock burst is the butterfly-shaped plastic zone shape caused by deviatoric stress field. This can also be demonstrated from the actual size of TSF; in fact, TSF is unlikely to become the dominant factor in most cases, which has been already demonstrated in [52]. In addition, this study also preliminarily discovered the fractal characteristics of acoustic emission events distribution before sample failure.

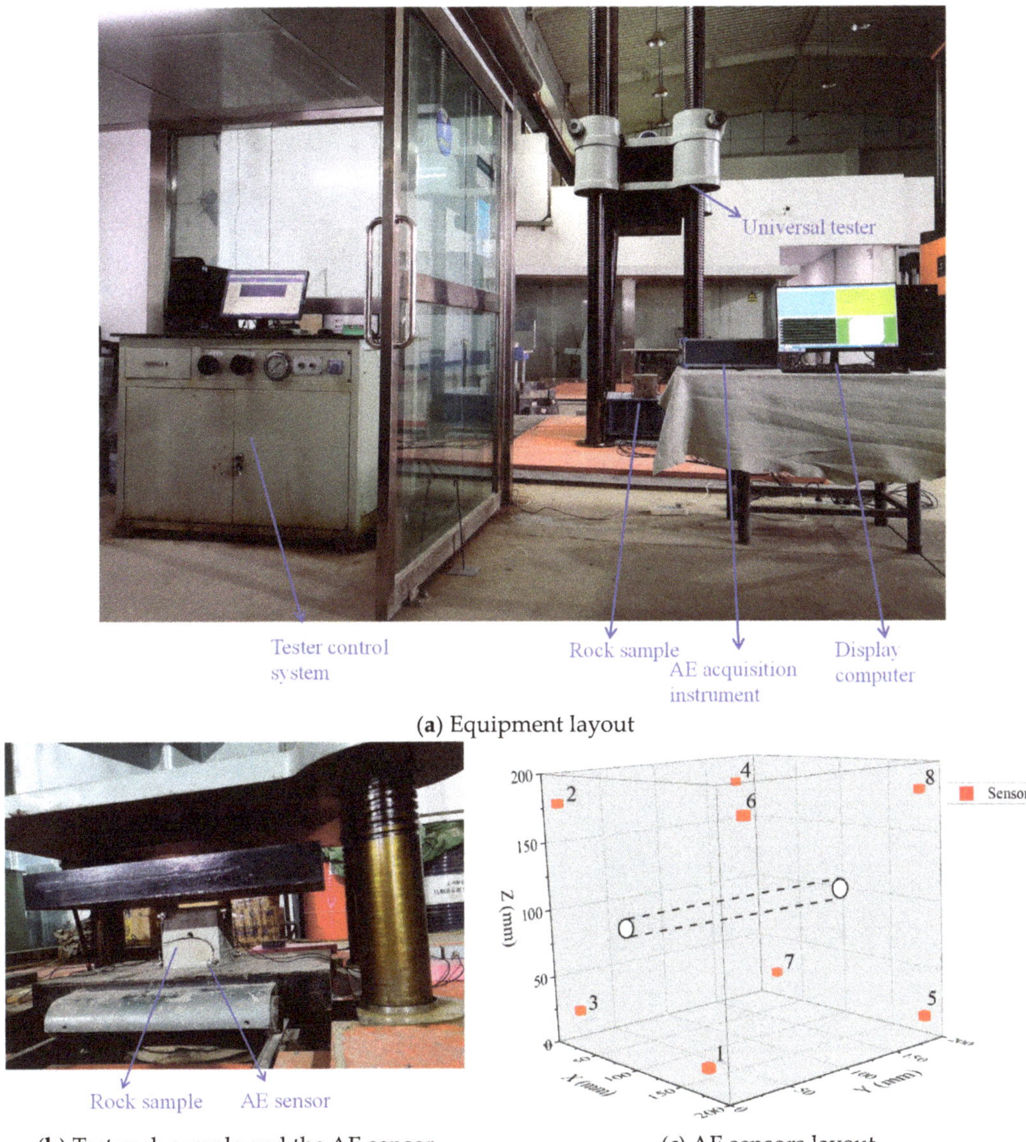

(**a**) Equipment layout

(**b**) Test rock sample and the AE sensor

(**c**) AE sensors layout

Figure 10. Equipment used for experimental verification.

The results of this study are based on a calculation method of energy to obtain the energy value before and after the change in the rock mass mechanical state, and the specific expression of dangerous stress state is obtained; that is, when the ratio of maximum principal stress to minimum principal stress reaches the 2.5/2.75/2.93 critical point. In addition, an experimental test was used to preliminarily verify the shape of the butterfly-shaped plastic zone on the eve of sample failure. Although the form of uniaxial compression was used for the convenience of the test, this form also showed butterfly shape in numerical simulation and theoretical calculation, which proved to be basically effective, but the butterfly shape under triaxial loading should be obtained more in later experiments. In addition, it is necessary to try measuring vibration signals in more directions, such as three axes [53]. The experimental verification has been preliminarily completed, and the on-site verification should be carried out later, and a warning method using the three-dimensional stress field or butterfly plastic zone state can also be studied in future. Furthermore, more detailed three-dimensional stress states [54,55] and anisotropy [56], or particle breakage [57] should also be considered, and the effects of deformation and cracks [58,59] can also be comprehensively considered.

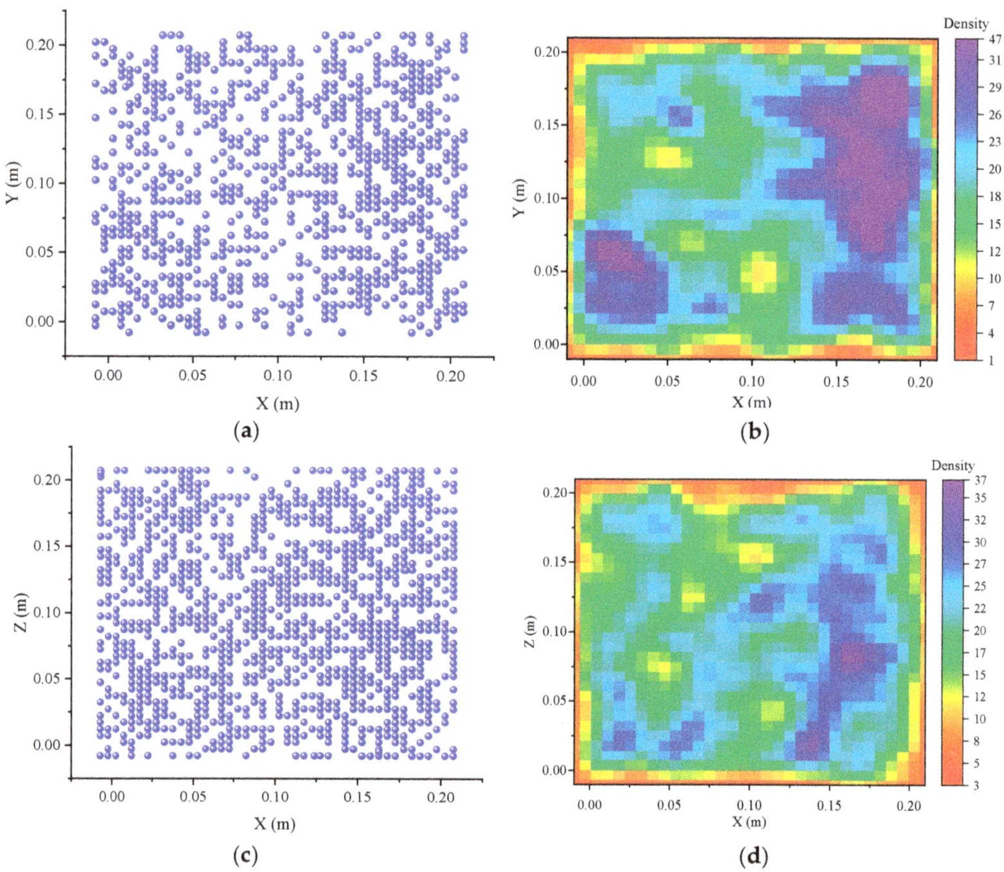

Figure 11. Scatter plot and heat map of AE events occurred during the period of 1990~1992s and 1993~1995s. (**a**) Scatter plot of AE events occurred during the period of 1990~1992s. (**b**) Heat map of AE events occurred during the period of 1990~1992s. (**c**) Scatter plot of AE events occurred during the period of 1993~1995s. (**d**) Heat map of AE events occurred during the period of 1993~1995s.

This study obtained the energy values before and after the occurrence of rock burst through the innovative energy calculation method proposed. By comparing them with the actual accident energy, the energy cloud maps of different mechanical states and the relationship between released energy were obtained. From an energy perspective, the formation factors of rock bursts were clarified, and the importance of deviatoric stress in

PSSF was emphasized. The minimum triggering stress value required to reach the critical value under different PSSFs was obtained, and the butterfly-shaped plastic zone formed by a deviatoric stress field was preliminarily verified at the laboratory scale. This study provides a new approach for studying the mechanism of rock dynamic disasters and lays a certain foundation for monitoring and warning of butterfly-shaped plastic zones.

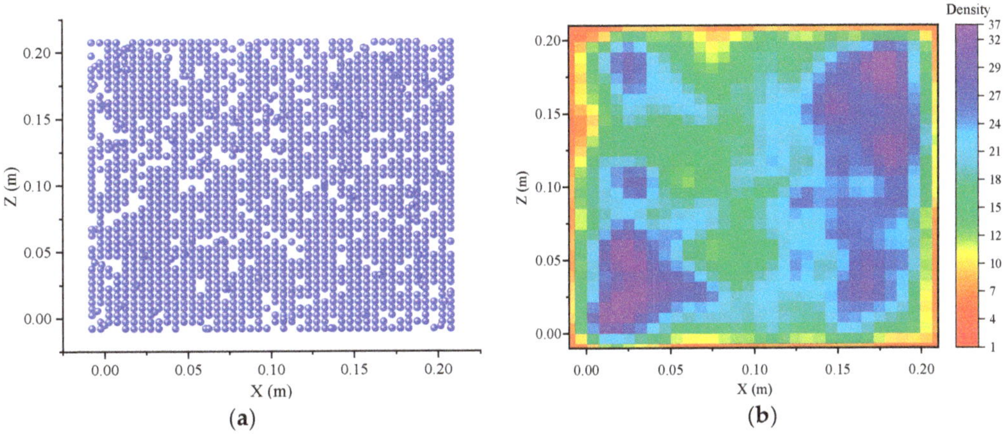

Figure 12. Scatter plot and heat map of AE events occurred during the whole period of 1990~1995s. (**a**) Scatter plot of AE events occurred during the whole period of 1990~1995s. (**b**) Heat map of AE events occurred during the whole period of 1990~1995s.

5. Conclusions

In this study, the released energy of rock burst accident is obtained by means of energy calculation, the formation factors of rock bursts are obtained by means of the calculation method, and the formation factors of butterfly-shaped plastic zone are obtained and verified by means of laboratory tests. It is easier to achieve quantification and accuracy from the perspective of energy, and the obtained factors of rock burst formations are clearer, which has certain scientific value. The main conclusions are as follows: (1) The energy conditions of rock burst accidents are obtained, which lays the foundation for determining the formation factors of rock bursts. (2) The formation factors of rock burst are the butterfly plastic zone caused by PSSF and appropriate TSF, and PSSF plays a leading role, which determines the possibility of rock burst accidents. The PSSF that leads to the butterfly-shaped plastic zone represents a dangerous state of deviatoric stress field, in which a small triggering stress can lead to large-scale energy release and rock failure. (3) The butterfly failure mode has been preliminarily verified at the laboratory scale through the location of AE events, and the rock sample shows a butterfly-shaped plastic zone before uniaxial loading failure. (4) This study also preliminarily discovered the fractal characteristics of acoustic emission events distribution before the sample failure.

Author Contributions: Conceptualization, W.Z.; methodology, J.M.; software, J.F.; validation, W.Z.; formal analysis, W.Z.; investigation, J.R.; resources, J.S.; data curation, J.R.; writing—original draft preparation, W.Z.; writing—review and editing, J.Z. All authors have read and agreed to the published version of the manuscript.

Funding: The authors gratefully acknowledge the financial support by Qingdao Huanghai University Doctoral Research Fund Project (2022boshi02) and National Natural Science Foundation of China (52174111).

Data Availability Statement: The data presented in this study are available on request from the corresponding author.

Conflicts of Interest: Author Junfeng Zhang was employed by Shandong Lansheng Mining Technology Co., Ltd. The remaining authors declare that the research was conducted in the absence of any commercial or financial relationships that could be construed as a potential conflict of interest.

References

1. Yang, J.H.; Jiang, Q.H.; Zhang, Q.B.; Zhao, J. Dynamic stress adjustment and rock damage during blasting excavation in a deep-buried circular tunnel. *Tunn. Undergr. Space Technol.* **2018**, *71*, 591–604. [CrossRef]
2. Feng, X.-T.; Liu, J.; Chen, B.; Xiao, Y.; Feng, G.; Zhang, F. Monitoring, Warning, and Control of Rockburst in Deep Metal Mines. *Engineering* **2017**, *3*, 538–545. [CrossRef]
3. Mark, C. Coal bursts in the deep longwall mines of the United States. *Int. J. Coal Sci. Technol.* **2016**, *3*, 1–9. [CrossRef]
4. Mark, C. Coal bursts that occur during development: A rock mechanics enigma. *Int. J. Min. Sci. Technol.* **2018**, *28*, 35–42. [CrossRef]
5. Zhang, Q.B.; He, L.; Zhu, W.S. Displacement measurement techniques and numerical verification in 3D geomechanical model tests of an underground cavern group. *Tunn. Undergr. Space Technol.* **2016**, *56*, 54–64. [CrossRef]
6. Zhang, W.; Qu, X.; Li, C.; Xu, X.; Zhang, S.; Jin, G.; Wang, Y. Fracture analysis of multi-hard roofs based on microseismic monitoring and control techniques for induced rock burst: A case study. *Arab. J. Geosci.* **2019**, *12*, 784. [CrossRef]
7. Carri, A.; Valletta, A.; Cavalca, E.; Savi, R.; Segalini, A. Advantages of IoT-Based Geotechnical Monitoring Systems Integrating Automatic Procedures for Data Acquisition and Elaboration. *Sensors* **2021**, *21*, 2249. [CrossRef]
8. Valletta, A.; Carri, A.; Savi, R.; Segalini, A. Algorithms for the Near-Real Time Identification and Classification of Landslide Events Detected by Automatic Monitoring Tools. In *International Scientific Conference Environmental Challenges in Civil Engineering*; Springer International Publishing: Cham, Switzerland, 2022; pp. 74–84.
9. Drusa, M.; Kais, L.; Dubovan, J.; Markovic, M.; Bahleda, F.; Mecar, M. Measurement of Axial Strain of Geogrid by Optical Sensors. *Sensors* **2021**, *21*, 6404. [CrossRef]
10. Rybak, J.; Khayrutdinov, M.M.; Kuziev, D.A.; Kongar-Syuryun, C.B.; Babyr, N.V. Prediction of the geomechanical state of the rock mass when mining salt deposits with stowing. *J. Min. Inst.* **2022**, *253*, 61–70. [CrossRef]
11. Liu, J.K.; Luan, H.J.; Zhang, Y.C.; Sakaguchi, O.; Jiang, Y.J. Prediction of unconfined compressive strength ahead of tunnel face using measurement-while-drilling data based on hybrid genetic algorithm. *Geomech. Eng.* **2020**, *22*, 81–95. [CrossRef]
12. Ren, J.; Zhang, W.; Wu, Z.; Li, J.; Shen, Y.; Zhang, G. Microseismic Signals in Heading Face of Tengdong Coal Mine and Their Application for Rock Burst Monitoring. *Shock Vib.* **2021**, *2021*, 6650446. [CrossRef]
13. Zhang, W.; Huo, T.; Li, C.; Wang, C.; Qu, X.; Xin, C.; Berardengo, M. Characteristics of Valuable Microseismic Events in Heading Face of an Underground Coal Mine Using Microseismic System. *Shock Vib.* **2021**, *2021*, 6683238. [CrossRef]
14. Shen, P.; Tang, H.; Zhang, B.; Ning, Y.; He, C. Investigation on the fracture and mechanical behaviors of simulated transversely isotropic rock made of two interbedded materials. *Eng. Geol.* **2021**, *286*, 106058. [CrossRef]
15. Cook, N.G.W.; Hoek, E.; Pretorius, J.P.; Ortlepp, W.D.; Salamon, M.D.G. Rock mechanics applied to the study of rock burst. *J. S. Afr. Inst. Min. Metall.* **1966**, *66*, 435–528.
16. Bieniawski, Z.T.; Denkhaus, H.G.; Vogler, U. Failure of fractured rock. *Int. J. Rock Mech. Min. Sci. Geomech. Abstr.* **1969**, *6*, 323–330. [CrossRef]
17. Brady, B.H.G.; Bnown, E.T. *Rock Mechonics for Underground Mining*; Springer Science & Business Media: Berlin/Heidelberg, Germany, 1985.
18. Bieniawski, Z.T. Mechanism of brittle of rocks. Part I, II and III. *Int. J. Rock Mech. Min. Sci. Geomech. Abstr.* **1967**, *4*, 395–406. [CrossRef]
19. Pan, Y. Disturbance response instability theory of rockburst in coal mine. *J. China Coal Soc.* **2018**, *43*, 2091–2098.
20. Gao, F.; Stead, D.; Kang, H. Simulation of roof shear failure in coal mine roadways using an innovative UDEC Trigon approach. *Comput. Geotech.* **2014**, *61*, 33–41. [CrossRef]
21. Li, Y. Mechanism of rock burst and its preliminary application. *J. China Inst. Min.* **1985**, *3*, 37–43.
22. Pan, J. Theory of rockburst start up and its complete technology system. *J. China Coal Soc.* **2019**, *44*, 173–182.
23. Zhao, Z.; Ma, N.; Guo, X.; Zhao, X.; Xia, Y.; Ma, Z. Mechanism conjecture of butterfly rock burst in coal seam roadway. *J. China Coal Soc.* **2016**, *11*, 2689–2697.
24. Ma, N.; Guo, X.; Zhao, Z.; Zhao, X.; Liu, H. Occurrence mechanisms and judging criterion on circular tunnel butterfly rock burst in homogeneous medium. *J. China Coal Soc.* **2016**, *11*, 2679–2688.
25. Jiang, L.; Kong, P.; Zhang, P.; Shu, J.; Wang, Q.; Chen, L.; Wu, Q. Dynamic Analysis of the Rock Burst Potential of a Longwall Panel Intersecting with a Fault. *Rock Mech. Rock Eng.* **2019**, *53*, 1737–1754. [CrossRef]
26. Wang, P.; Jiang, L.; Jiang, J.; Zheng, P.; Li, W. Strata Behaviors and Rock Burst–Inducing Mechanism under the Coupling Effect of a Hard, Thick Stratum and a Normal Fault. *Int. J. Geomech.* **2018**, *18*, 04017135. [CrossRef]
27. He, H.; Dou, L.; Gong, S.; Zhou, P.; Xue, Z. Rock burst rules induced by cracking of overlying key stratum. *Chin. J. Geotech. Eng.* **2010**, *32*, 1260–1265.

28. Lu, C.-P.; Liu, G.-J.; Liu, Y.; Zhang, N.; Xue, J.-H.; Zhang, L. Microseismic multi-parameter characteristics of rockburst hazard induced by hard roof fall and high stress concentration. *Int. J. Rock Mech. Min. Sci.* **2015**, *76*, 18–32. [CrossRef]
29. Xie, J.; Xu, J.; Wang, F. Mining-induced stress distribution of the working face in a kilometer-deep coal mine—A case study in Tangshan coal mine. *J. Geophys. Eng.* **2018**, *15*, 2060–2070. [CrossRef]
30. Cao, A.; Dou, L.; Cai, W.; Gong, S.; Liu, S.; Zhao, Y. Tomographic imaging of high seismic activities in underground island longwall face. *Arab. J. Geosci.* **2016**, *9*, 232. [CrossRef]
31. Lu, C.-P.; Dou, L.-M.; Zhang, N.; Xue, J.-H.; Wang, X.-N.; Liu, H.; Zhang, J.-W. Microseismic frequency-spectrum evolutionary rule of rockburst triggered by roof fall. *Int. J. Rock Mech. Min. Sci.* **2013**, *64*, 6–16. [CrossRef]
32. Xie, H.; Ju, Y.; Li, L. Criteria for strength and structural failure of rocks based on energy dissipation and energy release principles. *Chin. J. Rock Mech. Eng.* **2005**, *24*, 3003–3010.
33. Kong, P.; Jiang, L.; Shu, J.; Sainoki, A.; Wang, Q. Effect of Fracture Heterogeneity on Rock Mass Stability in a Highly Heterogeneous Underground Roadway. *Rock Mech. Rock Eng.* **2019**, *52*, 4547–4564. [CrossRef]
34. Zhu, S.; Jiang, F.; Wang, X.; Jiang, Y.; Ning, T.; Sun, S. Energy accumulation characteristics and rockburst mechanism of surrounding rock at heading face of extra-thick coal seam. *Chin. J. Geotech. Eng.* **2019**, *41*, 2071–2078.
35. Hao, S.; Huilin, Z.; Guansheng, Q.; Wei, L.; Mingjun, W. Preparation and CO_2 adsorption properties of TEPA-functionalized multi-level porous particles based on solid waste. *Colloids Surf. A Physicochem. Eng. Asp.* **2022**, *653*, 130004.
36. Liu, H.; Guo, L.; Zhao, X. Expansionary Evolution Characteristics of Plastic Zone in Rock and Coal Mass Ahead of Excavation Face and the Mechanism of Coal and Gas Outburst. *Energies* **2020**, *13*, 984. [CrossRef]
37. Guo, X.; Zhao, Z.; Gao, X.; Wu, X.; Ma, N. Analytical solutions for characteristic radii of circular roadway surrounding rock plastic zone and their application. *Int. J. Min. Sci. Technol.* **2019**, *29*, 263–272. [CrossRef]
38. Gao, F.; Kang, H.; Yang, L. An Experimental Investigation into the Strainburst Process Under Quasi-static Loading. *Rock Mech. Rock Eng.* **2020**, *53*, 5617–5629. [CrossRef]
39. Wu, Y.; Gao, F.; Chen, J.; He, J. Experimental Study on the Performance of Rock Bolts in Coal Burst-Prone Mines. *Rock Mech. Rock Eng.* **2019**, *52*, 3959–3970. [CrossRef]
40. Gao, F.; Kaiser, P.K.; Stead, D.; Eberhardt, E.; Elmo, D. Strainburst phenomena and numerical simulation of self-initiated brittle rock failure. *Int. J. Rock Mech. Min. Sci.* **2019**, *116*, 52–63. [CrossRef]
41. Gao, F.; Kang, H.; Li, J. Numerical simulation of fault-slip rockbursts using the distinct element method. *Tunn. Undergr. Space Technol.* **2021**, *110*, 103805. [CrossRef]
42. Haitjema, H. Occam's Razor. *Ground Water* **2019**, *57*, 349. [CrossRef]
43. Hsiung, S.M.; Blake, W.; Chowdhury, A.H.; Williams, T.J. Effects of mining-induced seismic events on a deep underground mine. *Pageoph* **1994**, *139*, 741–762. [CrossRef]
44. Xie, L.X.; Yang, S.Q.; Gu, J.C.; Zhang, Q.B.; Lu, W.B.; Jing, H.W.; Wang, Z.L. JHR constitutive model for rock under dynamic loads. *Comput. Geotech.* **2019**, *108*, 161–172. [CrossRef]
45. Zhu, F.; Zhao, J. Peridynamic modelling of blasting induced rock fractures. *J. Mech. Phys. Solids* **2021**, *153*, 104469. [CrossRef]
46. Gibowicz, S.J.; Kijko, A. *An Introduction to Mining Seismology*; Earthquake Publish House: San Diego, CA, USA, 1998.
47. Xie, L.X.; Lu, W.B.; Zhang, Q.B.; Jiang, Q.H.; Chen, M.; Zhao, J. Analysis of damage mechanisms and optimization of cut blasting design under high in-situ stresses. *Tunn. Undergr. Space Technol.* **2017**, *66*, 19–33. [CrossRef]
48. Wu, H.; Zhao, J.; Liang, W. Pattern transitions of localized deformation in high-porosity sandstones: Insights from multiscale analysis. *Comput. Geotech.* **2020**, *126*, 103733. [CrossRef]
49. Li, X.; Li, X.-F.; Zhang, Q.-B.; Zhao, J. A numerical study of spalling and related rockburst under dynamic disturbance using a particle-based numerical manifold method (PNMM). *Tunn. Undergr. Space Technol.* **2018**, *81*, 438–449. [CrossRef]
50. Zhao, T.-b.; Guo, W.-y.; Tan, Y.-l.; Lu, C.-p.; Wang, C.-w. Case histories of rock bursts under complicated geological conditions. *Bull. Eng. Geol. Environ.* **2017**, *77*, 1529–1545. [CrossRef]
51. Guo, W.; Gu, Q.; Tan, Y.; Hu, S. Case Studies of Rock Bursts in Tectonic Areas with Facies Change. *Energies* **2019**, *12*, 1330. [CrossRef]
52. Liu, B.; Zhang, H.; Zhang, B.; Lian, Z.; Yang, L.; Liu, T. Investigating the Characteristic of Weak Magnetic Stress Internal Detection Signals of Long-Distance Oil and Gas Pipeline Under Demagnetization Effect. *IEEE Trans. Instrum. Meas.* **2021**, *70*, 6011013. [CrossRef]
53. Herbut, A.; Rybak, J.; Brząkała, W. On a Sensor Placement Methodology for Monitoring the Vibrations of Horizontally Excited Ground. *Sensors* **2020**, *20*, 1938. [CrossRef]
54. He, S.H.; Goudarzy, M.; Ding, Z.; Sun, Y.F. Strength, Deformation, and Particle Breakage Behavior of Calcareous Sand: Role of Anisotropic Consolidation. *J. Geotech. Geoenviron. Eng.* **2023**, *149*, 17. [CrossRef]
55. Sun, Y.F.; Sumelka, W.; He, S.H.; Gao, Y.F. Enhanced Fractional Model for Soil-Structure Interface Considering 3D Stress State and Fabric Effect. *J. Eng. Mech.* **2022**, *148*, 16. [CrossRef]
56. Kong, B.; Dai, C.-X.; Hu, H.; Xia, J.; He, S.-H. The Fractal Characteristics of Soft Soil under Cyclic Loading Based on SEM. *Fractal Fract.* **2022**, *6*, 423. [CrossRef]
57. He, S.-H.; Ding, Z.; Sun, Y.; Chen, W.-Y.; Xia, T.-D. Cumulative deformations and particle breakage in calcareous sand subjected to drained high-cyclic loading: Experimental investigation. *Soil Dyn. Earthq. Eng.* **2022**, *161*, 107417. [CrossRef]

58. Ding, Z.; Zhang, M.-B.; Zhang, X.; Wei, X.-J. Theoretical analysis on the deformation of existing tunnel caused by under-crossing of large-diameter slurry shield considering construction factors. *Tunn. Undergr. Space Technol.* **2023**, *133*, 104913. [CrossRef]
59. Wang, R.; Ding, Z.; Zhang, Y.; Xu, Y. Self-healing of high-performance engineered cementitious materials with crystalline admixture in the seawater environment. *J. Build. Eng.* **2023**, *63*, 105472. [CrossRef]

fractal and fractional

MDPI

Article

Dilatancy Equation Based on the Property-Dependent Plastic Potential Theory for Geomaterials

Xuefeng Li [1,2,*], Houying Zhu [1,2] and Qi Yuan [1,2]

[1] School of Civil and Hydraulic Engineering, Ningxia University, Yinchuan 750021, China
[2] Solid Mechanics Institute, Ningxia University, Yinchuan 750021, China
* Correspondence: lixuefeng@nxu.edu.cn

Abstract: The dilatancy equation ignores the noncoaxiality of granular soil for the coaxial assumption of the direction of the stress and strain rate in conventional plastic potential theory, which is inconsistent with extensive laboratory tests. To reasonably describe the noncoaxial effects on dilatancy, the energy dissipation of plastic flow is derived based on the property-dependent plastic potential theory for geomaterials and integrates the noncoaxiality, the potential theory links the plastic strain of granular materials with its fabric, and the noncoaxiality is naturally related to the mesoscopic properties of materials. When the fabric is isotropic, the dilatancy equation degenerates into the form of the critical state theory, and when the fabric is anisotropic, it naturally describes the effects of noncoaxiality. In the plane stress state, a comparison between a simple shear test and prediction of the dilatancy equation shows that the equation can reasonably describe the effect of noncoaxiality on dilatancy with the introduction of microscopic fabric parameters, and its physical significance is clear. This paper can provide a reference for the theoretical description of the macro and micro mechanical properties of geomaterials.

Keywords: geomaterials; plastic potential theory; microscopic fabric; noncoaxiality; dilatancy

1. Introduction

Dilatancy is the variation of soil porosity due to the particle rearrangement under shear stress and is an essential mechanical property to distinguish geomaterial from other non-granular materials and to establish the constitutive model. Based on the energy principle, studying dilatancy is reasonable, and energy function at the critical state is usually used to describe dilatancy, such as Rowe's dilatancy equation in previous research [1]. Researchers gradually realized that the dilatancy of granular soil is also related to the material state in a later study [2]. Although Rowe also pointed out that the influence of material state should be considered in the dilatancy theory [1], after many scholars performed similar research, the stress dilatancy theory was successfully applied to the constitutive model of granular materials [3–5], and many scholars studied the particle breakage of rockfill from the perspective of energy [6–8]. Stress-dilatancy theory is widely used in the study of mechanical properties and the constitutive model of soil.

With the deepening of research, noncoaxiality has gradually attracted attention, which was first found by De Josselin de Jong [9] and has a significant effect on the mechanics and deformation of geomaterials. Therefore, based on the conventional plastic potential theory to derive dilatancy equations, there are some limitations in describing noncoaxiality due to the theory implying the coaxial assumption of direction between the stress and plastic strain rate. Researchers have conducted deeper research on noncoaxiality in subsequent theoretical and experimental studies [10,11]. Several scholars conducted simple shear tests to investigate noncoaxiality, such as the simple shear test of sand [12] and aluminum rod accumulation [13]. Lade [14] also found noncoaxial phenomena on the deviatoric plane in the true triaxial test of sand, where the direction of the stress and strain rate is coaxially under triaxial compression and tension tests and noncoaxial under other stress paths. Li [15] designed a similar true triaxial test, and although ideal spherical glass beads with a single particle size were used, obvious noncoaxiality was still observed. Xiong [16] found that noncoaxiality would cause the dilatancy curve to deviate from Rowe's line, which has a significant effect on sand dilatancy. Ignoring the noncoaxiality to deduce the dilatancy equation is unreasonable. Therefore, some scholars have integrated noncoaxiality into the research on dilatancy. Gutierrez [17] introduced the noncoaxial constant into the plastic theory and further analyzed plastic work and dilatancy. However, the noncoaxial angle seems to always exist in the whole stress space after the noncoaxial constant is set, which is inconsistent with the test results of Lade and Ducan [14]. Rudnicki [18] proposed a noncoaxial model for calculating the plastic deformation of the fractured rock mass. Xiong [16] modified Rowe's dilatancy equation by introducing Gutierrez's noncoaxial coefficient. Lashkari [19] proposed a dilatancy equation in the noncoaxiality constitutive model, and the noncoaxial coefficient is consistent with Gutierrez's previous research [20]. Tsegaye [21] established noncoaxial stress-dilatancy frames for axisymmetric, plane-strain, and general stress states, and presented a mechanism for establishing noncoaxial angular development in axisymmetric and plane-strain states. Pouragha [22] explored dilatancy aspects. The above research introduces corresponding parameters to improve the dilatancy equation based on the noncoaxial test phenomenon. At the microscopic level, noncoaxiality is a result of anisotropy [23] from the perspective of fabric properties described as a reasonable method.

To reasonably describe the stress–strain of granular materials, several researchers turned their theoretical studies to microscopic soil mechanics. Oda [24] used fabric tensors to describe the microscopic structure of granular materials and explored the connection between the initial fabric and the mechanical properties. Experiments by Wong [25] confirmed the connection between fabric anisotropy and the noncoaxiality of granular materials. Li and Dafalias [26] studied the anisotropy of sand from a microscopic perspective. In a study on microscopic soil mechanics, some researchers used fractal theory to describe the microstructure of granular materials [27]. The fractal dimension is used to characterize the particle breakage, and the fractal dimension before and after shearing is introduced into the constitutive model to simulate the influence of the change in the microscopic morphology of the particles on the stress–strain relationship [28,29]. Based on consideration of the microscopic characteristics of materials, Li [30] carried out relevant research work, established the anisotropic failure criterion of sand from the perspective of the combination of macro and micro, analyzed the noncoaxial characteristics of the simple shear test [31], and proposed the property-dependent plastic potential theory for geomaterials (potential theory for short) [32]. The theory connects the plastic deformation of materials with its microscopic fabric and describes the plastic strain rate related to material properties.

Dilatancy equations are usually derived according to the plastic flow rule and energy relation of materials. Therefore, from the perspective of the macro–micro combination, based on the potential theory, the dilatancy equation will be modeled combined with the idea of the energy transformation relation and the critical state of the soil. In the dilatancy equation, the microstructure of particles will be described by the long axis of particles and employed to define the noncoaxial coefficients. The noncoaxial coefficients will be introduced into the dilatancy equation to describe the effect of noncoaxiality on the geomaterials' dilatancy from a microscale perspective. Finally, the results of a simple shear test will be used to verify the rationality of the dilatancy equation.

2. Dilatancy Equation Based on the Potential Theory

2.1. Establishment Method

The dilatancy equation is established according to the energy transformation relation in plastic deformation and the critical state of the soil. In the potential theory, the plastic strain rate is shown in Equation (1), where g represents the plastic potential function to describe the direction of plastic flow, $d\lambda$ represents the plastic scalar factor, the derivation process can be referred to in reference [32], $\overline{F}_{lj}$ is the direction tensor of the fabric tensor F_{lj}, P_F is the size of the fabric tensor, and $P_F = 1/(3 + a_1 + a_2 + a_3)$. According to the characteristics of the direction tensor, $\overline{F}_{il}$ can be decomposed into two parts, such as Equation (2) and Equation (3). δ_{lj} in the equation is the Kronecker tensor, and when $i = j$, $\delta_{11} = \delta_{22} = \delta_{33} = 1$, it represents isotropy. $\widetilde{s}_{lj}$ represents the anisotropic part, and $a_{i_{(i=1,2,3)}}$ represents anisotropy in the main direction of the fabric. When the material is isotropic, $a_1 = a_2 = a_3 = 0$ and $\overline{F}_{lj} = \delta_{lj}$, the expression of plastic strain increment is consistent with the conventional plastic potential theory. When the material is anisotropic, the anisotropic parameter $a_{i_{(i=1,2,3)}}$ is nonzero, and the plastic strain rate is naturally decomposed into two parts, isotropic and anisotropic, as shown in Equation (4).

$$\mathrm{d}\varepsilon_{lj}^{p} = \mathrm{d}\lambda \frac{\partial g}{\partial \sigma_{il}} \overline{F}_{il} \tag{1}$$

$$F_{lj} = P_F\left(\delta_{lj} + \widetilde{s}_{lj}\right) = P_F\left(\overline{F}_{lj}\right) \tag{2}$$

$$\delta_{ij} = \begin{bmatrix} 1 & 0 & 0 \\ 0 & 1 & 0 \\ 0 & 0 & 1 \end{bmatrix}, \widetilde{s}_{ij} = \begin{bmatrix} a_1 & 0 & 0 \\ 0 & a_2 & 0 \\ 0 & 0 & a_3 \end{bmatrix} \tag{3}$$

$$\mathrm{d}\varepsilon_{lj}^{p} = \mathrm{d}\lambda \left(\frac{\partial g}{\partial \sigma_{ij}} + \frac{\partial g}{\partial \sigma_{il}} \widetilde{s}_{il}\right) \tag{4}$$

According to Equations (1)–(4), the plastic dissipated work can be written in the form of Equation (5):

$$\mathrm{d}W^{p} = \sigma_{lj}\mathrm{d}\varepsilon_{lj}^{p} = \sigma_{lj}\left(\mathrm{d}\lambda \frac{\partial g}{\partial \sigma_{ij}} + \mathrm{d}\lambda \frac{\partial g}{\partial \sigma_{il}} \widetilde{s}_{il}\right) \tag{5}$$

where $\mathrm{d}W^p$ is the plastic dissipated work, $\mathrm{d}\lambda \frac{\partial g}{\partial \sigma_{ij}}$ is the isotropic part of the plastic strain rate, and $\mathrm{d}\lambda \frac{\partial g}{\partial \sigma_{il}} \widetilde{s}_{il}$ is the anisotropic part of the plastic strain rate. The plastic dissipated work can be written as Equation (6) in the principal stress space, and stress σ_{lj} and strain increments $\mathrm{d}\varepsilon_{lj}^{p}$ are σ_k and $\mathrm{d}\varepsilon_k^p$ instead, respectively. σ_k is the principal stress expressed by stress invariants, which can be written as Equation (7).

$$\mathrm{d}W^{p} = \sigma_{lj}\, \mathrm{d}\varepsilon_{lj}^{p} = \sigma_k\, \mathrm{d}\varepsilon_k^{p} = \sigma_k\left(\mathrm{d}\lambda \frac{\partial g}{\partial \sigma_k} + \mathrm{d}\lambda \frac{\partial g}{\partial \sigma_k} s_k\right) \tag{6}$$

$$\sigma_k = p + \frac{2}{3} q \sin\left[\theta_{\sigma} + \frac{2\pi(2-k)}{3}\right] \tag{7}$$

According to Equation (1), the principal plastic strain increment $\mathrm{d}\varepsilon_k^p$ can be written by strain rate invariants as shown in Equation (8), where $\mathrm{d}\gamma^p$ is the plastic shear strain rate. The stress invariants in Equation (7) and strain increment invariants in Equation (8) are substituted into Equation (6) to obtain the energy relation based on the potential theory, as shown in Equation (9). In Equation (7), $p = (\sigma_1 + \sigma_2 + \sigma_3)/3$ represents the average principal stress, $q = \frac{1}{\sqrt{2}}\sqrt{(\sigma_1 - \sigma_2)^2 + (\sigma_2 - \sigma_3)^2 + (\sigma_1 - \sigma_3)^2}$ is the generalized shear stress, and the stress lode angle θ_σ represents the stress direction, which is defined on the deviatoric plane; the calculation formula is shown in Equation (12). $\mathrm{d}\varepsilon_v^p$ and $\mathrm{d}\gamma^p$ are the increments of the plastic volumetric strain and shear strain, respectively, and the angle $\theta_{\mathrm{d}\varepsilon}$ represent the direction of the strain rate, $\theta_{\mathrm{d}\varepsilon} = arctan\left[\frac{1}{\sqrt{3}}\left(2\frac{\mathrm{d}\varepsilon_2 - \mathrm{d}\varepsilon_3}{\mathrm{d}\varepsilon_1 - \mathrm{d}\varepsilon_3} - 1\right)\right]$.

$$\mathrm{d}\varepsilon_k^{p} = \frac{1}{3}\mathrm{d}\varepsilon_v^{p} + \mathrm{d}\gamma^{p} \sin\left[\theta_{\mathrm{d}\varepsilon} + \frac{2(2-k)\pi}{3}\right] \tag{8}$$

$$\begin{aligned} dW^p &= p\, d\varepsilon_v^p + \tfrac{2q\, d\gamma^p}{3} \sum_{k=1}^{3} \sin\left[\theta_\sigma + \tfrac{2\pi(2-k)}{3}\right] \sin\left[\theta_{d\varepsilon} + \tfrac{2\pi(2-k)}{3}\right] \\ &= p\, d\varepsilon_v^p + q\, d\gamma^p \cos(\theta_\sigma - \theta_{d\varepsilon}) \end{aligned} \quad (9)$$

According to the critical state theory, when the soil reaches the critical state, the stress and volumetric strain rate hold. Under the action of shear stress, shear deformation occurs continuously, and the stress ratio reaches the critical stress ratio η_c. At this point, the energy relation based on the potential theory can be simplified to Equation (10) in the critical state, and the three-dimensional dilatancy equation shown in Equation (11) based on the potential theory can be obtained.

$$dW^p = p\, d\varepsilon_v^p + q\, d\gamma^p \cos(\theta_\sigma - \theta_{d\varepsilon}) = \eta_c\, p\, d\gamma^p \quad (10)$$

$$\frac{d\varepsilon_v^p}{d\gamma^p} = \eta_c - \cos(\theta_\sigma - \theta_{d\varepsilon})\frac{q}{p} \quad (11)$$

where the stress lode angle is shown in Equation (12), which is consistent with Equation (7). The strain increments lode angle can be redefined in Equation (13) according to the potential theory and a detailed analysis process can be found in reference [32].

$$\theta_\sigma = \arctan\left[\frac{1}{\sqrt{3}}\left(2\frac{\sigma_2 - \sigma_3}{\sigma_1 - \sigma_3} - 1\right)\right] \quad (12)$$

$$\theta_{d\varepsilon} = \arctan\left[\frac{1}{\sqrt{3}}\left(2\frac{d\varepsilon_2 - d\varepsilon_3}{d\varepsilon_1 - d\varepsilon_3} - 1\right)\right] = \arctan\left[\frac{1}{\sqrt{3}}\left(2\frac{d\lambda \frac{\partial g}{\partial \sigma_2}\overline{F}_2 - d\lambda \frac{\partial g}{\partial \sigma_3}\overline{F}_3}{d\lambda \frac{\partial g}{\partial \sigma_1}\overline{F}_1 - d\lambda \frac{\partial g}{\partial \sigma_3}\overline{F}_3} - 1\right)\right] \quad (13)$$

According to Equation (10), when the stress lode angle is equal to the strain lode angle, it is coaxial. The analysis combined with Figure 1 and Equation (13) shows that when the material with the transversely isotropic fabric, $F_2 = F_3$, and under triaxial compression, $\sigma_1 > \sigma_2 = \sigma_3$, the principal stress acts in the three directions of F_1, F_2, and F_3, respectively, $\theta_\sigma = \theta_{d\varepsilon}$, and the stress and strain increments are coaxial. In the triaxial tensile state, $\sigma_1 = \sigma_2 > \sigma_3$, the principal stress acts on F_2, F_3, and F_1, respectively, the stress lode angle is equal to the strain lode angle, and the stress and strain increment are coaxial. In other stress states of the deviatoric plane, $\theta_\sigma \neq \theta_{d\varepsilon}$. Accordingly, the description of noncoaxiality under different stress states is associated with the meso-fabric properties of materials, which is different from the noncoaxial coefficients commonly used in previous research, has clearer physical significance, and is consistent with the experimental phenomenon in reference [14].

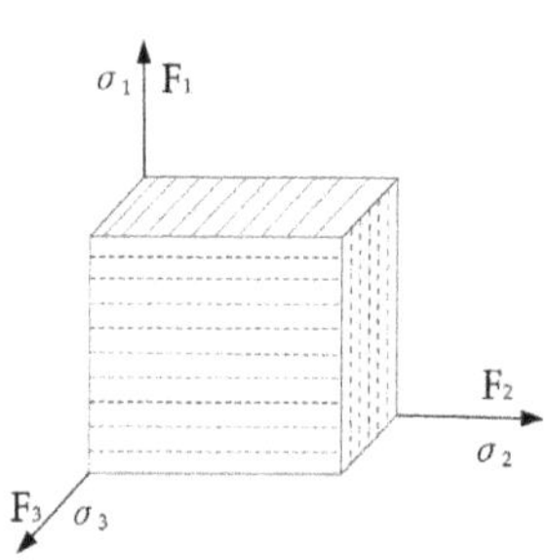

Figure 1. Stress and fabric relation.

2.2. Description of Dilatancy under Plane Stress State

To verify the rationality of the dilatancy equation in this paper, we first derived the dilatancy equation under the plane stress state. Mohr's circle under plane stress is shown in Figure 2, where ϕ_{dss} and ϕ_{p} are the friction angle and the peak friction angle, respectively. The relation between stress components σ_x, σ_y, σ_z, normal stress s, and shear stress t is shown in Equation (14). Similarly, the relation between the strain component increments $\mathrm{d}\varepsilon_x$, $\mathrm{d}\varepsilon_y$, and $\mathrm{d}\varepsilon_z$, volumetric strain, and shear strain in the strain Mohr's circle can be written as Equation (15).

$$\sigma_x = s - t\cos(2\alpha),\ \sigma_y = s + t\cos(2\alpha),\ \tau_{xy} = t\sin(2\alpha) \tag{14}$$

$$\left.\begin{array}{c} \mathrm{d}\varepsilon_x^p = \frac{1}{2}\mathrm{d}\varepsilon_v^p - \frac{1}{2}\mathrm{d}\gamma^p\cos(2\beta) \\ \mathrm{d}\varepsilon_y^p = \frac{1}{2}\mathrm{d}\varepsilon_v^p + \frac{1}{2}\mathrm{d}\gamma^p\cos(2\beta) \\ \mathrm{d}\varepsilon_{xy}^p = \frac{1}{2}\mathrm{d}\gamma^p\sin(2\beta) \end{array}\right\} \tag{15}$$

where α is the stress direction angle (as shown in Figure 2) and β is the strain increment direction angle in the strain space. Assuming the elastic strain is negligible, stress invariants and strain invariants can be written as:

$$\begin{array}{c} s = \frac{\sigma_x+\sigma_y}{2} = \frac{\sigma_1+\sigma_3}{2},\ t = \sqrt{\left(\frac{\sigma_x-\sigma_y}{2}\right)^2 + \tau_{xy}^2} = \frac{\sigma_1-\sigma_3}{2}, \\ \mathrm{d}\varepsilon_v^p = \mathrm{d}\varepsilon_1^p + \mathrm{d}\varepsilon_2^p,\ \mathrm{d}\gamma^p = \mathrm{d}\varepsilon_1^p - \mathrm{d}\varepsilon_2^p \end{array} \tag{16}$$

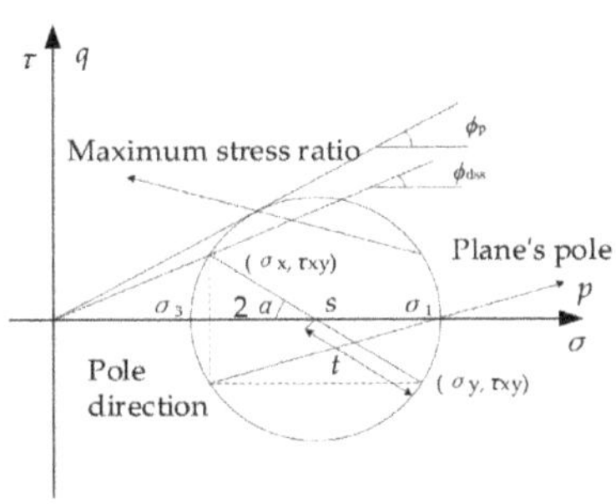

Figure 2. Stress Mohr circle.

In the plane stress state, the plastic dissipation work can be written as Equation (17). The stress component and strain component of Equation (14) and Equation (15) are substituted into Equation (17), respectively, and the incremental expression of the plastic dissipation work shown in Equation (18) is further determined. According to critical state soil mechanics, the volumetric strain increment of soil reaches zero and the stress ratio reaches the critical stress ratio η_c under the plane stress state, and the plastic dissipation work increment described in Equation (18) can be written the form of Equation (19) while the dilatancy equation under plane stress can be obtained through further modifications as Equation (20). Here, $\mathrm{d}\varepsilon_v^p$ and $\mathrm{d}\gamma^p$ are the volume strain and shear strain increments, and s and t are the average principal stress and shear stress, respectively.

$$\mathrm{d}W^p = \sigma_{ij}\mathrm{d}\varepsilon_{ij}^p = \sigma_x\mathrm{d}\varepsilon_x^p + \sigma_y\mathrm{d}\varepsilon_y^p + 2\tau_{xy}\mathrm{d}\varepsilon_{xy}^p \tag{17}$$

$$\mathrm{d}W^p = s\ \mathrm{d}\varepsilon_v^p + t\ \mathrm{d}\gamma^p\cos(2\Delta) \tag{18}$$

$$\mathrm{d}W^p = s\ \mathrm{d}\varepsilon_v^p + t\ \mathrm{d}\gamma^p\cos(2\Delta) = s\ \eta_c\mathrm{d}\gamma^p \tag{19}$$

$$\frac{d\varepsilon_v^p}{d\gamma^p} = \eta_c - c\frac{t}{s} \tag{20}$$

In Equation (17), $c = \cos(2\Delta) = \cos(2\alpha - 2\beta)$, as shown in Figure 2, α can be written as Equation (21) according to the geometric relation, and β can also be written as Equation (22) in the strain space. η_c can be regarded as different parameters for sand with different densities.

$$\alpha = \frac{1}{2}\arctan\frac{2\tau_{xy}}{\sigma_y - \sigma_x} \tag{21}$$

$$\beta = \frac{1}{2}\arctan\frac{2d\gamma^p}{d\varepsilon_y^p - d\varepsilon_x^p} \tag{22}$$

In the strain space, according to the idea of strain distribution in the potential theory, the strain increment in Equation (22) is expressed by Equation (1) and resubstituted into Equation (22), and the expression of strain rate direction angle Equation (23) based on the potential theory is obtained. It should be noted that in most cases, geomaterials are transversely isotropic, so the transversely isotropic fabric is adopted in the strain distribution by Equation (1). The specific components of the fabric are shown in Equation (24), and two components of the fabric direction tensor in Equation (4) can be written as Equation (25) where g represents the plastic potential function and a represents the anisotropic parameter of transversely isotropic fabric. The value of a is measured by the method proposed by Li [33], where $a_{i_{(i=1,2,3)}}$ represents the amplitude parameter of the orthorhombic anisotropy as shown in Equation (26) and the physical meanings of $\theta^{(K)}$ and $\alpha^{(K)}$ are shown in Figure 3. When $\alpha^{(K)} = \pi/4$, $\sin(\alpha^{(K)}) = \cos(\alpha^{(K)})$, $a_2 = a_3$, which is transversely isotropic, and the value of a is shown in Equation (27). In Equations (26) and (27), N is the particle number of the sample, and it represents the number of contact fabrics when F_{ij} is employed to describe the contact of particles.

$$\beta = \frac{1}{2}arctan\frac{2\frac{\partial g}{\partial \tau_{xy}}(1+a)}{\frac{\partial g}{\partial \sigma_y}(1-a) - \frac{\partial g}{\partial \sigma_x}(1+a)} \tag{23}$$

$$F_{Ij} = \frac{1}{3+a}\begin{bmatrix} 1-a & 0 & 0 \\ 0 & 1+a & 0 \\ 0 & 0 & 1+a \end{bmatrix} \tag{24}$$

$$\delta_{ij} = \begin{bmatrix} 1 & 0 & 0 \\ 0 & 1 & 0 \\ 0 & 0 & 1 \end{bmatrix}, \tilde{s}_{ij} = \begin{bmatrix} -a & 0 & 0 \\ 0 & a & 0 \\ 0 & 0 & a \end{bmatrix} \tag{25}$$

$$\begin{aligned}
a_1 &= \tfrac{1}{2N}\left\{\left[\sum_{K=1}^{2N}\left(\cos^2\left(\theta_1^{(K)}\right) - \sin^2\left(\theta_1^{(K)}\right)\cos^2\left(\alpha^{(K)}\right)\right)\right]^2 + \left[\sum_{K=1}^{2N}\sin\left(2\theta_1^{(K)}\right)\cos\left(\alpha^{(K)}\right)\right]^2\right\}^{\frac{1}{2}} \\
a_2 &= \tfrac{1}{2N}\left\{\left[\sum_{K=1}^{2N}\left(\cos^2\left(\theta_1^{(K)}\right) - \sin^2\left(\theta_1^{(K)}\right)\sin^2\left(\alpha^{(K)}\right)\right)\right]^2 + \left[\sum_{K=1}^{2N}\sin\left(2\theta_1^{(K)}\right)\sin\left(\alpha^{(K)}\right)\right]^2\right\}^{\frac{1}{2}} \\
a_3 &= \tfrac{1}{2N}\left\{\left[\sum_{K=1}^{2N}\sin^2\left(\theta_1^{(K)}\right)\cos\left(2\alpha^{(K)}\right)\right]^2 + \left[\sum_{K=1}^{2N}\sin^2\left(2\theta_1^{(K)}\right)\sin\left(2\alpha^{(K)}\right)\right]^2\right\}^{\frac{1}{2}}
\end{aligned} \tag{26}$$

$$a = \frac{1}{2N}\left\{\left[\sum_{K=1}^{2N}\left(\cos^2\left(\theta_1^{(K)}\right) - \frac{\sin^2\left(\theta_1^{(K)}\right)}{2}\right)\right]^2 + \frac{1}{2}\left[\sum_{K=1}^{2N}\sin\left(2\theta_1^{(K)}\right)\right]^2\right\}^{\frac{1}{2}} \tag{27}$$

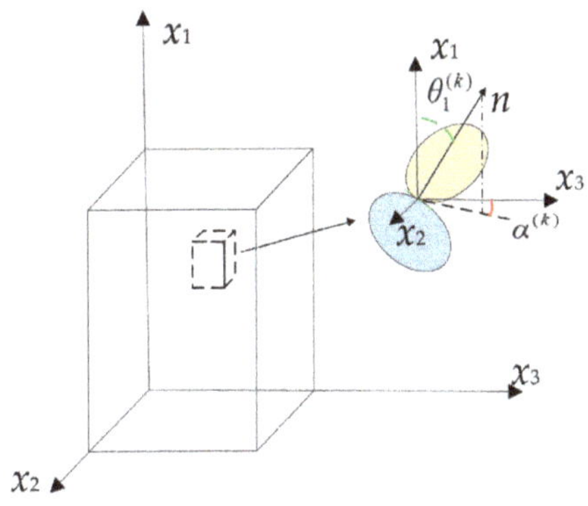

Figure 3. Particle direction diagram.

We then introduce two directional angles of Equations (21) and (23) into Equation (20), and the expression form of the dilatancy equation based on the potential theory under plane stress can be obtained. The plastic potential g can be obtained by integrating the equation $D = -\,\mathrm{d}t/\mathrm{d}s$; in the critical state, the plastic volumetric strain increment and shear stress are constant, which means $\mathrm{d}p\mathrm{d}\varepsilon_v^p + \mathrm{d}q\mathrm{d}\gamma^p = 0$. We let $D = \frac{\mathrm{d}\varepsilon_v^p}{\mathrm{d}\gamma^p} = -\frac{\mathrm{d}q}{\mathrm{d}p} = -\frac{\mathrm{d}t}{\mathrm{d}s}$, and introducing the dilatancy equation (Equation (28)) into the expression of the critical state obtains the plastic potential g by the integral as shown in Equation (29).

$$D = \frac{\mathrm{d}\varepsilon_v^p}{\mathrm{d}\gamma^p} = \mathrm{d}_0\left(\eta_c - c\frac{t}{s}\right) \tag{28}$$

$$g = t + \frac{\mathrm{d}_0\eta_c s}{\mathrm{d}_0 c - 1}\left[\left(\frac{s}{s_0}\right)^{\mathrm{d}_0 c - 1} - 1\right] \tag{29}$$

where s_0 represents the initial average principal stress. It can be seen that the dilatancy equation can reflect the noncoaxiality of stress and plastic strain increments in the plane stress state and also reflect the coaxiality when $a = 0$. According to Equation (23), the more obvious the anisotropy is, the more significant the noncoaxiality is. The dilatancy and stress ratio under plane stress are shown in Figure 4, which is drawn by setting different model parameters, in which d_0 was defined by Li [2] and anisotropic parameter a was under the same stress condition. With the gradual increase in the stress ratio, dilatancy presents as shear contraction followed by dilatancy. With the increase in the fabric parameter a, the difference in dilatancy is more significant with different anisotropy. With the decrease in material coefficient d_0, the volume contraction is lower, and the difference in dilatancy is lower.

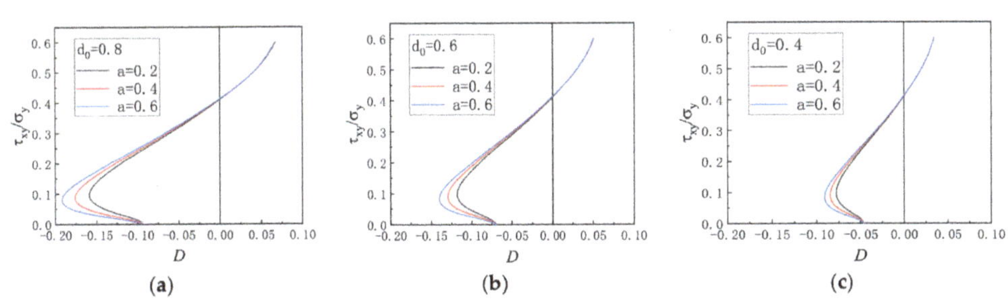

Figure 4. Stress dilatancy relation in plane stress state: **(a)** $\mathrm{d}_0 = 0.8$; **(b)** $\mathrm{d}_0 = 0.6$; **(c)** $\mathrm{d}_0 = 0.4$.

3. Model Verification

To verify the dilatancy equation, Equation (20) is compared with the results of the simple shear test [34]. First, the simple shear test is introduced. The principal stress rotation in the loading process is the noticeable feature of the simple shear test, and the position of the failure plane is uncertain, as shown in Figure 5. The black line is the initial stress unit, and the blue line is the stress unit after deformation. In the laboratory test, Cambridge University's Mark 5 DSS apparatus was used to shear the sample prepared by Leighton–Buzzard sand. The consolidation of the sample is loaded under the normal stress σ_y = 400 kPa. After the loading starts, shear is carried out at a constant shear rate until the peak value is reached. The test can reflect the rotation of principal stress and the noncoaxiality between the stress direction and strain increment direction, which is consistent with the research goal of this paper.

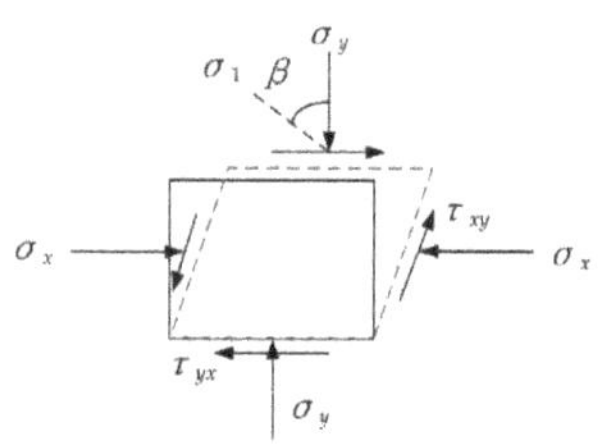

Figure 5. Stress-state diagram of a simple shear test.

3.1. Noncoaxiality Verification

The predicted and tested values of principal stress rotation are shown in Figure 6a. In this paper, the critical stress ratio sin φ_c of loose sand, medium dense sand, and dense sand in relevant tests are extracted from [17], which was performed by Cole [34]. The experimental results show an extremely small difference in the principal stress rotation angle among the samples, the simulation results are close to the experimental values, and the model's principal stress rotation angle is consistent with the experiment. Figure 6b–d shows the variation in noncoaxial angles, which decrease with the increase in the stress ratio (τ_{xy}/σ_y) under different densities, gradually close to the coaxial state. When the stress ratio of loose and medium-dense sand is close to 0.6, the noncoaxial angle is close to zero. When the stress ratio of dense sand reaches 0.8, the noncoaxial angle is approximately 5°. The dilatancy equation can aptly predict this law, which shows that the dilatancy equation in this paper is reasonable.

3.2. Verification of Dilatancy

Noncoaxiality has a significant effect on the dilatancy of sand [35] and the critical stress ratio with three densities is set to a fixed value in this section (sin φ_c= 0.5), the dilatancy coefficient D is calculated by different anisotropic parameters a, the effect of noncoaxiality on dilatancy is analyzed, and the predicted and experimental values($-\mathrm{d}v/\mathrm{d}\gamma$) are shown in Figure 7. When the fabric parameter a in the dilatancy equation takes different values, the dilatancy of prediction shows the same law with three densities, and the law of first contraction and then dilatancy is consistent with the experiment. The loose sand test results show significant volume contraction, which is closer to the predicted value and is consistent with the findings that noncoaxiality can lead to more significant volume contraction than in reference [16], indicating that the noncoaxial coefficient modified by the dilatancy equation from the mesoscopic perspective can reasonably reflect the influence of noncoaxiality on dilatancy. Meanwhile, under the low-stress ratio, the deformation is greatly affected by the material properties [36–38], and the different noncoaxiality values with different a values lead to the curves of dilatancy not coinciding. When reaching the high-stress ratio, the effect of material properties is weak, the stress–strain tends to be coaxial, and the dilatancy curves of different a values are less affected by noncoaxial and gradually coincide, which is consistent with the existing consensus.

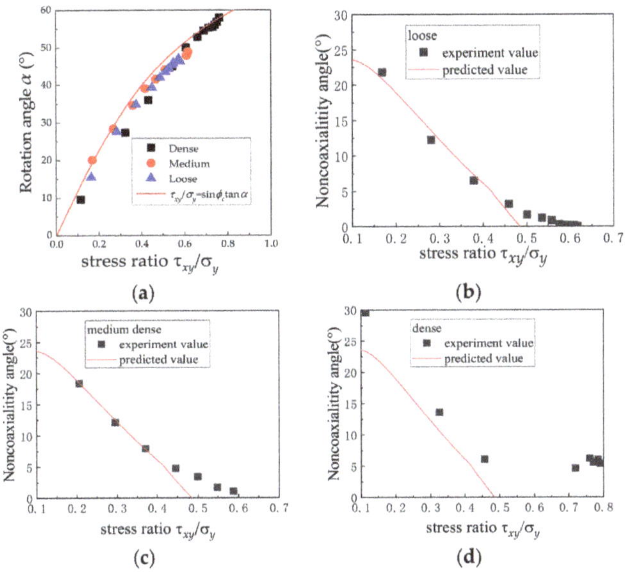

Figure 6. The predicted and tested values of principal stress direction and noncoaxial angle: (**a**) direction of principal stress; (**b**) loose sand; (**c**) medium dense sand; (**d**) dense sand.

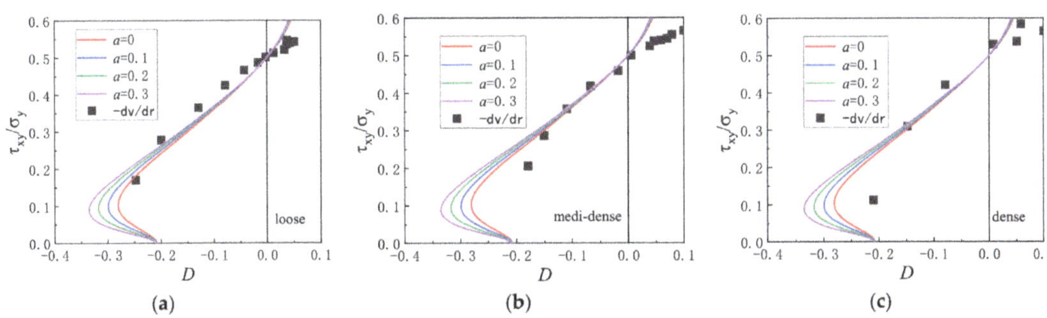

Figure 7. Comparison of predicted dilatancy values with test values: (**a**) loose sand; (**b**) medium dense sand; (**c**) dense sand.

4. Conclusions

To integrate the effect of noncoaxiality and describe the dilatancy of geomaterials reasonably, the strain increment lode angle $\theta_{d\varepsilon}$ defined in the potential theory by macro-micro combination is used to define the noncoaxial coefficient, introducing noncoaxiality to the dilatancy equation. It can reflect the influence of noncoaxiality on dilatancy and comprehensively considers the material properties. Finally, the rationality of the dilatancy equation is verified by a simple shear test, and the following conclusions are drawn:

(1) For noncoaxial conditions, calculation using stress invariants and strain increment invariants will overestimate the energy dissipated during loading. The energy transformation relation based on the potential theory introduces a new noncoaxial coefficient with values of 0–1, which can reasonably correct the influence of noncoaxiality

on energy dissipation. Meanwhile, the influence of material microscopic properties on energy dissipation is introduced, which is closer to the actual condition.

(2) The new noncoaxial coefficient is different from previous research, which is not only related to the stress level and stress direction but also related to the material microscopic fabric characteristics. The potential theory can be used to calculate the newly defined noncoaxial coefficient to provide a dilatancy equation considering noncoaxiality. When the microscopic fabric is isotropic, the noncoaxial coefficient is naturally 1, and the dilatancy equation can be reduced to the form of the critical state theory. When the fabric is anisotropic, the noncoaxial angle is related to the material anisotropy, the geometric relation between the fabric and the stress direction. The dilatancy equations can naturally describe noncoaxial effects, and the physical meaning is clearer.

(3) Under the simple shear stress state, after introducing the noncoaxial coefficient, the dilatancy equation can naturally reflect the influence of noncoaxiality on the dilatancy under the condition of principal stress rotation. At the low-stress ratio, the generation of noncoaxiality depends on the material properties and has a significant effect on dilatancy. When the stress ratio is high, the influence of material properties on stress and strain is not obvious, the stress and strain naturally tend to be coaxial, and the influence on dilatancy is weakened. The experimental results verify the effectiveness of the proposed dilatancy equation.

Author Contributions: Conceptualization, X.L. and H.Z.; methodology, X.L. and H.Z.; software, X.L. and H.Z.; validation, Q.Y.; formal analysis, H.Z. and Q.Y.; investigation, H.Z. and Q.Y.; resources, Q.Y.; data curation, Q.Y.; writing—original draft preparation, H.Z. and Q.Y.; writing—review and editing, X.L. and H.Z.; visualization, H.Z. and Q.Y.; supervision, X.L.; project administration, X.L.; funding acquisition, X.L. All authors have read and agreed to the published version of the manuscript.

Funding: This research was funded by the Projects for Leading Talents of Science and Technology Innovation of Ningxia (No. KJT2019001), the National Natural Science Foundation of China (No. 12162028), and the innovation team for multi-scale mechanics and its engineering applications of Ningxia Hui Autonomous Region (2021).

Data Availability Statement: The data used to support the findings of this study are available from the corresponding author upon request.

Conflicts of Interest: The authors declare no conflict of interest.

References

1. Rowe, P.W. The stress-dilatancy relation for static equilibrium of an assembly of particles in contact. *Proc. R. Soc. Lond.* **1962**, *269*, 500–527.
2. Li, X.S.; Dafalias, Y.F.; Wang, Z.L. State-dependant dilatancy in critical-state constitutive modelling of sand. *Can. Geotech. J.* **1999**, *36*, 599–611. [CrossRef]
3. Nova, R.; Wood, D.M. A constitutive model for sand intriaxial compression. *Int. J. Numer. Anal. Methods Geomech.* **1979**, *3*, 255–278. [CrossRef]
4. Jefferies, M.G. Nor-Sand: A simple critical state for sand. *Geotechnique* **1993**, *43*, 91–103. [CrossRef]
5. Wood, D.M.; Belkheir, K.; Liu, D.F. Strain softening and state parameter for sand modeling. *Geotechnique* **1994**, *44*, 335–339. [CrossRef]
6. Ueng, T.S.; Chen, T.J. Energy aspects of particle breakage in drained shear of sands. *Geotechnique* **2000**, *50*, 65–72. [CrossRef]
7. Guo, W.L.; Zhu, J.G. Particle breakage energy and stress dilatancy in drained shear of rockfills. *Géotech. Lett.* **2017**, *7*, 304–308. [CrossRef]
8. Salim, W.; Indraratna, B. A new elastoplastic constitutive model for coarse granular aggregates incorporating particle breakage. *Can. Geotech. J.* **2004**, *41*, 657–671. [CrossRef]
9. De Josselin de Jong, G. Statics and Kinematics in the Failable Zone of a Granular Material. Ph.D. Thesis, Technische Universiteit Delft, Delft, The Netherlands, 1959.
10. Spencer, A.J.M. Theory of the kinematics of ideal soils under plane strain conditions. *J. Mech. Phys. Solids* **1964**, *12*, 337–351. [CrossRef]
11. Tsutsumi, S.; Hashiguchi, K. General non-proportional loading behavior of soils. *Int. J. Plast.* **2005**, *21*, 1941–1969. [CrossRef]
12. Roscoe, K.H. The influence of strains in soil mechanics. *Géotechnique* **1970**, *20*, 129–170. [CrossRef]

13. Oda, M.; Konishi, J. Microscopic deformation mechanism of granular material in simple shear. *Soils Found.* **1974**, *14*, 25–38. [CrossRef] [PubMed]
14. Lade, P.V.; Duncan, J.M. Cubical triaxial tests on cohesionless soil. *J. Soil Mech. Found. Div.* **1973**, *99*, 793–812. [CrossRef]
15. Li, K.F.; Li, X.F.; Chen, Q.S.; Nimbalkar, S. Laboratory Analyses of Non-coaxiality and Anisotropy of Spherical Granular Media under True Triaxial State. *Int. J. Geomech.* **2023**, *23*, 04023150. [CrossRef]
16. Xiong, H.; Guo, L.; Cai, Y.Q. Effect of non-coaxiality on dilatancy of sand involving principal stress axes rotation. *Rock Soil Mech.* **2017**, *38*, 133–140.
17. Gutierrez, M.; Ishihara, K. Non-coaxiality and energy dissipation in granular materials. *Soils Found.* **2000**, *40*, 49–59. [CrossRef] [PubMed]
18. Rudnicki, J.W.; Rice, J.R. Conditions for the localization of deformation in pressure-sensitive dilatant materials. *J. Mech. Phys. Solids* **1975**, *23*, 371–394. [CrossRef]
19. Lashkari, A.; Latifi, M. A non-coaxial constitutive model for sand deformation under rotation of principal stress axes. *Int. J. Numer. Anal. Methods Geomech.* **2008**, *32*, 1051–1086. [CrossRef]
20. Gutierrez, M.; Ishihara, K.; Towhata, I. Flow theory for sand during rotation of principal stress direction. *Soils Found.* **1991**, *31*, 121–132. [CrossRef]
21. Tsegaye, A.B.; Benz, T.; Nordal, S. Formulation of non-coaxial plastic dissipation and stress-dilatancy relations for geomaterials. *Acta Geotech. Int. J. Geoengin.* **2020**, *15*, 2727–2739. [CrossRef]
22. Pouragha, M.; Kruyt, N.P.; Wan, R. Non-coaxial Plastic Flow of Granular Materials through Stress Probing Analysis. *Int. J. Solids Struct.* **2021**, *222–223*, 111015. [CrossRef]
23. Tian, Y.; Yao, Y.P.; Luo, T. Explanation and modeling of non-coaxiality of soils from anisotropy. *Rock Soil Mech.* **2018**, *39*, 2035–2042.
24. Oda, M. Initial fabrics and their relations to mechanical properties of granular materials. *Soils Found.* **1972**, *12*, 17–36. [CrossRef]
25. Wong, R.K.S.; Arthur, J.R.F. Sand sheared by stresses with cyclic variation in direction. *Géotechnique* **1986**, *2*, 215–226. [CrossRef]
26. Li, X.S.; Dafalias, Y.F. Constitutive Modeling of Inherently Anisotropic Sand Behavior. *J. Geotech. Geoenviron. Eng.* **2002**, *128*, 868–880. [CrossRef]
27. Tyler, S.W.; Wheatcraft, S.W. Fractal scaling of soil particle-size distributions: Analysis and limitations. *Soil Sci. Soc. Am. J.* **1992**, *56*, 362–369. [CrossRef]
28. Hou, H.; Pan, Z.; Jiang, P. Double Yield Surface Model of Calcareous Sand Considering Particle Breakage. *Adv. Eng. Sci.* **2021**, *53*, 132–141.
29. Xue, L.; Jiankun, L.; Jinze, L. Fractal dimension, particle shape, and particle breakage analysis for calcareous sand. *Bull. Eng. Geol. Environ.* **2022**, *81*, 106.
30. Li, X.F.; Huang, M.S.; Qian, J.G. Failure criterion of anisotropic sand with method of macro-meso incorporation. *Chin. J. Rock Mech. Eng.* **2010**, *29*, 1885–1892.
31. Li, X.F.; Huang, M.S.; Qian, J.G. Analysis of non-coaxial characters of sand for simple shear test with the method of macro-meso-incorporation. *Rock Soil Mech.* **2013**, *34*, 3417–3424.
32. Li, X.F.; Kong, L.; Huang, M.S. Property-dependent plastic potential theory for geomaterials. *Chin. J. Geotech. Eng.* **2013**, *35*, 1722–1729.
33. Li, X.F.; Wang, Q.; Liu, J.; Wu, W.; Meng, F. Quantitative Description of microscopic Fabric Based on Sand Particle Shapes. *China J. Highw. Transp.* **2016**, *29*, 29–36.
34. Cole, E. The Behaviour of Soils in the Simple-Shear Apparatus. Ph.D. Thesis, University of Cambridge, Cambridge, UK, 1967.
35. Xiong, H. Experimental Study on the Static and Dynamic Behavior of Anisotropic Sands Involving Rotation of Principal Stress Axes. Ph.D. Thesis, Zhejiang University, Hangzhou, China, 2015.
36. Yamada, Y.; Ishihara, K. Anisotropic Deformation Characteristics of Sand Under Three Dimensional Stress Conditions—ScienceDirect. *Soils Found.* **1979**, *19*, 79–94. [CrossRef] [PubMed]
37. Abelev, A.V.; Lade, P.V. Effects of cross anisotropy on three-dimensional behavior of sand. i: Stress—Strain behavior and shear banding. *J. Eng. Mech.* **2003**, *129*, 160–166. [CrossRef]
38. Lade, P.V.; Abelev, A.V. Effects of cross anisotropy on three-dimensional behavior of sand. ii: Volume change behavior and failure. *J. Eng. Mech.* **2003**, *129*, 167–174. [CrossRef]

MDPI AG
Grosspeteranlage 5
4052 Basel
Switzerland
Tel.: +41 61 683 77 34

Fractal and Fractional Editorial Office
E-mail: fractalfract@mdpi.com
www.mdpi.com/journal/fractalfract

www.ingramcontent.com/pod-product-compliance
Lightning Source LLC
Chambersburg PA
CBHW040313270926
51850CB00007BC/1203
9783725850839